Genotype
-by-
Environment
Interaction

Edited by
Manjit S. Kang
Hugh G. Gauch, Jr.

CRC Press
Taylor & Francis Group
Boca Raton   London   New York

CRC Press is an imprint of the
Taylor & Francis Group, an informa business

Published 1996 by CRC Press
Taylor & Francis Group
6000 Broken Sound Parkway NW, Suite 300
Boca Raton, FL 33487-2742

© 1996 by Taylor & Francis Group, LLC
CRC Press is an imprint of Taylor & Francis Group, an Informa business

First issued in paperback 2019

No claim to original U.S. Government works

ISBN 13: 978-0-367-44868-4 (pbk)
ISBN 13: 978-0-8493-4003-1 (hbk)

Visit the Taylor & Francis Web site at
http://www.taylorandfrancis.com

and the CRC Press Web site at
http://www.crcpress.com

Library of Congress Cataloging-in-Publication Data

Genotype-by-environment interaction / edited by Manjit S. Kang, Hugh G. Gauch, Jr.
   p. cm.
  Includes bibliographical references and index.
  ISBN 0-8493-4003-9 (alk. paper)
  1. Plant breeding. 2. Crops--Genetics. 3. Genotype-environment
interaction. I. Kang, Manjit S. II. Gauch, Hugh G., 1942- .
SB123.G427 1996
631.5'23—dc20
                                     96-5014
                                     CIP

Library of Congress Card Number 96-5014

PREFACE

Genotype-by-Environment Interaction (GEI) is a prevalent issue among farmers, breeders, geneticists, and production agronomists. This volume brings together contributions from knowledgeable plant breeders and quantitative geneticists to better understand the relationship between crop performance and environment. The information presented can help reduce the cost of extensive genotype evaluation by eliminating unnecessary testing sites and by fine-tuning breeding programs. The presence of a large GEI may necessitate establishment of additional regional breeding and testing sites, thus increasing the cost of developing a commercially usable product. The GEI relates to sustainable agriculture, as it affects efficiency of breeding programs and allocation of limited resources. Breeding for resistance or tolerance to biotic and abiotic stresses is crucial to develop genotypes with greater sustainability and stability.

In 1990, we focused attention on this vital issue by organizing an international symposium and producing a volume entitled *Genotype-By-Environment Interaction and Plant Breeding* (edited by M. S. Kang). Since then, fruitful associations among breeders, geneticists, and biometricians have emerged throughout the world and much new information has been generated.

Because of the fast pace of developments in the area of GEI and stability analyses during the past five years, new information needed to be properly organized and presented. This volume, *Genotype-by-Environment Interaction*, aims at providing an up-to-date compendium of techniques for analyzing GEI. The book contains authoritative articles by scientists from around the world and offers a first-time treatment of molecular approaches in analyzing GEI. It emphasizes: (i) how to select truly superior, adapted genotypes in diverse environments, (ii) methods that would effectively utilize GEI and help reduce breeding and testing costs, (iii) understanding the nature and causes of GEI and minimizing its deleterious implications and exploiting its beneficial potential through appropriate statistical and breeding methodologies, and (iv) identification of quantitative trait loci (QTL) interacting with environment.

A brief annotation of each chapter follows. Chapter 1 deals with incorporating stability into breeding programs and demonstrates why a greater emphasis on stability during the selection process would be beneficial to growers. The use of environmental covariates to remove heterogeneity from GEI is discussed in Chapters 1 and 2. In Chapter 2, two-way models are proposed that incorporate additional information on genotypic and environmental factors.

Chapter 3 deals with relationships among analytical and experimental methods used to study genotypic variation and GEI in plant breeding multi-environment trials. This chapter integrates the concepts of correlated response and selection. In Chapter 4, the authors expound on developments relative to AMMI analysis, emphasizing applications to yield trials.

As rapid developments in molecular genetics are occurring, it is important to integrate GEI and molecular mechanisms. Chapter 5 offers some of the statistical issues relative to the QTL and QTL-by-environment analyses.

These analyses help identify the underlying genetic mechanisms responsible for GEI. The main aims of Chapter 6 are to discuss strategies to obtain sustainable genotype performance and to assist the breeder in deciding whether observed stability differences are significant or merely due to sampling variation.

Chapter 7 is devoted to SHMM that separates genotypic effects from environmental effects; the emphasis here is on methods for clustering genotypes into groups to circumvent the effect of crossover interactions. In Chapter 8, the authors review multiplicative model methodology and summarize results on hypothesis testing for SHMM and other related model forms.

Chapter 9 treats GEI analyses by ranks. Emphasis is given to non-parametric approaches that use a concept of rank interaction. Simultaneous selection for yield and stability has implications in breeding.

Chapter 10 covers methods for estimating and testing hypotheses regarding probabilities (termed reliabilities) of cultivars of outperforming a check in a broad range of environments. Reliabilities can be useful to plant breeders, as demonstrated in Chapter 11, for characterizing target environments, selecting optimal testing sites, developing optimal population types, and determining optimal selection criteria to best allocate resources and identify the best breeding strategy. Chapter 12 describes identification of various genetic parameters affecting GEI, via a computer simulation approach, wherein GEI is incorporated into a traditional genetic model to simulate a trait.

Chapter 13 presents approaches that could help improve selection of superior genotypes. Experimental results are presented that relate to relationships between different stability statistics and means; also included are principal component and cluster analyses. This chapter highlights prediction of genotypes for specific locations. Chapter 14 deals with spatial analysis and application of AMMI analysis to wheat and tea fertility experiments. The results show the usefulness of AMMI to describe spatial patterns and clarify treatment effects.

The Appendix contains a bibliography of 83 references that should aid the reader in getting acquainted with some of the publications on GEI. It also explains computer access to a large, continually updated bibliography.

We trust the readers will find the volume informative and useful in dissecting GEI and developing breeding strategies. This project could not have been completed without the cooperation and interest of the authors; and CRC Press, especially Marsha Baker. Special thanks are due to Robert Magari who diligently assisted in ensuring that all chapters followed the correct format and formatted properly those chapters that did not conform to the style used. A word of thanks also goes to Jennifer Tullier and Michelle Averitt for their assistance.

Manjit S. Kang
Baton Rouge, Louisiana

Hugh G. Gauch, Jr.
Ithaca, New York

THE EDITORS

Manjit S. Kang received a B.Sc. (Agri. & A.H.) with Honors (1968) from the Punjab Agricultural University at Ludhiana, India; an M.S. in Biological Sciences, majoring in Plant Genetics (1971) from Southern Illinois University at Edwardsville; an M.A. in Botany (1977) from Southern Illinois University at Carbondale; and a Ph.D. in Crop Science (also 1977) from University of Missouri-Columbia.

Dr. Kang is a well-published author. He edited the volume *Genotype-By-Environment Interaction and Plant Breeding* (1990) that resulted from an international symposium that he organized at Louisiana State University in February 1990. He is the author/publisher of *Applied Quantitative Genetics* (1994) that resulted from eight years of teaching a graduate level course on Quantitative Genetics in Plant Improvement. He has published 70[+] refereed journal articles in international journals and 80 other scientific publications. His research interests are: genetics of resistance to aflatoxin, weevils, and herbicides in maize; genetics of grain dry-down rate and stalk quality in maize; genotype-by-environment interaction and stability analyses; and interorganismal genetics.

Dr. Kang taught Plant Breeding and Plant Genetics courses at Southern Illinois University-Carbondale (1972-74). He has been teaching a graduate level Applied Quantitative Genetics course since 1986 at Louisiana State University. He team-taught an Advanced Plant Genetics course (1993-95). He has advised several M.S. and Ph.D. students. He has been a Full Professor in the Department of Agronomy at LSU since 1990.

Dr. Kang's biographical sketches have been included in Marquis Who's Who in the South and Southwest (19th ed.), Who's Who in the Frontiers of Science and Technology (2nd ed.), and Who's Who in Science and Engineering (3rd ed). He belongs to Gamma Sigma Delta and Sigma Xi.

Hugh G. Gauch, Jr. received a B.S. in Botany (1964) from the University of Maryland and an M.S. in Plant Genetics (1966) from Cornell University. He is a Senior Research Specialist in the Department of Soil, Crop and Atmospheric Sciences at Cornell University. During the past 25 years, his research has been concerned with applications of multivariate statistics to ecological and agricultural data. This work has resulted in two books: *Multivariate Analyses in Community Ecology* (Cambridge University Press, 1982) and *Statistical Analysis of Regional Yield Trials: AMMI Analysis of Factorial Designs* (Elsevier, 1992). He has published several refereed, scientific journal articles.

Mr. Gauch's present research focuses on analysis of agricultural yield trials. The three primary goals are to model and understand complex genotype-environment interactions using AMMI and related analyses, to estimate yields accurately by partitioning the variation into pattern-rich model and a discarded noise-rich residual, and to design efficient experiments.

TABLE OF CONTENTS

Chapter 1

NEW DEVELOPMENTS IN SELECTING FOR PHENOTYPIC STABILITY IN CROP BREEDING

M. S. Kang and R. Magari [1]

I. INTRODUCTION

Genotype x environment (GE) interaction is a major concern in plant breeding for two main reasons; *first,* it reduces progress from selection, and *second,* it makes cultivar recommendation difficult because it is statistically impossible to interpret the main effects. GE interaction occurs in both short-term (3 to 4 yr testing at a location) and long-term (several years at several locations) crop performance trials. Several methods have been proposed to analyze GE interaction (Lin et al., 1986; Becker and Leon, 1988; Kang, 1990). Usually researchers ignore GE interaction encountered, especially in short-term trials, and base genotype selection solely on mean performance across environments. Only recently has attention been focused on usefully incorporating GE interaction into genotype selection in short-term trials (Kang and Pham, 1991; Kang, 1993; Magari and Kang, 1993).

An obstacle in adopting a method or selection criterion that simultaneously selects for yield and stability could be the misconception that there would be yield reduction when selection is based on yield and stability. Understanding that the main purpose of crop performance trials is to estimate or predict, using past performance data, genotype performance in future years, especially on growers' farms, may clarify this misconception. Especially when crossover type of GE interaction [i.e., one that causes genotype rank changes (Baker, 1990)] is present, mean yield of genotypes selected via a method that combines yield and stability would usually be lower than that of genotypes selected on the basis of yield alone (Kang et al., 1991; Bachireddy et al., 1992). However, the lower yield relates to past performance, and it would not necessarily translate into reduced yield on growers' farms, as pointed out later in this chapter.

Another way to clarify the misconception is by examining consequences to growers of researchers' committing Type I (rejecting the null hypothesis when it is true) and Type II errors (accepting the null hypothesis when it is false) relative to selection on the basis of yield alone (conventional method, CM) and that on the basis of yield and stability. Generally, Type II errors constitute the most serious risk for growers (Glaz and Dean, 1988; Johnson et al., 1992). Only for Kang's (1991) modified rank-sum (KMR) method can Type I and Type II error rates be determined for the stability component but not for the yield component. Therefore, a new statistic, designated as Yield-Stability statistic (YS), was proposed (Kang,

[1]Department of Agronomy, Louisiana Agricultural Experiment Station, Louisiana State University Agricultural Center, Baton Rouge, LA 70803-2110. Approved for publication by the director of the Louisiana Agricultural Experiment Station as manuscript no. 95-09-9050.

1993) and a computer program is available, free of charge, for calculating this statistic (Kang and Magari, 1995).

The stability component in YS is based on Shukla's (1972) stability-variance statistic (Sh-σ^2). Shukla (1972) partitioned GE interaction into components, one corresponding to each genotype, and termed each component as stability variance. Lin et al. (1986) classified Sh-σ^2 as Type 2 stability, meaning that it was a relative measure dependent on genotypes included in a particular test. Kang et al. (1987) reported on the relationship between Shukla's stability variance and Wricke's ecovalence (Wricke, 1962) and concluded that they were identical in ranking cultivars for stability (rank correlation coefficient=1.00). This measure should be acceptable and useful to breeders and agronomists, as it provides contribution of each genotype in a test to total GE interaction attributable to all genotypes.

In analyzing GE interactions, plant breeders often strive to grow all genotypes in all environments, thus producing balanced data. This, sometimes, is not possible, especially when a wide range of environments or long-term trials are considered. Also, the number of replications may not be equal for all genotypes owing to discarding of several experimental plots for various reasons. In such cases, plant breeders have to deal with unbalanced data that are more common in practice than considered in literature.

Searle (1987) classified unbalancedness into planned unbalanced data and missing observations. In studying GE interaction, both categories of unbalancedness may occur, but planned unbalancedness (a situation when, for different reasons, one does not have all genotypes in all environments) is more difficult to handle. Researchers have used different approaches for studying GE interaction in unbalanced data (Freeman, 1975; Pedersen et al., 1978; Zhang and Geng, 1986; Gauch and Zobel, 1990; Rameau and Denis, 1992). Usually environmental effects are considered as random and cultivar effects as fixed. Inference on random effects using least squares in the case of unbalanced data is not appropriate because information on variation among random effects is not incorporated (Searle, 1987). For this reason, mixed model equations (MME) are usually recommended (Henderson, 1975).

II. RELATIVE IMPORTANCE OF ERROR RATES

Growers would prefer to use a high-yielding cultivar that performs consistently from year to year. They may even be willing to sacrifice some yield if they are guaranteed, to some extent, that a cultivar would produce consistently from year to year (Kang et al., 1991). The guarantee that a cultivar would perform consistently would be in statistical terms, based on Type I and Type II error rates for a selection criterion that encompasses both yield and stability (Kang, 1993).

A. ERROR RATES FOR YIELD COMPARISONS

Let us consider the null hypothesis, H_0: $\mu_1 \geq \mu_0$, where μ_1 is mean yield of a genotype and μ_0 is overall mean yield of all genotypes. The alternative hypothesis is H_a: $\mu_1 < \mu_0$. If H_0 is rejected when it is true, a Type I error is committed, and the conclusion is that the yield of a given genotype is < overall mean yield. In this situation, the grower would miss using a superior genotype. This could mean an economic loss depending on alternative genotype chosen. On the other hand, if H_0 is accepted when it is false, a Type II error is committed, and the conclusion that the yield of a given genotype is $\geq$ overall mean yield could lead to recommendation of an inferior cultivar. The risk of concluding that an inferior genotype is superior is expressed as γ or Type III error rate. Under this scenario, the grower would use an inferior genotype and definitely suffer an economic loss. Thus, Type II errors would be more harmful to growers than Type I errors when selection is based on CM.

Table 1

Estimates of $\beta(1)$ (Type II error rate, one-tail) at several levels of $\alpha(1)$ (Type I error rate, one-tail) and δ (minimal detectable difference) for pairwise comparisons among overall yield means calculated from 16 environments.[a]

δ t ha[1]	$\alpha(1)$ level				
	0.001	0.01	0.05	0.10	0.25
0.2	0.993	0.956	0.845	0.743	0.515
0.4	0.967	0.857	0.649	0.505	0.278
0.5	0.940	0.777	0.527	0.382	0.181
0.6	0.890	0.674	0.401	0.271	0.111
0.8	0.723	0.425	0.189	0.108	0.032
1.0	0.486	0.206	0.066	0.030	0.006
1.2	0.251	0.074	0.016	0.006	0.001
1.4	0.097	0.019	0.003	0.001	0.000
1.6	0.027	0.003	0.000	0.000	0.000
1.8	0.005	0.000	0.000	0.000	0.000
2.0	0.001	0.000	0.000	0.000	0.000

[a] From Kang (1993).

The β values for several levels of $\alpha(1)$ [$\alpha(1)$ refers to one-tailed test] and δ (minimal detectable difference) are shown in Table 1. The comparisonwise Type II error rates (one-tailed β) were estimated using the following formula adapted from Zar (1981): $t_{\beta(1),v} \geq [\delta/(2EMS/n)^{\frac{1}{2}}] - t_{\alpha(1),v}$, where t is Student's t statistic, EMS=error mean square, and n=number of observations making up a genotype mean. Calculations are based on a corn (*Zea mays L.*) dataset of 16 hybrids grown in 16 environments (4 years, 4 locations and 4 replications per location) (Kang, 1993). These β estimates are based on an EMS of 1.741 with 686 df and n=64 (16 environments and four replications). It is obvious that for a specific δ, Type II error risk would be reduced by choosing a higher level of α. A higher level of α would not be as harmful to growers as a higher level of β. At a realistic δ value of about 0.5 t ha^{-1} ≈ 1LSD at α(2)=0.1 [α(2)=0.1 corresponds to α(1)=0.05 in Table 1], Type II error rate would be 0.527 and power of the test (1-β) 0.473. Carmer (1976) presented arguments for using α(2) in the range of 0.20 to 0.40 for LSD for making pairwise multiple comparisons between observed means from crop performance trials. Carmer's suggestion is equivalent to setting α(1) in the range of 0.10 and 0.20 when one-tailed hypotheses are tested. If α(1)=0.25 and δ were in the range of 1.0 to 1.4, Type II error rate would be almost zero.

B. ERROR RATES FOR STABILITY VARIANCE

This part of the discussion deals with hypothesis testing relative to GE interaction or stability-variance statistic. Here comparisons are made between GE interaction means. The null hypothesis would be that the difference between mean yield of two genotypes is the same for all environments. This is the same as testing that Sh-σ^2=0.

If H_0 is rejected when it is true (Type I error), the erroneous conclusion would be that genotype performance was inconsistent or unstable across environments. In this situation, the consequence of a Type I error would be that the grower might miss using a stable genotype and might or might not suffer an economic loss, depending on the alternative genotype chosen (Kang, 1993).

On the other hand, if H_0 is accepted when it is false (Type II error), the erroneous conclusion would be that genotype performance was stable or consistent across test environments. Consequently, an unstable genotype would not be penalized for instability. The consequence of this decision could be disastrous for growers, as they would expect high yield from the genotype on their farms, but actually the genotype could perform poorly. This implies that a higher penalty for unstable performance, i.e., a greater emphasis on stable performance, would not necessarily be harmful to growers (Kang, 1993).

The β values for three levels of α (0.01, 0.05, and 0.1) and different levels of δ (minimal detectable difference between GE interaction means) are shown in Table 2. Here, n=4 (number of replications) in the formula for estimating β. An LSD$_{\alpha=0.05}$ of 1.6 to 1.8 would be needed to declare differences between GE interaction means as significant. Therefore, at δ=1.8, to achieve power≈0.61, α(2)=0.1, i.e., α(1)=0.05 would be desirable for significance testing. If one desires to reduce Type II error rate to 0.26 (power=0.74), α(2) level would need to be increased to 0.20 (not shown in Table 2) to label a genotype as unstable. Researchers' placing more emphasis on stability of performance than currently done in the selection process would benefit growers. Growers would have a greater risk of suffering a real yield loss when a cultivar is chosen only on the basis of mean yield than when cultivar selection is based on both yield and performance stability.

Table 2

Estimates of $\beta(1)$ (Type II error rate, one-tail) at three levels of $\alpha(2)$ (Type I error rate, two-tail) and six levels of δ (minimal detectable difference) for pairwise comparisons among genotype × year interaction yield means calculated from four replications.[a]

δ t ha^{-1}	$\alpha(2)$ level		
	0.01	0.05	0.10
0.5	0.980	0.924	0.865
1.2	0.902	0.753	0.640
1.4	0.860	0.677	0.557
1.6	0.807	0.600	0.474
1.8	0.743	0.513	0.390
2.0	0.666	0.429	0.309

[a] From Kang (1993)

III. STABILITY VARIANCE FOR UNBALANCED DATA

Consider a simple case with "t" genotypes, "b" environments, and "n" replications. The general linear model for this analysis is:

$$Y_{ijk} = \mu + \alpha_i + \beta_k + \gamma_{ik} + \epsilon_{ijk}, \qquad (1)$$

where Y_{ijk} is a single observation, μ is the grand mean, α_i is an environmental effect, β_k is a cultivar effect, γ_{ik} is a GE interaction, and ϵ_{ijk} is experimental error. Genotypes are considered as fixed and environments as random. GE interaction effect will be considered as random because the interaction between a random factor and a fixed factor is random. In the above equation the response variable is assumed to be normally distributed with mean of $(\mu+\beta_k)$ and $(\sigma^2_e + \sigma^2_{ge} + \sigma^2_w)$, where σ^2_e is environmental variance, σ^2_{ge} is GE interaction variance, and σ^2_w is experimental error variance.

For calculating stability variances, Shukla (1972) partitioned GE interaction into "t" genotype x environment components. These variances are unbiased and have minimum variance among all possible quadratic unbiased estimators (Shukla, 1972). Sh-σ^2 can be negative because they are calculated as the differences of two statistically dependent sums of squares. Although negative variance estimates are not uncommon, this is a negative feature of this approach. Computation of Sh-σ^2 is impossible in the case of unbalanced data, but genotype x environment variance components for each genotype ($\sigma^2_{ge(k)}$) can be estimated using the restricted maximum likelihood (REML) approach.

For stability analysis, environmental and error effects of (1) are assumed to be independent identically normally distributed with zero mean and σ^2_e and σ^2_w variance, respectively, while GE interaction effects are assumed to be independent normally distributed with zero mean and $\sigma^2_{ge(k)}$ variances. Equation (1) can be modifed for experiments conducted with randomized complete block design or other designs without changing the model for GE interaction. This equation can be solved using Henderson's (1975) MME and REML effects calculated. Environmental effects are assumed not to be correlated, but a relationship matrix based on environmental similarities or other criteria may be added in the model.

To demonstrate this procedure, an unbalanced subset of Louisiana corn hybrid yield trials of 11 hybrids grown at 4 locations (Alexandria, Baton Rouge, Bossier City, St. Joseph) for 4 years (1985 - 1988) were analyzed. All hybrids were not included in all years or/and locations. Experiments were conducted using a randomized complete block design with 4 replications at a location. A few experimental plots were discarded during the course of the experiments for various reasons.

For calculating phenotypic stability variances the Expectation Maximization (EM)-type algorithm was used (Patterson and Thompson, 1971). All variances were computed by iterating on MME and tested for H_0: $\sigma^2_{ge(k)} = 0$. For the mixed model (1), and unbalanced data, the variances of random effects affect the estimation of fixed effects (hybrid mean yields) as well (Searle, 1987). For this reason, MME is used, which gives best linear unbiased estimators (BLUE) for hybrids (fixed effects) and best linear unbiased predictors (BLUP) for random effects (Henderson, 1975). Hybrid yields, presented in the form of BLUE (Table 3), were the solutions for hybrid means from the MME equations. Hybrid 6 was the highest yielding, although not statistically different ($p > 0.23$) from hybrid 9 and hybrid 3. For stability interpretation, $\sigma^2_{ge(k)}$ has the same properties as Sh-σ^2. It can be regarded as a Type 2 stability statistic (Lin et al., 1986) or as a statistic related to agronomic concept of stability (Becker, 1981). High values of $\sigma^2_{ge(k)}$ indicate that the genotypes are not stable or that they interact with the environments. The values of $\sigma^2_{ge(k)}$ (Table 3) were statistically not different from zero for both 0.05 and 0.01 probability levels except $\sigma^2_{ge(8)} = 1.0962$ that was statistically different from zero at 0.05 probability level. Kang (1993) also reported hybrid 8 (SB1827) to be unstable with a probability > 99%.

For calculating stability variances, we used REML because it can be effectively adapted to unbalanced data (Searle, 1987). Although the procedure requires that data have a multivariate normal distribution (Patterson and Thompson, 1971), Searle et al. (1992) indicated that this assumption in many circumstances is not seriously incorrect. The method is translation invariant and

REML estimators of $\sigma^2_{ge(k)}$ will always be positive if positive starting values are used (Harville, 1977). Piepho (1993) demonstrated by Monte Carlo simulation that for a given sufficient number of environments, maximum likelihood estimates (MLE) are superior to Sh-σ^2 for ranking genotypes. REML estimates are generally preferred to MLE because the degrees of freedom for fixed effects are considered for calculating error.

Table 3
Best linear unbiased estimators (BLUE) and stability variances for considered corn hybrids form the Louisiana Corn Hybrid Yield Trials.

Hybrid	BLUE ($\mu + \beta_k$)	Stability variance ($\sigma^2_{ge(k)}$)
	t ha^{-1}	t ha^{-1}
1	6.8953	0.2878
2	6.3860	0.2363
3	7.1219	0.2814
4	6.5325	0.3081
5	6.6149	0.6851
6	7.3808	0.2886
7	6.5333	0.7613
8	6.8055	1.0962*
9	7.2765	0.4017
10	6.1882	0.4214
11	5.6819	0.4631

* Statistically different from zero at the 0.05 probability level.

Calculation of REML stability variances for unbalanced data allows one to obtain a reliable estimate of stability parameters as well as to overcome the difficulties for manipulating unbalanced data.

IV. SIMULTANEOUS SELECTION FOR YIELD AND STABILITY

Several methods for simultaneous selection for yield and stability are discussed by Kang and Pham (1991). Simultaneous selection for yield and stability reduces the probability of committing Type II error. The combined rate of committing a Type II error in this case will be the product of β for comparisons of overall yield mean (Table 1) and β for comparisons of GE interaction means (Table 2). For example, for $\delta=0.5$ for overall mean yield comparisons and $\delta=1.6$ for GE interaction means, the combined probability of Type II error would be 0.627, 0.316, and 0.181 at $\alpha=0.01$, 0.05, and 0.10, respectively. The combined probability of committing a Type II error for both yield and stability would be negligible (0.030 x 0.309=0.01) if δ=1.0 at α=0.10 for mean yield comparisons and δ=2.0 at α=0.1 for GE interaction means.

Table 4
Example showing the operation of a new method (calculation of yield-stability statistic, YS) for simultaneous selection for yield and stability in corn performance trials.[a]

Genotype	Yield	Rank (Y')	Adjust. to Y'[†]	Adjusted (Y)	Stability variance (Sh-σ^2)	Stability rating (S)	YS
		t ha^{-1}					
M8172	7.58	16	+2	18	3.8**	-8	10 ✓§
3147	7.41	15	+2	17	2.8‡	-2	15 ✓
SB1860	7.31	14	+2	16	3.2*	-4	12 ✓
3165	7.11	13	+1	14	6.2**	-8	6 ✓
G-4673A	6.94	12	+1	13	2.2	0	13 ✓
Coker 21	6.88	11	+1	12	2.5	0	12 ✓
SB1827	6.84	10	+1	11	5.7**	-8	3
M8150	6.61	9	-1	8	5.9**	-8	0
G-4765	6.59	8	-1	7	1.8	0	7 ✓
G-4733	6.47	7	-1	6	2.3	0	6 ✓
3389	6.46	6	-1	5	1.2	0	5
3320	6.42	5	-1	4	3.1*	-4	0
SB1802	6.36	4	-1	3	2.6	0	3
G-4522	6.26	3	-1	2	2.6	0	2
8951	6.15	2	-2	0	3.4*	-4	-4
8990	5.74	1	-2	-1	15.1**	-8	-9
Mean	6.69						5.1
LSD$_{0.05}$	0.46						

[a] From Kang (1993)
[†]Adjustment of +1 for mean yield > overall mean yield (OMY), +2 for mean yield ≥ 1LSD above OMY, +3 for mean yield ≥ 2LSD above OMY, -1 for mean yield < OMY, -2 for mean yield ≤ 1LSD below OMY, and -3 for mean yield ≤ 2LSD below OMY.
‡, *, ** Denote significance at the 0.1, 0.05, and 0.01 levels of probability, respectively and also indicate that genotype was judged to be unstable.
§✓ indicates genotypes selected on the basis of YS.

To integrate yield and stability in one index, Kang (1988) proposed the rank-sum method. Based on this method, ranks are assigned for mean yield, with the genotype with the highest yield receiving the rank of 1. Similarly, ranks are assigned for Sh-σ^2, with the lowest estimated value receiving the rank of 1. The two ranks for each genotype are summed and the lowest rank-sum is regarded as most desirable. This method assumes equal weight for yield and stability. However, plant breeders may prefer to assign more weight to either yield or stability.

Statistical significance of $Sh-\sigma^2$ is considered when the modified rank-sum (KMR) is calculated (Kang, 1991). **Ratings of -8, -4, and -2 are assigned** to $Sh-\sigma^2$ according to statistical significance at 0.01, 0.05, and 0.1, respectively, and 0 when $Sh-\sigma^2$ is not significant.

Alternatively, to consider statistical significance for both yield and stability, Kang (1993) proposed the yield-stability statistic (YS). Illustration of the calculation is based on the corn dataset of 16 hybrids and 16 environments used in section II. The calculations of all these statistics are based on $Sh-\sigma^2$ as an indicator of stability. **Considering its equivalency with $Sh-\sigma^2$** (Kang et al., 1987), ecovalence also may be used in calculating sum of ranks but not KMR and/or YS because ecovalence does not provide a test of significance. REML variance component estimates of GE interaction for each genotype ($\sigma^2_{ge(k)}$) may be useful for calculating simultaneous selection statistics, especially when the data are not balanced.

Following the detection of a significant GE interaction, steps to calculate the YS statistic for the *i*th genotype are listed below and relate to Table 4:

Determine contribution of each genotype to GE interaction by calculating $Sh-\sigma^2$ (Shukla, 1972) .

Arrange genotypes from highest to lowest yield and assign yield rank (Y'), with the lowest yielding genotype receiving the rank of 1.

Calculate $LSD_{\alpha(2)}$ for mean yield comparisons [$\alpha(2)$ refers to a two-tailed test] as [$t_{\alpha(2)},\nu$ $(2EMS/(b \times n))^{1/2}$], where EMS=error mean square, ν=df associated with EMS, b=number of environments, and n=number of replications.

Adjust Y' according to LSD, and determine adjusted yield rank (Y) [See Table 4].

Assign respective $Sh-\sigma^2$ values to genotypes and determine whether or not $Sh-\sigma^2$ is significant at $\alpha(2)$=0.10, 0.05, 0.01, using an approximate F test with (b-1),ν degrees of freedom.

Assign stability rating (S) as follows: -8, -4, and -2 for $Sh-\sigma^2$ significant at α=0.01, 0.05, and 0.1, respectively; and 0 for nonsignificant $Sh-\sigma^2$. The stability ratings of -8, -4, and -2 were chosen because they changed genotype ranks from those based on yield alone (Y').

Sum adjusted yield rank (Y) and stability rating (S) for each genotype to compute YS statistic.

Calculate mean YS as $\sum YS/t$, where t=number of genotypes. Select genotypes with YS > mean YS.

If more emphasis on stability is desired, ranks may be assigned to genotypes in accordance with stability variances and adjusted for stability ratings. The adjusted YS' in this case is calculated by adding up adjusted yield rank and adjusted stability rank.

Relationships between several statistics for simultaneous selection for yield and stability are discussed by Kang and Pham (1991). A summary of their relationships with yield is given in Table 5.

Table 5
Rank-correlation coefficients between yield and rank-sum (RS), S_i^3 and S_i^6, and superiority measure (P_i) for five yield datasets.[a]

Dataset	Number of entries	RS	S_i^3	S_i^6	P_i
1	14	-0.26	-0.28	-0.62*	-0.97**
2	23	-0.72**	-0.79**	-0.92**	-0.99**
3	16	-0.68**	-0.37	-0.64**	-0.90**
4	18	-0.73**	-0.57*	-0.87**	-0.96**
5	12	-0.65*	-0.34	-0.77**	-1.00

[a] Adapted from Kang and Pham (1991)
*, ** Significant at the 0.05 and 0.01 probability levels, respectively.

RS selects more for stability than S_i^6, and P_i, which would be safer for farmers because of the lower rate of committing a Type II error when stability is emphasized. In Table 6 are given relationships between yield and statistics that are modified to consider significance levels for stability or/and yield based on corn yield trials data (Magari and Kang, 1993). KMR depends on the number of entries per trial. The fewer the entries per trial, the greater the emphasis on stability. YS and YS' do not depend on the number of entries, but YS' obviously selects more for stability. According to the breeding objectives, base germplasm involved, and the range of target environments, either YS or YS' can be used. If adaptation for a specific or a narrow range of environments is desired, YS is more useful. On the other hand, YS' may give better results when selecting for a wider range of adapted genotypes.

Table 6

Rank-correlations coefficients between yield and rank-sum (RS), Kang's modified rank-sum (KMR), YS and YS' for three sets of corn yield trials.

Set	Number of entries	RS	KMR	YS	YS'
1	9	-0.62*	0.54	0.84**	0.57
2	12	-0.45	0.69**	0.78**	0.18
3	16	-0.69**	0.76**	0.84**	0.55

*, ** Significant at the 0.05 and 0.01 probability levels, respectively.

V. CONTRIBUTION OF ENVIRONMENTAL VARIABLES TO GENOTYPE-BY-ENVIRONMENT INTERACTION

Yield stability or GE interaction for yield is very complex. There is evidence that yield stability is genetically controlled, but the amount of genetic variation is related to the statistics used for stability evaluation as well as to the range of environments. Yield stability depends on yield components and other plant characteristics, such as resistance to pests, tolerance to stress factors and environmental variables. All of these contribute in different ways to GE interaction. If relative importance of these components to GE interaction is evaluated and the limiting one(s) is(are) determined, then its(their) manipulation might increase stability. For example, a certain genotype yields less in some environments because it is susceptible to a given disease (low stability), then breeding for resistance to the disease will reduce losses in those environments and increase its stability. The same thing applies when, for example, low temperature is the most important variable that contributes to GE interaction and low stability for a certain genotype. Increasing tolerance to low temperatures would result in greater stability for that genotype.

Contributions of different environmental variables to GE interaction have been reported in several articles (Saeed and Francis, 1984; Kang and Gorman, 1989; Kang et al., 1989; Gorman et al., 1989; Rameau and Denis, 1992). Based on corn hybrid yield trials conducted in multi-years and multi-locations in Louisiana, Kang and Gorman (1989) removed from GE interaction heterogeneity due to maximum and minimum temperatures, rainfall, and relative humidity. In the linear model used in this case, GE interaction is explained in terms of the covariate used, as shown by Shukla (1972):

$$Y_{ijk} = \mu + \alpha_i + \theta_{ij} + \beta_k + b_k z_i + \epsilon_{ijk}, \qquad (2)$$

where θ_{ij} is blocks within environment effect, b_k is a regression coefficient of the kth genotype's yield in different environments and z_i is an environmental covariate. Much of the GE interaction was removed by the environmental index (difference between mean of all genotypes in ith environment and grand mean) although not statistically significant. This accounted for 9.61% of total GE interaction. Heterogeneity removed from using other covariates was low (Table 7). Rainfall during the growing season removed 1.4% of the total GE interaction, whereas the amounts removed by minimum and maximum temperature and relative humidity were negligible. Considering the complexity of GE interaction, the relative contribution of a single environmental variable might be very small in comparison with the total number of variables affecting it. Similar results have been reported for sorghum (Gorman et al., 1989), soybean (Kang et al., 1989), and other corn trials (Magari and Kang, 1993).

Table 7
Heterogeneity removed from genotype x environment interaction via four environmental covariates in corn yield trials.[a]

		Mean squares		
Source	df	Environmental index	Minimum temperature	Rainfall
GE interaction	176	2.8**	2.8**	2.8**
Heterogeneity	16	2.8	0.01	0.4
Residual	160	2.8**	3.1**	3.0**
Pooled error	561	1.8	1.8	1.8

[a] Adapted from Kang and Gorman (1989)
** Significant at 0.01 probability level.

When a number of environmental variables are considered, the combination of two or more variables would remove more heterogeneity from GE interaction than individual variables do. Methods developed in Chapter 2 (van Eeuwijk et al., 1995) may be helpful for this purpose.

REFERENCES

Bachireddy, V.R., R. Payne, Jr., K.L. Chin, and M.S. Kang. 1992. Conventional selection versus methods that use genotype x environment interaction in sweet corn trials. *HortScience* 27(5):436-438.
Baker, R.J. 1990. Crossover genotype-environmental interaction in spring wheat. p. 42-51. In M. S. Kang (ed.) *Genotype-by-environment interaction and plant breeding.* Louisiana State University Agricultural Center, Baton Rouge, Louisiana, USA.
Becker, H.C. 1981. Correlations among some statistical measures of phenotypic stability. *Euphytica* 30:835-840.
Becker, H.C., and J. Leon. 1988. Stability analysis in plant breeding. *Plant Breeding* 101:1-23.

Carmer, S.G. 1976. Optimal significance levels for application of the least significant difference in crop performance trials. *Crop Science* 16:95-99.

Freeman, G.H. 1975. Analysis of interactions in incomplete two-way tables. *Applied Statistics* 24:46-55.

Gauch, H.G., and R.W. Zobel. 1990. Imputing missing yield trial data. *Theoretical and Applied Genetics* 70:753-761.

Glaz, B., and J.L. Dean. 1988. Statistical error rates and their implications in sugarcane clone trials. *Agronomy Journal* 80:560-562.

Gorman D.P., M.S. Kang, and M.R. Milam. 1989. Contribution of weather variables to genotype x environment interaction in grain sorghum. *Plant Breeding* 103:299-303.

Harville, D.A. 1977. Maximum-likelihood approaches to variance component estimation and to related problems. *Journal of the American Statistical Association* 72:320-340.

Henderson, C.R. 1975. Best linear unbiased estimation and prediction under a selection model. *Biometrics* 31:423-447.

Johnson, J.J., J.R. Alldredge, S.E. Ullrich, and O. Dangi. 1992. Replacement of replications with additional locations for grain sorghum cultivar evaluation. *Crop Science* 32:43-46.

Kang, M.S. 1988. A rank-sum method for selecting high-yielding, stable corn genotypes. *Cereal Research Communications* 16:113-115.

Kang, M.S. (ed) 1990. *Genotype by environment interaction and plant breeding*. Louisiana State University Agricultural Center, Baton Rouge, Louisiana, USA.

Kang, M.S. 1991. Modified rank-sum method for selecting high yielding, stable crop genotypes. *Cereal Research Communications* 19:361-364.

Kang, M.S. 1993. Simultaneous selection for yield and stability in crop performance trials: Consequences for growers. *Agronomy Journal* 85:754-757.

Kang, M.S., and D.P. Gorman. 1989. Genotype x environment interaction in maize. *Agronomy Journal* 81:662-664.

Kang, M.S., D.P. Gorman, and H.N. Pham. 1991. Application of a stability statistic to international maize yield trials. *Theoretical and Applied Genetics* 81:162-165.

Kang, M.S., B.G. Harville, and D.P. Gorman. 1989. Contribution of weather variables to genotype x environment interaction in soybean. *Field Crops Research* 21:297-300.

Kang, M.S., and R. Magari. 1995. STABLE: Basic program for calculating yield-stability statistics. *Agronomy Journal* 87:276-277.

Kang, M.S., and J.D. Miller. 1984. Genotype x environment interactions for cane and sugar yield and their implications in sugarcane breeding. *Crop Science* 24:435-440.

Kang, M.S., J.D. Miller, and L.L. Darrah. 1987. A note on relationship between stability variance and ecovalence. *Journal of Heredity* 78:107.

Kang, M.S., and H.N. Pham. 1991. Simultaneous selection for high yielding and stable crop genotypes. *Agronomy Journal* 83:161-165.

Lin, C.S., M.R. Binns, and L.P. Lefkovitch. 1986. Stability analysis: Where do we stand? *Crop Science* 26:894-900.

Magari, R., and M.S. Kang. 1993. Genotype selection via a new yield-stability statistic in maize yield trials. *Euphytica* 70:105-111.

Patterson, H.D., and R. Thompson. 1971. Recovery of inter-block information when block sizes are unequal. *Biometrika* 58:545-554.

Pedersen, A.R., E.H. Everson, and J.E. Grafius. 1978. The gene pool concept as basis for cultivar selection and recommendation. *Crop Science* 18:883-886.

Piepho, H.P. 1993. Use of maximum likelihood method in the analysis of phenotypic stability. *Biometrical Journal* 35:815-822.

Rameau, C., and J.-B. Denis. 1992. Characterization of environments in long-term multi-site trials in asparagus, through yield of standard varieties and use of environmental covariates. *Plant Breeding* 109:183-191.

Saeed, M., and C.A. Francis. 1984. Association of weather variables with genotype x environment interaction in grain sorghum. *Crop Science* 24:13-16.

Searle, S.R. 1987. Linear models for unbalanced data. John Wiley & Sons, New York, New York.

14

Searle, S.R., G. Casella, and C.E. McCulloch. 1992. *Variance components.* John Wiley & Sons, New York, New York.

Shukla, G.K. 1972. Some statistical aspects of partitioning genotype-environmental components of variability. *Heredity* 29:237-245.

van Eeuwijk, F.A., J.-B. Denis, and M.S. Kang. 1995. Incorporating additional information on genotypes and environments in models for two-way genotype-by-environment tables. Chapter 2, this volume.

Wricke, G. 1962. Uber eine Methode zur Erfassung der okologischen Streubreite in Feldversuchen. *Z. Pflanzenzucht* 47:92-96.

Zar, J. H. 1981. Power of statistical testing: Hypotheses about means. *American Laboratory (Fairfield, Connecticut.)* 12(b):102-107.

Zhang, Q., and S. Geng. 1986. A method of estimating varietal stability for long-term trials. *Theoretical and Applied Genetics* 71:810-814.

Chapter 2

INCORPORATING ADDITIONAL INFORMATION ON GENOTYPES AND ENVIRONMENTS IN MODELS FOR TWO-WAY GENOTYPE BY ENVIRONMENT TABLES

F.A. van Eeuwijk[1], J.-B. Denis[2] and M.S. Kang[3].

I. INTRODUCTION

For genotypes and environments making up the factor levels in two-way genotype by environment tables, often substantial additional information is either available or easily obtainable. For genotypes, additional quantitative information may be present from laboratory and greenhouse tests bearing on the physiology of the plants, while additional qualitative information may be present from various categorizations, like those on basis of genealogy. For environments, quantitative information can consist in edaphic and climatological data, whereas a minimum in qualitative information consists in year and location groupings. The additional information on genotypes and environments includes more than direct measurements. More remotely, statistics calculated from previous or comparable trials, concerning the variable under study as well as other variables, may be used.

For the additional information, more or less clear-cut hypotheses may be entertained regarding its relation with the structure of the genotype by environment interaction in the variable to be analyzed. To test these hypotheses statistically, models are necessary that allow the incorporation of that information. In plant breeding, the emphasis has long been on models not offering this opportunity. We feel that for a broad spectrum of hypotheses, between exploratory and inferential, it is imperative to pay more attention to regression based statistical methods. As a consequence, more parsimonious models may be built, providing more accurate tools to decide and act on. Similar ideas have been expressed by Hinkelmann (1974), Denis and Vincourt (1982), Tai (1990), van Eeuwijk (1993), and Federer and Scully (1993).

The main point of this paper is to give a survey of the most important regression based models for two-way tables and to illustrate the interpretation of their interaction parameters. Three families of models will be presented.

[1] DLO - Centre for Plant Breeding & Reprod. Research, P.O. Box 16, 6700 AA Wageningen, The Netherlands.
[2] Laboratoire de Biométrie, I.N.R.A., route de Saint-Cyr, F-78026 Versailles, France.
[3] Dep. of Agronomy, 104 Madison Sturgis Hall, Louisiana State Univ., Baton Rouge, Louisiana 70803-2110, United States of America.

After some details on notation (Section II), fixed factorial regression is introduced first (Section III), and an account is given of how quantitative as well as qualitative covariates may be included. Secondly, reduced rank factorial regression, based on bilinear descriptions of the interaction, will be dealt with (Section IV). Lastly, mixed factorial regression is presented, in which either the genotypes or the environments are supposed to represent a random sample from a population (Section V). Estimation and testing, together with software availability, are briefly discussed in Section VI. Considerations playing a role in model choice form the subject of Section VII. Finally, Section VIII presents alternatives and further extensions to the models presented in the sections III, IV and V.

II. DATA AND NOTATION

A. THE DATA TABLE TO BE INTERPRETED

The topic of the paper will be restricted to the interpretation of the joint effects of the factors 'genotype' and 'environment' on a continuous variable Y. The subscript i $(1,...,I)$ will be used to indicate the genotype, and j $(1,...,J)$ to indicate the environment. Typically, Y_{ij} represents yield, but many other quantitative variables are equally substitutable. Averaging over replicates makes Y_{ij} to comply closer with distributional assumptions.

For a number of statistical tests with regard to the structure of interaction an estimate of error is required. This estimate may be obtained from the mean intra-block error, or from the part of the interaction not modeled. We will not consider the question of which estimate to use, but assume that a non-controversial estimate of error is available.

B. ADDITIONAL INFORMATION

Besides the data, Y_{ij}, additional information is assumed to be present at the levels of the genotypes and/or environments. The value of the k-th covariate $(k = 1,...,K)$ for the i-th genotype will be denoted by x_{ik}. Covariates are either quantitative, as in multiple regression, or qualitative, as in analysis of variance. Examples of quantitative genotypic covariates are physiological characterizations, such as earliness, and disease susceptibilities. Examples of qualitative genotypic covariates are genetic and geographic origin. When only one covariate is considered, the subscript in question is dropped, i.e., x_i instead of x_{ij}. Quantitative covariates are throughout supposed to be centered.

Similarly, we will denote the h-th covariate $(h = 1,...,H)$ for the j-th environment by z_{jh}. For environments, we can think of humidity and soil pH as quantitative covariates, and location (region, country) and cultivation regimes as qualitative covariates.

To indicate that the covariate values are considered known, they are written with lower case letters. This does not mean that the corresponding factor may not be random, but only that the analyses are done conditionally on the values of the covariates.

A popular environmental covariate is y_j, the mean over genotypes (Finlay and Wilkinson, 1963). y_j can be treated as any other covariate (Mandel, 1961), because of its statistical independence of the interaction estimates. Also the genotypic main effect, y_i, can be used in that way. A further possibility is to use the product of y_j and y_i, as a covariate for the entire table. A simple extension uses covariates of the type o_i and o_j, i.e., main effects of another variable on the same set of genotypes and environments. For even further extensions, see Baril (1992). Though mostly covariates are used in a linear fashion only, higher order terms as squares, cubes, and cross-products can be considered equally well.

Table 1

Hypothetical yields for 17 genotypes (rows) in 12 environments (columns).

	E1	E2	E3	E4	E5	E6	E7	E8	E9	E10	E11	E12
G1	10	6	6	7	9	7	2	7	8	6	5	11
G2	9	6	5	7	8	8	2	6	7	7	5	9
G3	11	7	6	7	9	6	4	6	7	7	6	8
G4	8	7	4	6	8	8	5	7	7	9	5	7
G5	9	7	6	7	9	6	3	6	8	9	6	9
G6	8	8	4	7	8	7	3	8	7	8	6	8
G7	11	8	5	7	9	7	4	8	9	9	7	9
G8	10	8	5	8	11	9	4	9	9	11	8	10
G9	8	8	6	7	9	7	3	7	9	9	7	9
G10	8	6	6	7	9	6	3	7	6	10	6	10
G11	8	7	4	5	9	7	3	5	7	9	4	9
G12	8	6	4	6	8	7	3	8	6	8	5	9
G13	8	8	4	6	8	9	4	7	7	8	5	8
G14	9	6	6	6	10	7	3	7	8	12	5	9
G15	9	7	5	9	11	9	5	8	6	10	4	8
G16	7	7	4	6	8	6	4	6	7	8	6	7
G17	8	6	3	7	9	5	3	6	6	8	4	8

In some models, pseudo covariates are estimated. When they are defined as linear combinations of measured covariates, they will be designated as synthetic covariates. When they are only subject to statistical/numerical construction rules, they will be called artificial covariates.

C. DESCRIPTION OF THE MODELS

The models presented below consist of sums of model terms. Terms are related to the expectation or the variance. Fixed parameters are represented by lower case Greek letters, random parameters and variates by standard upper case letters, and observations on (co)variates by standard lower case letters. The error term is written E_{ij}. Unless stated otherwise, E_{ij}'s are assumed to have zero mean, constant variance, and to be uncorrelated. For fixed effects models, the decomposition of the degrees of freedom (parametric dimension) corresponding to the different model terms is displayed via recapitulative tables. These tables

Table 2

Genotypic covariates. Rstc represents resistance, a centred measurement. The covariates $\rho_{(2)}$, $\rho_{(12)}$ and $\xi_{(16)}$ are artificial covariates corresponding to models (2), (12) and (16).

	Rstc	$\rho_{(2)}$	$\rho_{(12)}$	$\xi_{(16)}$
G1	-41	-0.0082	0.5525	0.5820
G2	-13	-0.0038	0.2614	0.2320
G3	-37	-0.0053	0.3335	0.3073
G4	34	0.0041	-0.2899	-0.3620
G5	-5	-0.0023	0.1635	0.1991
G6	25	0.0002	-0.0297	-0.0792
G7	-32	-0.0025	0.1780	0.1522
G8	-7	0.0017	-0.1037	-0.1038
G9	-24	-0.0024	0.1538	0.1169
G10	2	0.0014	-0.0795	0.0564
G11	16	0.0010	-0.0347	-0.0748
G12	8	0.0011	-0.0737	-0.0438
G13	-13	0.0012	-0.0900	-0.2335
G14	11	0.0043	-0.2351	-0.1182
G15	39	0.0069	-0.4850	-0.4306
G16	14	0.0008	-0.0840	-0.1441
G17	23	0.0019	-0.1373	-0.0558

are two-dimensional depictions showing the composition of the models, in which each model term corresponds with a zone in the table, and where the area of this zone is proportional to the associated degrees of freedom (Denis, 1991).

D. EXAMPLE DATA SET

To enhance understanding, for some of the presented models a numerical example will be used. The data set used is a modified and rounded version of the data used by Kang and Gorman (1989), including yield figures for 17 genotypes in 12 environments (Table 1). For each genotype, an associated fictitious resistance measure, Rstc, is available (Table 2). The 12 environments are characterized by four climatological variables (Table 3). An independent estimate for the error variance is also present.

Table 3

Environmental covariates. Maximum temperature (MaxT), minimum temperature (MinT), Rain, and relative humidity (RH) are measurements that were averaged over the growing season. The covariates $\tau_{(3)}$ and $\zeta_{(16)}$ are artificial covariates corresponding to models (3) and (16).

	MaxT	MinT	Rain	RH	$\tau_{(3)}$	$\zeta_{(16)}$
E1	5	-3	-119	-63	-0.0228	0.3628
E2	11	7	-39	-10	0.0028	-0.1212
E3	15	1	-121	61	-0.0164	0.3370
E4	16	12	41	65	0.0055	0.0145
E5	-2	3	21	20	0.0092	-0.0839
E6	-10	-10	41	18	0.0112	-0.3009
E7	0	14	111	13	0.0202	-0.3695
E8	-8	1	91	4	0.0083	-0.1516
E9	-7	-5	-99	-37	-0.0174	0.2259
E10	-18	-18	291	-6	0.0308	-0.4894
E11	13	13	-119	-24	-0.0152	0.1739
E12	-15	-15	-99	-41	-0.0160	0.4025

III. FIXED FACTORIAL REGRESSION

A general formal treatment of the models presented in this section is given by Denis (1980, 1988).

A. THE ADDITIVE MODEL AS BASE LINE

It is common to define interaction in two-way tables relative to the two-way additive model,

$$Y_{ij} = \mu + \alpha_i + \beta_j + E_{ij}. \tag{1}$$

The additive model provides a first, rough approximation to the data. Analysis of variance (ANOVA) on our example data showed that 82% of the total sum

Table 4

Two-way ANOVA, and interaction as described by various models. Numbers for models correspond to numbers for model formulations in text.

Model	Source	Df	SS	MS
	Genotypic main effect	16	67.3	4.208
	Environmental main effect	11	550.6	50.053
	Interaction (non-additivity)	176	136.7	0.853
(6)	Rstc.v.MaxT	1	4.2	4.185
(6)	Rstc.v.MinT	1	0.05	0.049
(6)	Rstc.v.Rain	1	28.0	27.954
(6)	Rstc.v.RH	1	4.9	4.913
(8)	Rstc.$(\Sigma v_{1h}.z_{jh})$	4	29.2	7.294
(3)	Rstc.τ_j	11	31.0	2.816
(2)	ρ_i.Rain	16	38.3	2.395
(12)	$\rho_i.(\Sigma\lambda_h.z_{jh})$	19	38.8	2.042
(16)	$\xi_i\zeta_j$	26	42.6	1.640
	Error	561	252.4	0.450

of squares could be explained by only 13% of the degrees of freedom (Table 4). Nevertheless, interaction was highly significant and could not be omitted.

For many purposes it is useful to express (1) as a double regression model with the constant covariates $1_i=1_j=1$:

$$Y_{ij}=\mu+\alpha_i 1_j+1_i\beta_j+E_{ij}. \qquad (1')$$

The main effects (α_i,β_j) thus are the regression coefficients for these non-informative constant covariates. The structure of the model is displayed in Table 5.

Table 5
Recapitulative table associated with model (1).

	1	
1	μ 1	β_j J-1
	α_i I-1	(I-1).(J-1)

B. INCLUDING ONE QUANTITATIVE ENVIRONMENTAL COVARIATE

Perhaps the simplest way of introducing a covariate associated with the environments is to write the interaction as a regression on this covariate with the coefficient depending on the genotype. Early applications of this type of model include Knight (1970), and Freeman and Perkins (1971). More recent applications are Fakorede and Opeke (1986), and McGraw et al. (1986). The model can be written as

$$Y_{ij} = \mu + \alpha_i + \beta_j + \rho_i z_j + E_{ij}. \tag{2}$$

For illustration we choose the rainfall data from Table 3 as environmental covariate. The estimates for the genotypic regression coefficients, ρ_i, are given in Table 2. These coefficients can be interpreted as underlying a differential

Table 6
Recapitulative table associated with model (2).

	1	z_j	
1	μ 1		β_j J-1
	α_i I-1	$\rho_i.z_j$ I-1	(I-1).(J-2)

genotypic response to rainfall. For example, for *G15* yield increases with rainfall relative to what might have been expected on the basis of an additive model. Under dry circumstances, this yield decreases. Recall that the covariates were all centered.

The partitioning of the interaction sum of squares according to model (2) is given in Table 4. The sum of squares due to heterogeneity of genotypic slopes amounted to 38.3, with 16 degrees of freedom. The corresponding mean square, 2.395, is clearly greater than the mean square for the total interaction, 0.853. Therefore, rainfall can be considered to be a good explanatory covariate.

Table 6 reveals the structure of the model by showing its recapitulative table. Because the environmental main effect was fitted before the regression on z_j, the $\rho_i z_j$ term corresponds to $I\text{-}1$ degrees of freedom for I parameters.

Table 7

Recapitulative table associated with model (3).

	1	
1	μ *1*	β_j *J-1*
x_i		$x_i \cdot \tau_j$ *J-1*
	α_i *I-1*	*(I-2).(J-1)*

C. INCLUDING ONE QUANTITATIVE GENOTYPIC COVARIATE
The counterpart of model (2), including one genotypic covariate, is

$$Y_{ij} = \mu + \alpha_i + \beta_j + x_i \tau_j + E_{ij}. \tag{3}$$

An untimely use of this model can be found in Freeman and Crisp (1979). For our example, x_i is a resistance measure for genotype i, as given in Table 2. Now, τ_j can be interpreted as the potential of environment j to favor the spread of the disease. If the environment is beneficial to the spread of the disease, i.e., τ_j is large and positive, and if the genotype is susceptible, i.e., x_i is large and negative, then the correction term $x_i \tau_j$ will be large and negative, implying a decrease in yield. Estimates for τ_j are given in Table 3, and the explained sum of squares is given in Table 4. The mean square amounted to 2.816, again much greater than the total interaction mean square. The number of degrees of

Table 8
Recapitulative table associated with model (4).

1	$\{z_{j1} \oplus z_{j2}\}$	
μ *1*	β_j *J-1*	
α_i *I-1*	$\rho_{i1} \cdot z_{j1} + \rho_{i2} \cdot z_{j2}$ *(I-1).2*	*(I-1).(J-3)*

freedom attributed to a term is not determined by the factor with which the covariate is associated, but by the opposite factor (Table 7).

D. INCLUDING SEVERAL QUANTITATIVE ENVIRONMENTAL COVARIATES

A generalization of (2), including two environmental covariates, leads to

$$Y_{ij} = \mu + \alpha_i + \beta_j + \rho_{i1}z_{j1} + \rho_{i2}z_{j2} + E_{ij}, \tag{4}$$

where z_{j1} and z_{j2} can be rainfall and average maximum temperature over the growing season in environment j. The structure of the model is given in Table 8. When covariates are correlated, inclusion of more than one covariate complicates interpretation of the coefficients, just as for multiple regression.

Table 9
Recapitulative table associated with model (5).

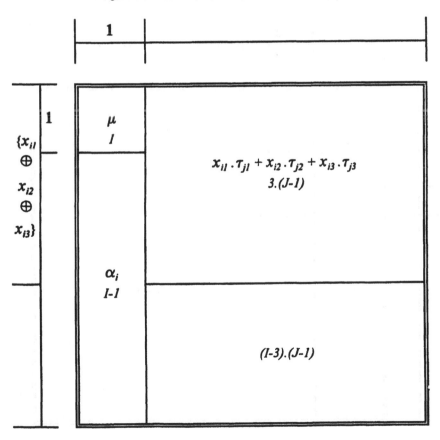

Coefficients are conditional upon the values of the other included covariates; so one should be cautious in interpretations. Examples of the application of model (4) can be found in Hardwick (1972), Hardwick and Wood (1972), Rameau and Denis (1992), and van Eeuwijk and Elgersma (1993).

E. INCLUDING ONE QUALITATIVE GENOTYPIC COVARIATE

Qualitative genotypic covariates attribute group membership to genotypes. Let x_i be a qualitative variable that indicates to which of three groups with a common ancestor a genotype belongs. For example, $x_i=3$ would mean that genotype i belongs to the third group of genotypes. Including this variable x_i in model (3) does not result in anything sensible, because the numbering of the groups is arbitrary and does not refer to something inherent to that group of genotypes. What we can do is replace the qualitative variable x_i by indicator

Table 10
Recapitulative table associated with model (6).

	1	z_j	
1	μ *1*		β_j *J-1*
x_i		$x_i.v.z_j$ *1*	
	α_i *I-1*		*(I-1).(J-1) - 1*

variables (valued 0 or 1), just as when ANOVA models are presented in multiple regression form. Now x_{i1}, x_{i2}, and x_{i3} attribute membership when they have value 1, e.g., $x_{i3}=1$ means genotype i belongs to group 3. Of course, if $x_{i3}=1$, then $x_{i1}=x_{i2}=0$, therefore $x_{i1}+x_{i2}+x_{i3}$ is always one. This redundancy can be removed by leaving out one of the indicator variables, or imposing an additional constraint. Another possibility is to remove the environmental main effect, as is done in

$$Y_{ij}=\mu+\alpha_i+x_{i1}\tau_{j1}+x_{i2}\tau_{j2}+x_{i3}\tau_{j3}+E_{ij}. \tag{5}$$

The parameters τ_{jk} represent the environmental 'main' effects for each of the three groups of genotypes separately. Table 9 displays the structure of the model.

Table 11
Recapitulative table associated with model (7).

	1	z_j	
1	μ 1	β_j $J-1$	
x_i		$x_i.v.z_j$ 1	$x_i.\tau_j$ $J-2$
	α_i $I-1$	$\rho_i.z_j$ $I-2$	$(I-2).(J-2)$

F. INCLUDING GENOTYPIC AND ENVIRONMENTAL COVARIATES
1. Quantitative-quantitative

The simplest extension of the additive model including one genotypic and one environmental covariate is

$$Y_{ij} = \mu + \alpha_i + \beta_j + x_i \nu \, z_j + E_{ij}. \tag{6}$$

It can be derived from models (2) or (3) by imposing the restriction of $\rho_i = x_i \nu$, or $\tau_j = \nu z_j$, respectively. In Table 10 it is shown how the single parameter ν represents one degree of freedom in the interaction space.

The model has been fitted for all combinations of genotypic and environmental covariates at our disposal (Table 4). The combination of genotypic resistance and environmental rainfall produced the highest mean square, as might have been expected from the previous results.

One step further than model (6), a simple combination of (2), (3), and (6) gives

$$Y_{ij} = \mu + \alpha_i + \beta_j + x_i \nu \, z_j + x_i \tau_j + \rho_i z_j + E_{ij}. \tag{7}$$

The recapitulative table (Table 11) for this model shows how $x_i \nu z_j$ is common to both $x_i \tau_j$ and $\rho_i z_j$. To estimate ν, supplementary constraints have to be imposed on τ_j and ρ_i. The ANOVA table (Table 12) shows that a significant

Table 12
Two-way ANOVA with decomposition of the interaction according to model (7), with Rstc for x_i and Rain for z_j.

Source	Df	SS	MS
Genotype	16	67.3	4.208
Environment	11	550.6	50.053
Interaction	176	136.7	0.853
$Rstc_i.v.Rain_j$	1	28.0	27.954
$Rstc_i.\tau_j$	10	3.0	0.301
$\rho_i.Rain_j$	15	10.4	0.691
Remainder	150	108.7	0.725
Error	561	252.4	0.450

amount of interaction was left unexplained by model (7). For a more telling example, see Paul et al. (1993).

Model (6) can straightforwardly be extended to include several genotypic, as well as environmental covariates, to give

$$Y_{ij} = \mu + \alpha_i + \beta_j + \sum_{k=1}^{K} \sum_{h=1}^{H} x_{ik} v_{kh} z_{jh} + E_{ij}. \tag{8}$$

Good illustrations of applications of model (8) are given by Charmet et al. (1993), and Baril et al. (1995).

Table 13
Recapitulative table associated with model (9).

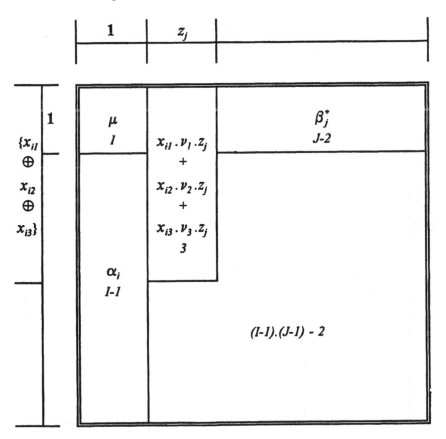

2. Qualitative-quantitative

Taking x_i qualitative in model (6) leads to the model

$$Y_{ij} = \mu + \alpha_i + \beta_j^* + x_{i1}\nu_1 z_j + x_{i2}\nu_2 z_j + x_{i3}\nu_3 z_j + E_{ij}, \tag{9}$$

and Table 13. The parameters β_j^* represent the environmental main effects after adjustment for the general mean and the regressor z_j. The β_j^*'s may be interpreted as a type of residual. For applications see Saeed and Francis (1984), and Royo et al. (1993).

3. Qualitative-qualitative

When in model (9) both the genotypic covariate and the environmental covariate are qualitative, we arrive at

Table 14
Recapitulative table associated with model (10).

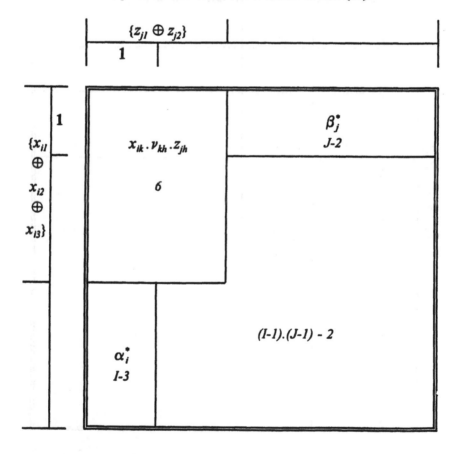

$$Y_{ij} = \alpha_i^* + \beta_j^* + x_{i1}\nu_1 z_{j1} + x_{i2}\nu_{21} z_{j1} + x_{i3}\nu_{31} z_{j1} + \tag{10}$$
$$x_{i1}\nu_{12} z_{j2} + x_{i2}\nu_{22} z_{j2} + x_{i3}\nu_{32} z_{j2} + E_{ij},$$

and Table 14. The environmental covariate z_j may indicate one of two regions, and is represented in the model by two indicator variables, z_{j1} and z_{j2}. In addition to the β_j^*'s of model (9), α_i^*'s appear, representing the genotypic main effects after adjustment for cross-product terms involving x_i. The parameters ν_{kh} represent the mean for the genotypes of the genotypic group k (descendance) in the environments of the environmental group h (region).

Table 15
Recapitulative table associated with model (10').

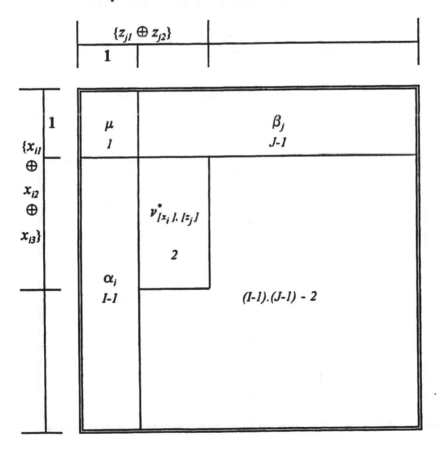

Model (10) can be reparametrized giving

$$Y_{ij}=\mu+\alpha_i+\beta_j+v^{\cdot}_{[x_i,],[z_j,]}+E_{ij}, \tag{10'}$$

and Table 15. In (10') the usual main effects are included, and the $v^{\cdot}_{[x_i,],[z_j,]}$'s have to sum to zero over genotypes (sum over i) and environments (sum over j). Being adjusted for the main effects, they are interaction parameters. Interaction is exclusively of the 'between by between' type. One might think of classifying the original data in the six groups following from the intersection of the three genotypic groups with the two environmental groups. Interaction is present only between these six groups.

Many authors have studied models of the type exemplified by (10'). Although the use of a priori groupings is inferentially superior over the use of

Table 16
Recapitulative table associated with model (12).

1	$\{z_{j1} \oplus ... \oplus z_{jH}\}$	
1		
μ	β_j	
1	$J-1$	
	$\rho_i.\lambda_h.z_{jh}$	
	$1+H-2$	
α_i		
$I-1$	$(I-1).(J-1-H) + (I-2).(H-1)$	

a posteriori groupings, most references relate to the latter (Horner and Frey, 1957; Abou-el-Fittouh et al., 1969; Lin and Thompson, 1975; Byth et al., 1976; Denis, 1979; Seif et al., 1979; Berbigier et al., 1980; Brennan et al., 1981; Brown et al., 1983; Lefkovitch, 1985; Lin and Butler, 1988; Corsten and Denis, 1990; Crossa et al., 1990; Arntzen and van Eeuwijk, 1992; Muir et al., 1992; Oliveira and Charmet, 1992). With a priori grouping, the procedure is fully inferential; otherwise it is more exploratory. The inferential value when using a posteriori groupings remains a point of discussion. Certainly, the type of testing needs more consideration in these cases.

IV. REDUCED RANK FACTORIAL REGRESSION

Theory on general reduced rank regression models has been developed over time by a number of authors belonging to very different disciplines. Among the major contributions we list Rao (1964), Izenman (1975), van den Wollenberg (1977), Gabriel (1978), Obadia (1978), Tso (1981), Davies and Tso (1982), Sabatier et al. (1989), van der Leeden (1990), and Velu (1991). As a solution to genotype by environment interaction problems in plant breeding, reduced rank *factorial* regression models have been proposed. Important contributions are due to Wood (1976), Denis (1991), van Eeuwijk (1992a), and van Eeuwijk (1995a).

A. ONE-WAY REDUCED RANK REGRESSION WITH ONE TERM
Considering model (4) and Table 8, we see that up to J-1 covariates are conceivable. For the case of J-1 covariates, the interaction described would be equal to the total non-additivity remaining from the additive two-way model (1). A number of covariates would then very likely be modeling mere noise, as in most situations with large numbers of covariates. A method allowing the incorporation of substantial amounts of covariates, while using fewer degrees of freedom than a comparable factorial regression model, is reduced rank (factorial) regression. Basically, a so-called synthetic covariate is formed as a linear combination of the available covariates, i.e., the most explanatory linear combination that can be constructed according to a least squares criterion. A synthetic covariate can be incorporated in a model like (2) without further complications. Define the synthetic covariate

$$\zeta_j = \sum_{h=1}^{H} \lambda_h z_{jh}. \tag{11}$$

The coefficients λ_h are unknown parameters to be estimated from the data.

The model becomes:

$$Y_{ij} = \mu + \alpha_i + \beta_j + \rho_i \left[\sum_{h=1}^{H} \lambda_h z_{jh} \right] + E_{ij}. \tag{12}$$

Table 16 shows the distribution of the degrees of freedom over the various terms. It is obvious that substantial amounts of degrees of freedom can be won. As an illustration, compare Table 16 with Table 8. Table 8 gives the degrees of freedom for the interaction in a factorial regression model with 2 environmental covariates ($H=2$), $2(I-1)$. The comparable reduced rank regression model (12) uses I degrees of freedom. In general, the difference

Table 17
Recapitulative table associated with model (13).

1	$\{z_{j1} \oplus ... \oplus z_{jH}\}$	

1	μ 1	β_j $J-1$	
		$\rho_{i1} \cdot \lambda_{h1} \cdot z_{jh}$ $I+H-2$	
			$\rho_{i2} \cdot \lambda_{h2} \cdot z_{jh}$ $I+H-4$
α_i $I-1$			$(I-1).(J-H-1) +$ $(I-3).(H-2)$

R=2

between a reduced rank model as (12) and the corresponding full rank model amounts to $(I\text{-}2)(H\text{-}1)$ degrees of freedom. The difference increases with I and H. This increase in parsimony can express itself in greater accuracy and stability. For interpretational purposes, one should try to integrate the synthetic covariate in subject matter knowledge about the environments. For the genotypic sensitivities, ρ_i, a physiological basis should be sought.

Model (12) is not linear in its parameters, but bilinear. Least squares estimates are no longer linear combinations of the observations, but can come from a singular value decomposition of the fitted values matrix of the factorial regression model including the same set of covariates.

Wood (1976) contains an example of a reduced rank factorial regression model with one synthethic covariate.

Table 18
Recapitulative table associated with model (15).

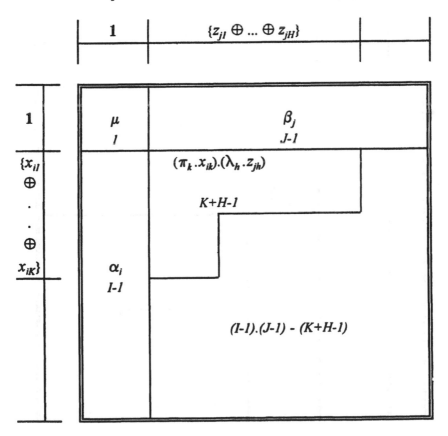

B. ONE-WAY REDUCED RANK REGRESSION WITH SEVERAL TERMS

There is no need to restrict the number of synthetic covariates in reduced rank models to just one,

$$Y_{ij} = \mu + \alpha_i + \beta_j + \sum_{r=1}^{R} \rho_{ir} \left[\sum_{h=1}^{H} \lambda_{hr} z_{jh} \right] + E_{ij}. \tag{13}$$

Model (12) follows from (13) by taking $R=1$. Table 17 represents the recapitulative table.

Illustrations of the use of model (13) are presented in van Eeuwijk (1992a), and van Eeuwijk et al. (1995).

Table 19
Recapitulative table associated with model (16).

C. TWO-WAY REDUCED RANK REGRESSION

Synthetic covariates may be used on both the genotypic as well as the environmental dimension of the table. We define the genotypic synthetic covariate as

$$\xi_i = \sum_{k=1}^{K} \pi_k x_{ik},$$ (14)

with the π_k as unknown parameters to be estimated. This leads to the model

$$Y_{ij} = \mu + \alpha_i + \beta_j + \left[\sum_{k=1}^{K} \pi_k x_{ik}\right] \left[\sum_{h=1}^{H} \lambda_h z_{jh}\right] + E_{ij},$$ (15)

with the recapitulative table given in Table 18.

D. REDUCED RANK REGRESSION INDEPENDENT OF COVARIATES

When I-1 linearly independent genotypic covariates are used to create a synthetic covariate, there is no restriction on the ξ_i's of having to be a linear combination of the x_k's. The same holds true for the ζ_j's when there are J-1 environmental covariates. Model (15) can thus be defined without reference to covariates as

$$Y_{ij} = \mu + \alpha_i + \beta_j + \xi_i \zeta_j + E_{ij}.$$ (16)

Model (16) is known under various names, like AMMI model (Gauch, 1988) and bilinear model (Denis, 1991). Recently it was placed in the biadditive model family by Denis and Gower (1992, 1994a), in an attempt to create a more unified nomenclature for models for two-way tables. Table 19 gives the distribution of the degrees of freedom over the model terms.

For completeness we give the extension of (16) to more than one term,

$$Y_{ij} = \mu + \alpha_i + \beta_j + \sum_{r=1}^{R} \xi_{ir} \zeta_{jr} + E_{ij}.$$ (17)

The recapitulative table of model (17) is shown in Table 20.

Models (16) and (17) are extensively used in plant breeding. A good review that emphasizes prediction can be found in Gauch (1992). A brief exposition emphasizing interpretation is given by van Eeuwijk (1992b). Generalized bilinear models are described in van Eeuwijk (1995b).

38

V. MIXED FACTORIAL REGRESSION

General presentations of the models proposed in the subsections A and B can be found in Goldstein and McDonald (1988), and Denis and Dhorne (1989).

A. GENOTYPES FIXED AND ENVIRONMENTS RANDOM

Sometimes, the environments included in an experiment can be assumed to represent a random sample from a population of environments, thereby fulfilling a sufficient condition for a mixed model approach. Model (2) can be

Table 20
Recapitulative table associated with model (17).

changed into a mixed model by replacing the fixed environmental parameters, indexed by j, by random parameters:

$$Y_{ij} = \mu + \alpha_i + B_j + \rho_i z_j + E_{ij}. \qquad (18)$$

Although the environments are considered random, z_j is not considered to be random, as the analysis proceeds conditional on the value of z_j. The random parameters B_j can be obtained as best linear unbiased predictions, after estimation of the variance component $\sigma_{BB} = \mathrm{var}(B_j)$ (Searle et al., 1992).

Table 21
Genotypic variances when no covariates are included, when rain has been included, and when maximum and minimum temperature, rain and relative humidity have been included. The asterisks indicate P<0.05 for the χ^2 test proposed by Shukla (1972b).

	No cov.	Rain	MaxT+MinT +Rain+RH
G1	1.9104 *	0.8121 *	0.6109
G2	0.6472	0.4469	0.3895
G3	1.2316 *	0.8261 *	0.6176
G4	0.8452 *	0.6169	0.7042
G5	0.4048	0.3559	0.3828
G6	0.5306	0.5990	0.6612
G7	0.5785	0.5249	0.1611
G8	0.5104	0.5193	0.5838
G9	0.5987	0.5660	0.8117
G10	0.9164 *	0.9870 *	1.2004 *
G11	0.6593	0.7235	0.6811
G12	0.4134	0.4481	0.4988
G13	0.8578 *	0.9308 *	1.2673 *
G14	1.4336 *	1.2328 *	0.8737
G15	1.8593 *	1.1239 *	1.0126 *
G16	0.5851	0.6479	0.4165
G17	0.5139	0.5131	0.6351

B. ENVIRONMENTS FIXED AND GENOTYPES RANDOM

In the early phases of the selection process, plant breeders tend to work with groups of genotypes that are considered to be samples from larger populations, whose performance needs to be estimated in a number of well defined environments. By inserting a random genotype in model (2) we obtain

$$Y_{ij} = \mu + A_i + \beta_j + R_i z_j + E_{ij}. \tag{19}$$

Variance components to be estimated are $\sigma_{AA} = \mathrm{Var}(A_i)$, $\sigma_{RR} = \mathrm{Var}(R_i)$, and $\sigma_{AR} = \mathrm{Cov}(A_i, R_i)$. For individual genotypic performances, again best linear unbiased predictions can be calculated. A noticeable feature of model (19) is that the variance of Y_{ij} depends on j, the environment,

$$Var(Y_{ij}) = \sigma_{AA} + 2\sigma_{AR} z_j + \sigma_{RR}(z_j)^2 + \sigma_{EE}. \tag{20}$$

When one wants to use model (19) to predict future genetic gain, one should be aware that the gain depends on the particular environment j.

C. GENOTYPES FIXED, ENVIRONMENTS RANDOM, AND RANDOM INTERACTION DEPENDING ON GENOTYPE

Shukla (1972a,b) introduced a model that included fixed genotypes and random environments, besides a genotypic specific error component. Interesting applications are present in Kang and Miller (1984), Gorman et al. (1989), Kang and Gorman (1989), Gravois et al. (1990), Helms (1993), Magari and Kang (1993), and Kang (1993). The model formulation is very similar to (18):

$$Y_{ij} = \mu + \alpha_i + B_j + \rho_i z_j + E_{ij}, \tag{21}$$

with the variance of E_{ij} depending on the genotype,

$$Var(E_{ij}) = \sigma_{EE}(i). \tag{22}$$

The variance $\sigma_{EE}(i)$ is usually interpreted as a stability associated with genotype i. Inclusion of more than one covariate is straightforward:

$$Y_{ij} = \mu + \alpha_i + B_j + \sum_{h=1}^{H} \rho_{ih} z_{jh} + E_{ij}. \tag{23}$$

Results of the application of models (21) and (23) to the example data are presented in Table 21.

Deleting the regression term $\rho_i z_j$ from model (21) produces a well-known, particularly simple type of heteroscedastic model, often discussed in literature (Russel and Bradley, 1958; Shukla, 1982; Snee, 1982; Denis, 1983; Vincourt et al., 1984; Longford, 1987; Searle et al., 1992; Mudholkar and Sarkar, 1992).

VI. ESTIMATION AND TESTING

A. FIXED FACTORIAL REGRESSION

Fixed factorial regression models fall in the class of fixed linear models and therefore no special problems arise with regard to estimation of parameters and testing of hypotheses.

B. REDUCED RANK FACTORIAL REGRESSION

The inclusion of bilinear terms complicates estimation and testing. Closed form least squares estimators and asymptotic variances are only known for orthogonal cases, i.e., without missing values and with proportional numbers of replications (Denis and Gower, 1992, 1994b). In non-orthogonal cases, numerical approaches are inevitable, and tests and confidence intervals will be approximate.

C. MIXED FACTORIAL REGRESSION

With the exception of the models developed by Shukla, which seem to need a specific procedure, estimation for mixed factorial regressions can be done using restricted maximum likelihood.

D. SOFTWARE

Most of the models presented can be processed with the main statistical packages that include programming facilities, for example Genstat (1993), SAS (1992), and S-plus (1994). The most important Genstat statements for fixed full and reduced rank factorial regression have been added as an Appendix to van Eeuwijk et al. (1995). Special purpose packages also have been developed. We mention first MatModel (Gauch, 1990), which deals mainly with AMMI models. An attractive feature of this package is the cross-validation procedure for assessing the number of interaction terms, when replicates are present. INTERA (Decoux and Denis, 1991) offers facilities for a wide range of fixed factorial regression models and AMMI models, applicable to balanced and unbalanced data. Furthermore, INTERA can fit models combining features of both factorial regression and AMMI. Computer programs to calculate the ecovalence (Wricke, 1962) and Shukla's stability statistics, σ_i^2 and s_i^2 (Shukla, 1972a), are described in Kang (1988, 1989). Presently a new program is available that calculates, in addition to the above mentioned statistics, the YS_i statistic, which combines yield and stability into a single selection criterion (Kang, 1993).

VII. SOME CONSIDERATIONS WITH RESPECT TO MODEL CHOICE

The basic question for the experimenter is, which model to choose out of all the possibilities enumerated above? No definite answer is possible. The choice strongly depends on the desired goal. Various choices accompany the model selection process. We briefly address four issues.

A. CONSTRAINTS

All the models described are overparameterized. Supplementary constraints can be imposed to solve this indeterminacy. Various possibilities exist. Natural extensions of the sum-to-zero constraints were proposed by Denis (1991). These lead to orthogonal decompositions that are convenient for the construction of recapitulative tables. However, we feel that mathematical convenience should always be made subordinate to biological knowledge, also in choosing identification constraints.

B. COVARIATE SELECTION

The most difficult point in the application of factorial regression models seems to be the choice of a good subset of covariates for genotypes as well as environments. It is a variable selection problem having the square of the complexity of that of variable selection in the standard 'one-way' multiple regression context. It is important to keep in mind that the size of the sample for factorial regressions is not IJ, but I-1 for regressions with genotypic covariates, and J-1 for regressions with environmental covariates.

Denis (1988) contains a discussion of variable selection strategies for factorial regression models. It is shown how nesting relationships between models can be used to test for the inclusion of covariates and the possibility of rank reduction.

In the absence of subject matter knowledge, exhaustive variable searches may be used as exploratory analyses. One should then be cautious against over-interpretation, and correct for selection bias by using an appropriate experiment-wise error rate. If possible, it is, however, always preferable to work inferentially, i.e., test specific hypotheses following from subject matter knowledge about the interaction of physiological processes in the plant with defined environmental factors. The relevancy of selected covariates can be further investigated in future trials, as a safeguard against conclusions based on chance correlations.

C. FIXED OR RANDOM

Another important question is the choice of terms as fixed or random. Two main types of arguments can be distinguished. A first type of argument is based on sampling considerations. Do the genotypes and/or environments in the experiment constitute a sample from a population to which the inference is directed? The second kind of arguments is more pragmatic, and involves the desirability of shrinkage and recovery of information, and the convenience of choosing a model term random when many parameters are associated with the term. With regard to shrinkage we may question whether it is reasonable to shrink estimates deviating from the mean of the sample back towards that mean. Or, should relatively good genotypes pay for being an element of a relatively bad sample, while relatively bad genotypes benefit from being an element of a relatively good sample? Considerations concerning recovery of information play a role when data are unbalanced. At all times it must be possible to assess whether the random effects indeed could have come from the assumed distribution. For example, for the estimation of a variance component, at least 10 degrees of freedom should be available, otherwise it is preferable to take the term fixed. The same remark applies to Shukla's approach; many environments are needed for accurate estimates of individual genotypic variances.

D. PARSIMONY

In model building and model choice, one should always take into account the parsimony principle (Gauch, 1988), i.e., avoid over fitting. By including ever more covariates, the amount of interaction described will keep on increasing. However, as a consequence, more noise will be fitted, leading to less robust models. From this perspective, reduced rank regression models are attractive as they allow more covariates for the same number of degrees of freedom. A word of caution should be given for too uncritically accepting the degrees of freedom attributed to synthetic and artificial covariates. When pattern does not clearly dominate noise, these degrees of freedom will be too low, thus declaring the influence of synthetic and artificial covariates significant, when it is not (Gauch, 1992; Williams and Wood, 1993; Cornelius, 1993).

VIII. ALTERNATIVES AND EXTENSIONS

Despite the long enumeration of models given above, the possibilities of modeling interaction using additional information are not exhausted. In this final section, we give some further ideas on the subject.

A. DECOMPOSING MAIN EFFECTS

Use of covariates for decomposition of variation need not be kept restricted to interaction effects. Main effects can be split into a part due to regression on a covariate and a residual

$$Y_{ij}=\mu+x_i\alpha_0+\alpha_i^{*}+\beta_0 z_j+\beta_j^{*}+x_i\nu z_j+E_{ij}. \qquad (6')$$

The residual from a main effects regression almost always is strongly significant because of the dominant role of the main effects in the description of the total variation.

B. ANALYSIS OF COVARIANCE

Some authors (Snedecor and Cochran, 1976; Searle, 1979) have used the following model under the name of analysis of covariance:

$$Y_{ij}=\mu+\alpha_i+\beta_j+\rho\, o_{ij}+E_{ij}, \qquad (24)$$

where o_{ij} is a covariate whose value depends specifically on the cell (i,j). As previously indicated, covariates defined on cell level can be subsumed under the factorial regression models by defining a genotypic covariate $x_i=o_{i.}-o_{..}$ and an environmental covariate $z_j=o_{.j}-o_{..}$, where a dot means averaging.

C. PARTIAL LEAST SQUARES REGRESSION

Multivariate partial least squares regression models have been proposed to model interaction in dependence on covariates (Aastveit and Martens, 1986; Talbot and Wheelwright, 1989). These models can be interpreted in a way reminiscent of reduced rank regression. Partial least squares can be viewed as a robust estimation procedure.

D. BIADDITIVE MIXED MODELS

An interesting conjunction of model classes is given by allowing the multiplicative covariates in biadditive models, to which the Finlay-Wilkinson and AMMI model belong, to be random. Some preliminary work has been done here by Oman (1991).

E. PIECEWISE REGRESSION

Genotypic responses to many environmental factors will reach an upper limit. A simple way to model this kind of response is

$$Y_{ij}=\mu+\alpha_i+\beta_j+\rho_i Min(\phi_i, z_j)+E_{ij}. \qquad (25)$$

In model (25) the covariate z_j is replaced by the minimum of a threshold ϕ_i and the covariate z_j. Each genotype has its own threshold after which the response cannot increase any more.

F. GENERALIZED LINEAR AND BILINEAR MODELS

All fixed factorial regression models dealt with so far assume that the expectation can be modeled linearly in the parameters and that the variance is constant. Deviations from these assumptions sometimes can be cured by transformation of the response. However, the optimal transformation for achieving linearity need not be the same as the optimal transformation for achieving homogeneity of variance. For the models in the class of generalized linear models it is not necessary to find a transformation of the response as a compromise between first (expectation linear in the parameters) and second order (homogeneous variance) requirements. In generalized linear models a suitable transformation of the expectation can be combined with a convenient choice for a variance function, expressing the dependence of the variance on the mean (McCullagh and Nelder, 1989). Generalized factorial regression models extend considerably the range of application for factorial regression models. A further elaboration in the form of generalized bilinear models is discussed in van Eeuwijk (1995b).

G. HIGHER WAY FACTORIAL REGRESSION

Genotype by environment problems often involve more than two factors. Environments are usually cross-classifiable by years and locations. This fact does not complicate the use of factorial regression models for the fixed and mixed cases. Somewhat harder to make are the extensions to the class of biadditive models, although progress is made also here. In van Eeuwijk and Kroonenberg (1995) quadri-additive models are introduced for three-way interaction.

ACKNOWLEDGMENT

Thanks are due to L.C.P. (Paul) Keizer for his help in the preparation of the manuscript.

REFERENCES

Aastveit, A.H., and H. Martens. 1986. ANOVA interactions interpreted by partial least squares regression. *Biometrics* 42:826-844.

Abou-el-Fittouh, H.A., J.O. Rawlings and P.A. Miller. 1969. Classification of environments to control genotype by environment interactions with an application to cotton. *Crop Sci* 9:135-140.

Arntzen, F.K., and F.A. van Eeuwijk. 1992. Variation in resistance levels of potato genotypes and virulence level of potato cyst nematode populations. *Euphytica* 62:135-143.

Baril, C.P. 1992. Factor regression for interpreting genotype-environment interaction in bread-wheat trials. *Theor. Appl. Genet.* 83:1022-1026.

Baril, C.P., J.-B. Denis, R. Wustman and F.A. van Eeuwijk. 1995. Analysing genotype by environment interaction in Dutch Potato Variety Trials using factorial regression. *Euphytica* 82:149-155.

Berbigier, A., J.-B. Denis and C. Dervin. 1980. Interaction variété-lieu: analyse du rendement d'orges de printemps. *Ann. Amélior. Plantes* 30:79-94.

Brennan, P.S., D.E. Byth, D.W. Drake, I.H. DeLacy and D.G. Butler. 1981. Determination of the location and number of test environments for a wheat cultivar evaluation program. *Aust. J. Agric. Res.* 32:189-201.

Brown, K.D., M.E. Sorrells and W.R. Coffman. 1983. Method for classification and evaluation of testing environments. *Crop Sci* 23:889-893.

Byth, D.E., R.L. Eisemann and I.H. DeLacy. 1976. Two-way pattern analysis of a large data set to evaluate genotypic adaptation. *Heredity* 37:215-230.

Charmet, G., F. Balfourier, C. Ravel and J.-B. Denis. 1993. Genotype x environment interactions in a core collection of French perennial rye grass populations. *Theor. Appl. Genet.* 86:731-736.

Cornelius, P.L. 1993. Statistical tests and retention of terms in the additive main effects and multiplicative interaction model for cultivar trials. *Crop Sci* 33:1186-1193.

Corsten, L.C.A., and J.-B. Denis. 1990. Structuring interaction in two-way tables by clustering. *Biometrics* 46:207-215.

Crossa, J., W.H. Pfeiffer, P.N. Fox and S. Rajaram. 1990. Multivariate analysis for classifying sites: application to an international wheat yield trial. p. 214-233. *In* M.S. Kang (ed.) *Genotype-by-environment interaction and plant breeding.* Louisiana State Univ., Baton Rouge.

Davies, P.T., and M.K.-S. Tso. 1982. Procedures for reduced-rank regression. *Appl. Statist.* 31:244-255.

Decoux, G., and J.-B. Denis. 1991. INTERA, version 3.3, Notice d'utilisation, logiciel pour l'interprétation statistique de l'interaction entre deux facteurs. Laboratoire de Biométrie, INRA, route de St-Cyr, F-78026 Versailles Cédex, 175pp.

Denis, J.-B. 1979. Structuration de l'interaction. *Biom. Praxim.* 19:15-34.

Denis, J.-B. 1980. Analyse de régression factorielle. *Biom.Praxim.* 20:1-34.

Denis, J.-B. 1983. Extension du modèle additif d'analyse de variance par modélisation multiplicative des variances. *Biometrics* 39:849-856.

Denis, J.-B. 1988. Two-way analysis using covariates. *Statistics* 19:123-132.

Denis, J.-B. 1991. Ajustement de modèles linéaires et bilinéaires sous contraintes linéaires avec données manquantes. *Rev. Statist. Appl.* 39:5-24.

Denis, J.-B., and T. Dhorne. 1989. Modelling interaction by regression with random coefficients. *Biuletyn Oceny Odmian* 21-22:63-73.

Denis, J.-B., and J.C. Gower. 1992. Biadditive models. Technical report of the Laboratoire de Biométrie, INRA, F-78026 Versailles, 33 pp.

Denis, J.-B., and J.C. Gower. 1994a. Biadditive models. Letter to the editor. *Biometrics* 50:310-311.

Denis, J.-B., and J.C. Gower. 1994b. Asymptotic covariances for the parameters of biadditive Models. *Utilitas Mathematica* 46: 193-205.

Denis, J.-B., et P. Vincourt. 1982. Panorama des méthodes statistiques d'analyse des interactions génotype x milieu. *Agronomie* 2:219-230.

Fakorede, M.A.B., and B.O. Opeke. 1986. Environmental indices for the analysis of genotype x environment interaction in maize. *Maydica* 31:233-243.

Federer, W.T., and B.T. Scully. 1993. A parsimonious statistical design and breeding procedure for evaluating and selecting desirable characteristics over environments. *Theor. Appl. Genet.* 86:612-620.

Finlay, K.W., and G.N. Wilkinson. 1963. The analysis of adaptation in a plant-breeding program. *Aust. J. Agric. Res.* 14:742-754.

Freeman, G.H., and P. Crisp. 1979. The use of related variables in explaining genotype-environment interactions. *Heredity* 42:1-11.

Freeman, G.H., and J.M. Perkins. 1971. Environmental and genotype-environmental components of variability. VIII. Relations between genotypes grown in different environments and measures of these environments. *Heredity* 27:15-23.

Gabriel, K.R. 1978. Least squares approximation of matrices by additive and multiplicative models. *J. R. Statist. Soc.* B 40:186-196.

Gauch, H.G.Jr. 1988. Model selection and validation for yield trials with interaction. *Biometrics* 44:705-715.

Gauch, H.G.Jr. 1990. *MATMODEL version 2.0, AMMI and related analyses for two-way data matrices.* Cornell University, Ithaca, New York 14853, USA, 69pp.

Gauch, H.G.Jr. 1992. *Statistical analysis of regional yield trials: AMMI analysis of factorial designs.* Elsevier, Amsterdam, 278pp.

Genstat 5 Committee. 1993. *Genstat 5 release 3, reference manual.* Clarendon Press, Oxford.

Goldstein, H., and R.P. McDonald. 1988. A general model for the analysis of multi level data. *Psychometrika* 53:455-467.

Gorman, D.P., M.S. Kang and M.R. Milam. 1989. Contribution of weather variables to genotype x environment interaction in grain sorghum. *Plant Breeding* 103:299-303.

Gravois, K.A., K.A.K. Moldenhauer and P.C. Rohman. 1990. Genotype-by-environment interaction for rice yield and identification of stable, high-yielding genotypes. p. 181-188. *In* M.S. Kang (ed.) *Genotype-by-environment interaction and plant breeding.* Louisiana State Univ., Baton Rouge.

Hardwick, R.C. 1972. Method of investigating genotype environment and other two factor interactions. *Nature New Biology* 236:191-192.

Hardwick, R.C., and J.T. Wood. 1972. Regression methods for studying genotype-environment interactions. *Heredity* 28:209-222.

Helms, T.C. 1993. Selection for yield and stability among oat lines. *Crop Sci* 33:423-426.

Hinkelmann, K. 1974. Genotype-environment interaction: Aspects on statistical design, analysis and interpretation. Invited paper to the 8th International Biometric Conference, Constanta, Romania.

Horner, T.W., and K.J. Frey. 1957. Methods for determining natural areas for oat varietal recommendations. *Agron. J.* 49:313-315.

Izenman, A.J. 1975. Reduced-rank regression models for the multivariate linear model. *J. Mult. Anal.* 5:248-264.

Kang, M.S. 1988. Interactive BASIC program for calculating stability-variance parameters. *Agron. J.* 80:153.

Kang, M.S. 1989. A new SAS program for calculating stability-variance parameters. *J. Hered.* 80:415.

Kang, M.S. 1993. Simultaneous selection for yield and stability in crop performance trials: consequences for growers. *Agron. J.* 85:754-757.

Kang, M.S., and D.P. Gorman. 1989. Genotype x environment interaction in maize. *Agron. J.* 81:662-664.

Kang, M.S., and J.D. Miller. 1984. Genotype x environment interactions for cane and sugar yield and their implications in sugar cane breeding. *Crop Sci* 24:435-440.

Knight, R. 1970. The measurement and interpretation of genotype-environment interactions. *Euphytica* 19:225-235.

Lefkovitch, L.P. 1985. Multi-criteria clustering in genotype-environment interaction problems. *Theor. Appl. Genet.* 70:585-589.

Lin, C.S., and G. Butler. 1988. A data-based approach for selecting locations for regional trials. *Can. J. Plant. Sci.* 68:651-659.

Lin, C.S., and B. Thompson. 1975. An empirical method of grouping genotypes based on a linear function of the genotype-environment interaction. *Heredity* 34:255-263.

Longford, N.T. 1987. A fast algorithm for maximum likelihood estimation in unbalanced mixed models with nested random effects. *Biometrika* 74:817-827.

Magari, R., and M.S. Kang. 1993. Genotype selection via a new yield-stability statistic in maize yield trials. *Euphytica* 70:105-111.

48

Mandel, J. 1961. Non-additivity in two-way analysis of variance. *J. Am. Statist. Ass.* 56:878-888.

McCullagh, P., and J.A. Nelder. 1989. *Generalized linear models*. Chapman and Hall, London.

McGraw, R.L., P.R. Beuselinck and R.R. Smith. 1986. Effect of latitude on genotype x environment interactions for seed yield in birdsfoot trefoil. *Crop Sci* 26:603-605.

Mudholkar, G.S., and I.C. Sarkar. 1992. Testing homoscedasticity in a two-way table. *Biometrics* 48:883-888.

Muir, W., W.E. Nyquist and S. Xu. 1992. Alternative partitioning of the genotype-by-environment interaction. *Theor. Appl. Genet.* 84:193-200.

Obadia, J. 1978. L'analyse en composantes explicatives. *Rev. Statist. Appl.* 26:5-28.

Oliveira, J.A., and G. Charmet. 1992. Genotype by environment interaction in Lolium perenne: grouping wild populations by cluster analysis. *Invest. Agr.: Prod. Prot. Veg.* 7:117-128.

Oman, S.D. 1991. Multiplicative effects in mixed model analysis of variance. *Biometrika* 78:729-739.

Paul, H., F.A. van Eeuwijk and W. Heijbroek. 1993. Multiplicative models for cultivar by location interaction in testing sugar beets for resistance to beet necrotic yellow vein virus. *Euphytica* 71:63-74.

Rameau, C., and J.-B. Denis. 1992. Characterization of environments in long-term multi-site trials in Asparagus, through yield of standard varieties and use of environmental covariates. *Plant Breeding* 109:183-192.

Rao, C.R. 1964. The use and interpretation of principal component analysis in applied research. *Sankhya* B 26:329-358.

Royo, C., A. Rodriguez and I. Romogosa. 1993. Differential adaptation of complete and substituted triticale. *Plant Breeding* 111:113-119.

Russel, T.S., and R.A. Bradley. 1958. One-way variances in a two-way classification. *Biometrika* 45:111-129.

Sabatier, R., J.-D. Lebreton and D. Chessel. 1989. Principal component analysis with instrumental variables as a tool for modelling composition data. p. 341-352. *In* R. Coppi and S. Bolasco (eds.) *Multiway Data Analysis*. Elsevier, North-Holland.

Saeed, M., and C.A. Francis. 1984. Association of weather variables with genotype x environment interactions in grain sorghum. *Crop Sci* 24:13-16.

SAS Institute Inc., SAS*. 1992. *Software: changes and enhancements, release 6.07*. Technical report P-229, SAS/STAT*, Cary, NC: SAS Institute Inc., 620 pp.

Searle, S.R. 1979. Alternative covariance models for the 2-way crossed classification. *Comm. St.* A 8:799-818.

Searle, S.R., G. Casella and C.E. McCulloch. 1992. *Variance Components*. Wiley, New York.

Seif, E., J.C. Evans and L.N. Balaam. 1979. A multivariate procedure for classifying environments according to their interaction with genotypes. *Aust. J. Agric. Res.* 30:1021-1026.

Shukla, G.K. 1972a. An invariant test for the homogeneity of variances in a two-way classification. *Biometrics* 28:1063-1072.

Shukla, G.K. 1972b. Some statistical aspects of partitioning genotype-environmental components of variability. *Heredity* 29:237-245.

Shukla, G.K. 1982. Testing the homogeneity of variances in a two-way classification. *Biometrika* 69:411-416.

Snedecor, G.W., and W.G. Cochran. 1976. *Statistical Methods* (6th edn., 8th printing). Iowa State Univ. Press, Ames.

Snee, R.D. 1982. Nonadditivity in a two-way classification: is it interaction or nonhomogeneous variance? *J. Am. Statist. Ass.* 77:515-519.

S-plus. 1994. *S-plus for Windows version 3.2 supplement*. Seattle: StatSci, a division of MathSoft, Inc.

Tai, G.C.C. 1990. Path analysis of genotype-environment interactions. p. 273-286. *In* M.S. Kang (ed.) *Genotype-by-environment interaction and plant breeding*. Louisiana State Univ., Baton Rouge.

Talbot, M., and A.V. Wheelwright. 1989. The analysis of genotype x environment interactions by partial least squares regression. *Biuletyn Oceny Odmian* 21-22:19-25.

Tso, M.K.-S. 1981. Reduced-rank regression and canonical analysis. *J. R. Statist. Soc.* B 43:183-189.

van den Wollenberg, A.L. 1977. Redundancy analysis an alternative for canonical correlation analysis. *Psychometrika* 42:207-219.

van der Leeden, R. 1990. Reduced rank regression with structured residuals. DSWO Press, Leiden.

van Eeuwijk, F.A. 1992a. Interpreting genotype-by-environment interaction using redundancy analysis. *Theor. Appl. Genet.* 85:89-100.

van Eeuwijk, F.A. 1992b. Multiplicative models for genotype-environment interaction in plant breeding. *Statistica Applicata* 4:393-406.

van Eeuwijk, F.A. 1993. Genotype by environment interaction; Basic ideas and selected topics. p. 91-104. *In* Johan H.L. Oud and Rian A.W. van Blokland-Vogelesang (eds.) *Advances in longitudinal and multivariate analysis in the behavioral sciences.* ITS, Nijmegen.

van Eeuwijk, F.A. 1995a. Linear and bilinear models for the analysis of multi-environment trials: I. An inventory of models. *Euphytica* 84:1-7.

van Eeuwijk, F.A. 1995b. Multiplicative interactions in generalized linear models. *Biometrics* 51:1017-1032.

van Eeuwijk, F.A., and A. Elgersma. 1993. Incorporating environmental information in an analysis of genotype by environment interaction for seed yield in perennial ryegrass. *Heredity* 70:447-457.

van Eeuwijk, F.A., L.C.P. Keizer and J.J. Bakker. 1995. Linear and bilinear models for the analysis of multi-environment trials: II. An application to data from the Dutch Maize Variety Trials. *Euphytica* 84:9-22.

van Eeuwijk, F.A., and P.M. Kroonenberg. 1995. Multiplicative decompositions of interactions in three-way analysis of variance, with applications to plant breeding. Submitted.

Velu, R.P. 1991. Reduced rank models with two sets of regressors. *Appl. Statist.* 40:159-170.

Vincourt, P., M. Derieux and A. Gallais. 1984. Quelques méthodes de choix des génotypes à partir d'essais multilocaux. *Agronomie* 4:843-848.

Williams, E.R., and J.T. Wood. 1993. Testing the significance of genotype-environment interaction. *Aust. J. Agric. Res.* 35:359-362.

Wood, J.T. 1976. The use of environmental variables in the interpretation of genotype-environment interaction. *Heredity* 37:1-7.

Wricke, G. 1962. Über eine Methode zur Erfassung der ökologischen Streubreite in Feldversuchen. *Z. Pflanzenzücht.* 47:92-96.

Chapter 3

RELATIONSHIPS AMONG ANALYTICAL METHODS USED TO STUDY GENOTYPE-BY-ENVIRONMENT INTERACTIONS AND EVALUATION OF THEIR IMPACT ON RESPONSE TO SELECTION

I.H. DeLacy, M. Cooper and K.E. Basford[1]

I. INTRODUCTION

The focus of this book is on new perspectives for the study and accommodation of genotype by environment (GxE) interactions in plant breeding. Their ubiquitous nature (DeLacy and Cooper, 1990) and impact on genetic improvement were discussed in a number of the papers included in the precursor to this volume (Kang, 1990).

Many statistical methods have been advocated as a basis for the analysis of genotypic variation to accommodate GxE interactions and provide guidance for exploiting positive interactions. These techniques have often been discussed as competing methods and usually are compared with the view to identify the "best" method. Generally the criteria for defining the "best" method are based on the efficacy with which the alternative methods represent the genotypic variation intrinsic to the data set under study rather than in terms of the more important criterion of realised response to selection. Implicit in the former is that the method which more effectively describes the current data set is more likely to provide a basis for predicting performance in future environments. This strategy is based on two assumptions. Firstly, that the data set being analysed is adequate for prediction of response to selection and secondly, that the description of the data characterises differences among the genotypes which can be exploited by a selection strategy. The plant breeder, interested in response to selection and not directly involved in the development and evaluation of these methods, must select an analytical strategy from the competing plethora to optimise selection among genotypes. Selection among genotypes is the primary objective of conducting multi-environment trials (METs) and must guide the selection of analytical methods.

Why so many alternatives? Each appears to have some merit, particularly when discussed by advocates. Clearly there are situations where application of any of the techniques will provide the user with some insight into the nature and causes of GxE interactions. Instead of viewing the methods as competing strategies our approach is to examine them for their inter-relationships. The models used in quantitative genetic theory provide an appropriate framework for such an examination.

---

[1]Department of Agriculture, The University of Queensland, Brisbane, 4072, Queensland, Australia.

The advantage of this framework is that an understanding of the relationships among the alternative analytical methods enables judicious joint application of the methods to exploit any complementary strengths. This would avoid the weaknesses inherent in rigid adherence to a particular choice. However, there are two disadvantages. Firstly, proficiency in many methods is required rather than mastery of one. Secondly, extra time is required to conduct multiple analyses, a luxury which is often not available when selection decisions must be made in "real time". The first requires willingness to evaluate the alternatives in relation to the objectives of the breeding program. Once a decision has been made on the appropriate combination of analytical methods the second can be overcome by developing appropriate computer software for the breeding program.

In the next two sections of this chapter we discuss the principles which we have used to study GxE interaction in context with the genetic improvement of quantitative traits such as yield, yield indices and measures of quality. The next section outlines the framework for analysis of genotypic variation taking into account GxE interaction. In the section that follows we develop additional theory to link aspects of the framework and provide two examples of its application.

II. FRAMEWORK FOR ANALYSIS OF GENOTYPIC VARIATION

To outline a framework for the analysis of GxE interaction it is necessary to consider jointly both (1) the context within which the interactions exist in the crop production system targeted by a breeding program and how they are encountered in multi-environment trials (METs) (Figure 1) and (2) the objectives of selection in a breeding program and how an understanding of GxE interaction can impact upon selection strategies and response to selection (Table 1). To develop this larger picture we combine the concepts of a target population of environments (Comstock, 1977) and a germplasm pool accessed by the breeding program. The objective is to quantify the scope for genetic improvement of the crop within the target production system. This involves evaluating the relative genetic merit of the genotypes from the germplasm pool for their performance for various traits in the target population of environments (Figure 1). Schematically we may consider that there is a distribution of some index of merit of the genotypes and a target population of environments which presents a mixture of distributions of challenges (resources, constraints, hazards) to the expression of the merit of a genotype. From this population perspective the phenotypic performance of genotypes in combinations of environments can be analysed to quantify the amount of variation attributable to the effects of environment, genotype and GxE interactions (Table 1). To estimate these population parameters, samples or selected genotypes from the germplasm pool, are evaluated in samples of environments taken from the target population. The resultant experiments are the plant breeding METs (Figure 1). If a completely random model is adopted and "practised" then response to selection can be predicted from the estimates of components of

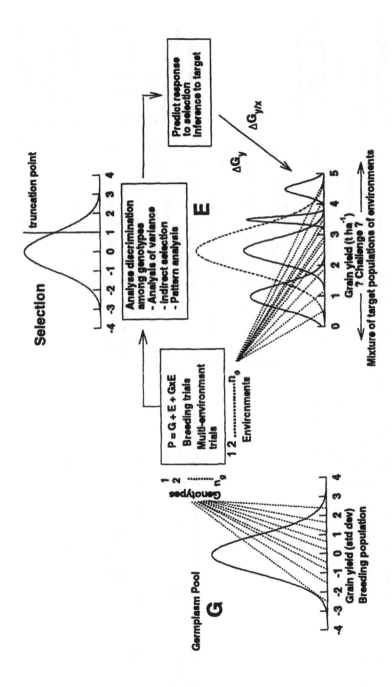

Figure 1. Schematic representation of the recurrent steps involved in the conduct of multi-environment trials where the objective is to sample test environments from a pre-defined target population of environments to test the relative merit of genotypes.

Table 1

Considerations for the analysis and understanding of the form of genotype by environment (GxE) interactions in terms of their application to selection in plant breeding

Form of GxE Interaction	Model assumptions	Analysis methods	Application in plant breeding	
			Objectives of analysis	Selection strategy
Non-repeatable	Environments: random Genotypes: random	Analysis of variance REML Best linear unbiased predictors (BLUPs) of genotype performance	1. Estimate components of variance to determine the relative sizes of sources of variation and estimate heritability. 2. Characterise the form of GxE interactions by examining them for both genotypes and environments for (a) Heterogeneity (HV) + Lack of correlation (LC) partition (this enables calculation of the pooled genetic correlation) (b) Rank change (RC) + No Rank Change (NRC) partition (c) The impact of rank change on the composition of the select group at a defined selection intensity.	Selection for broad adaptation. Decisions on sample size (*i.e.* how many test environments, replicates and genotypes to use?)
Mixture of non-repeatable and repeatable	Environments: a mixture of random & fixed Genotypes: random	Indirect selection Pattern analysis	3. Relationships among environments measured in terms of indirect response to selection. 4. Grouping, ordination and partitioning (size and shape) of GxE interactions for individual environments.	Selection for broad adaptation and specific adaptation to types of environments.
Mixture of non-repeatable and repeatable	Environments: a mixture of random & fixed Genotypes: a mixture of random & fixed	Pattern analysis	5. Grouping, ordination and partitioning of GxE interactions for environments and genotypes. 6. Investigation of causes of differences in patterns of adaptation.	Selection for specific adaptation and stability.
Repeatable	Environments: fixed Genotypes: fixed	Pattern analysis Biological models	7. Interpretation of causes of GxE interactions.	Decisions on breeding and selection strategies. How many and what types of test environments?

variance, heritability and pooled genetic correlations (Table 1). Response to selection can be viewed as either a direct (when the sample of environments is from the target set) or indirect (when the test environments predict advance in future or other target sets) genetic improvement. The performance prediction of the genotypes can then be used as a basis for truncating the population and implementing selection (Figure 1).

From the perspective of the random model, GxE interactions are viewed as a source of error in prediction of genotype performance in the target population of environments and selection is practised for broad adaptation. Where response is considered as a case of direct selection the GxE interaction component of variance is included in the denominator of the heritability estimate and therefore reduces the predicted response to selection. Alternatively where response is considered as a case of indirect selection, GxE interactions reduce the genetic correlation between the selection sample and target population of environments and thus reduce indirect response to selection. Where the random model is applicable to the target crop production system of the breeding program, a MET program which randomly samples the range of possible GxE interactions is necessary. Questions that need to be resolved include the size of the random samples of environments, replications and genotypes needed to optimise progress from selection, efficiency in experimental design and analytical estimation or prediction procedures (Table 1).

The alternative view is that the GxE interactions are not random and that there exist repeatable and heritable patterns within the interaction complex which can be exploited by selection, i.e., for specific adaptation. With the recognition that some of the components of the GxE interaction complex may be repeatable there has been interest in describing, understanding and exploiting GxE interactions. The strategies for accommodating repeatable GxE interactions in breeding programs are varied but a basic ingredient is a clear definition of the environmental factor(s) contributing to the interaction (the challenge) and then selection for adaptation to that challenge. Where this is possible the "E" from GxE becomes more explicitly defined. For example, for the yield of wheat in Queensland, Australia, repeatable GxE interactions have been attributed to the environmental factor root-lesion nematode (RLN) (*Pratylenchus thornei*) and genetic variation for tolerance to the nematode. Therefore, part of the GxE complex is now recognised as Gx(RLN) interaction and selection for tolerance and resistance to the nematode provides the basis for exploiting specific adaptation.

There is nothing new about this strategy; it has a long history of successful application in breeding for resistance to biotic stresses. However, its application to understanding GxE interactions in general, where the "E" factor is a result of uncharacterised and unknown mixtures of biotic and abiotic stresses, is recent (Eisemann et al., 1990; Cooper et al., 1993). We argue that these un-characterised, non-random interactions have a quantifiable degree of repeatability within the target production system. While this view is not uncommon there are few examples where such a quantification has been attempted. Two examples, with which we are familiar, that consider repeatability of GxE interactions on a temporal basis are those discussed by DeLacy et al. (1994) and Mirzawan et al. (1994). Notwithstanding this, to achieve progress in the understanding of the causes of GxE interactions and to enable a distinction between repeatable and random (non-repeatable, noise) components, it is desirable to combine a biological and a statistical model when analysing genotypic adaptation. As with the statistical models, the biological models can take many forms. It is beyond the scope of this paper to discuss this approach in detail and the reader is referred to Cooper and Hammer (1996) which contains numerous examples.

Clearly, the strategies for dealing with repeatable and non-repeatable GxE interactions are different. The traditional selection for broad adaptation across non-repeatable GxE interactions can be complemented by selection for specific adaptation where repeatable interactions are identified and of sufficient importance to warrant special attention (Table 1). This model of the GxE interaction complex enables the target population of environments to be viewed from a new perspective as a mixture of target populations, rather than one target population. For this model the strategy for dealing with GxE interaction evolves to one of resolving and defining components of the "E" in GxE to enable a stratified sample of environments which accommodates the structure identified in the target population of environments. The statistical theory appropriate for analysing data which are sampled from a mixture of underlying populations has been considered as mixture models (McLachlan and Basford, 1988) and goes back to original work by Karl Pearson (1894).

Where such a resolution of the GxE interaction is achieved the completely random model can and should be re-evaluated. Most plant breeders are familiar with such an approach where the environments are sampled as a cross-classification of years and locations. Where interactions are specifically associated with either of these factors they are explicitly sampled in the METs. Here we are advocating further resolution of the "E" as warranted. Where a particular environmental factor is identified as important in both the target population of environments and in generating GxE interactions the question arises of how best to sample that environmental factor. It can be argued that if it is important it will be appropriately sampled by a random sampling strategy. This applies where sufficiently large samples can be taken. At many stages of the breeding program resource limitations necessitate small samples, e.g. early generation selection for yield and quality where large numbers of lines are tested in few environments. In these cases specific environmental factors could be imposed as a fixed effect by stratified sampling or by management of the environments. This strategy is under investigation for improving yield and quality of wheat in Queensland, Australia, for the abiotic environmental factors water and nitrogen (Eisemann et al., 1990, Cooper et al., 1995). Where the target population has a known structure, theory indicates that a stratified sampling strategy is superior to a random sampling strategy. However, it is too early to tell whether this can be practically applied and a benefit realised for these two abiotic factors in our breeding work.

Once a set of environmental factors have been recognised as important and are known to contribute to GxE interactions, decisions must be made on how to analyse the data collected from METs (Figure 1) (Cooper and DeLacy, 1994). The analysis must enable selection for both broad and specific adaptation. Selection for broad adaptation on average performance across environments is the first priority. Where GxE interactions are significant, the associated change in relative performance across environments is examined. We use a combination of three analytical methods: (i) analysis of variance, (ii) direct and indirect selection and (iii) pattern analysis (Table 1). The combination of analytical methods used is influenced by two factors: (1) the degree with which the form of the GxE interactions are identified as being repeatable and (2) the model (random, fixed or mixture) used to represent the genotypes and environments. The three analytical methods and their relationships are discussed below. This is followed by a section which deals with their application in plant breeding.

A. ANALYSIS OF VARIANCE

The size of the GxE interaction can be judged in terms of the relative size of the genotypic and interaction components of variance. The larger the relative size of the interaction component the more complex the problem of identifying broadly adapted genotypes. It is necessary to consider the effect of any interactions in terms of the objectives of selection. To achieve this, distinction should be made between those interactions which impact on selection decisions and those which do not. This can be done in a number of ways, each useful for their intended purpose.

The GxE interaction component can be partitioned into a component due to heterogeneity of genotypic variance among environments and a component attributable to lack of genetic correlation (Robertson, 1959; Cockerham, 1963). Muir et al. (1992) gave the equivalent partition in terms of sums of squares. The component of GxE interaction due to heterogeneity does not intrinsically complicate selection and attention can be focussed on the component due to lack of correlation. The distinction between these two components of GxE interaction allows estimation of the pooled genetic correlation (Dickerson, 1962) (Table 1). This statistic is complementary to heritability as it distinguishes between GxE interaction due to heterogeneity and lack of correlation whereas the estimate of heritability traditionally does not.

Given that an interaction component attributable to lack of correlation exists, it is those interactions which result in a change in the rank of genotypes across environments (reranking) which directly affect selection decisions (Haldane, 1947). Therefore, it would be useful to partition the interaction into that component which contributes to reranking and that which does not. We provide a method for partitioning sums of squares into these two components in the second part of this paper. Baker (1988, 1990) discussed methods for testing whether specific rank change interactions were significant. At a different level of resolution the GxE interactions can be investigated in terms of whether they contribute to a change in the composition of the select group at a defined truncation level (Gilmour, 1988; Eisemann et al., 1990).

Even if a fully random model is assumed, which indicates that only pooled analyses over all genotypes and environments should be evaluated, it is common to examine the data for patterns. This enables the breeder to understand the nature of the interactions present. This may enable a better biological understanding when integrated with other information, but may also enable repeatable patterns in the interaction to be recognised.

To enable patterns to be described the pooled statistics mentioned above may be decomposed into parts attributable to either individual environments on one hand or to individual genotypes on the other. That is, either the pooled variance components or the pooled sums of squares can be partitioned into: variation among genotype means, GxE interaction variation, heterogeneity of variance, lack of correlation, reranking or non-reranking parts for individual environments or genotypes (Table 1). The incentive to do this for environments is to investigate their ability to provide discrimination among the genotypes grown in them. The incentive for genotypes is to provide a measure of their reliability or stability. The details of these partitions are discussed later.

Many of the statistics can be further partitioned into all pair-wise comparisons of environments or genotypes. Examples are covariance, correlation or squared Euclidean distance matrices among environments or genotypes, which are collectively known as proximity matrices. Pattern analysis procedures (classification and ordination) are powerful methods for examining these proximity matrices and are discussed later.

The analysis of variance can be applied in a number of ways to quantify the magnitude of GxE interaction and particular categories of interaction which impact upon selection decisions. The impact of these on response to selection can be further investigated by estimation of direct and indirect selection parameters.

B. DIRECT AND INDIRECT SELECTION

The response to selection can be investigated as a case of direct or indirect response to selection. With direct response to selection the environments sampled in a MET are viewed as a sample from the target population of environments and are used as the basis for predicting performance. For indirect response to selection the environments in the MET are used to predict indirect response to selection for future, other or production environments. While they provide different frameworks for analysing response to selection, the approaches are related.

For a given population of environments the genotypic component of variance from the combined analysis of variance is equivalent to the average of the genetic components of covariance over all pair-wise combinations of environments. These genetic covariance components give rise to the genetic correlation between environments. Falconer (1952) proposed that the influence of GxE interactions on response to selection can be examined by investigating the genetic correlations between environments. A high genetic correlation indicates little GxE interaction. Therefore, investigation of variation for average performance across environments can proceed through analysis of the form of indirect response to selection among the environments. However, we have found the indirect selection framework more useful in cases where there are clearly defined selection and target populations of environments. Where this applies, response to selection in the target population can be analysed in terms of the indirect response from selection in the selection population of environments.

This approach was used to assess the opportunity to select wheat lines for Australian conditions based on their performance in international trials conducted by CIMMYT (Cooper et al., 1993a, 1993b; and Cooper and Woodruff, 1993) and to evaluate managed-environments as a strategy for stratifying the sampling of environmental conditions responsible for generating GxE interactions (Cooper et al., 1995).

C. PATTERN ANALYSIS

Pattern analysis refers to the joint use of classification and ordination methods (Williams, 1976). Byth et al. (1976) applied this approach to the study of GxE interactions for the yield of wheat in an international trial conducted by CIMMYT. At that time the method was advocated as an exploratory tool for description of patterns of adaptation of genotypes observed in METs. Since that initial stimulus, further work has been conducted to relate the use of pattern analysis methodology to quantitative genetic theory and selection objectives (Fox and Rosielle, 1982; Cooper and DeLacy, 1994).

We have found that pre-treatment of data from METs by the standardisation proposed by Fox and Rosielle (1982) (centring by removing the environment main-effect and scaling by dividing by the phenotypic standard deviation of each environment), prior to pattern analysis, results in a more meaningful analysis of the data where the objective is to consider the impact of GxE interaction on selection. The impact of this transformation is to correct for two aspects of variation which do not impact on selection decisions, environmental main-effects and heterogeneity of variance among environments. From this work we now have a combination of pattern analysis methods which complement the information obtained by application of the analysis of variance and selection theory methods. The pattern analysis methods we employ are used to explore the correlation structure among the environments and are thus related to selection objectives by concentrating on the component of GxE interaction which contributes to lack of correlation. Classification is used to identify groupings of environments which produce similar discrimination among genotypes. Following the relationship given by Gower (1966, 1967), enabling complementary dissimilarity and similarity measures to be calculated, the ordination is conducted on the standardised data using the singular value decomposition method, and biplots (Gabriel, 1971; Bardu and Gabriel, 1978; Kempton, 1984) are used to graphically portray the relationships among genotypes and environments in low dimensional space.

Note that the ordination procedure described above gives a different decomposition of the data than does the decomposition of the untransformed data matrix recently called AMMI analysis (Gauch, 1992). It is also possible to use the mixture method of clustering (McLachlan and Basford, 1988) to resolve the target population of environments into a mixture of populations for which the response of the genotypes can be studied. This method of clustering explicitly assumes an underlying model of a mixture of target populations of environments.

D. APPLICATION OF GxE INTERACTION ANALYSIS IN PLANT BREEDING

Each of the above data analyses can be implemented by standard statistical computer packages. We have also developed a dedicated computer package to implement the analyses. Access to high performance computer capability ensures that all analyses can be conducted in "real time". Two forms of data analysis may be implemented: (i) analysis to assist routine selection decisions, and (ii) detailed analysis of research experiments.

In the first case the main objective in data analysis is to predict response to selection for broad adaptation and characterise how similarities and differences in broad adaptation arose by investigating the patterns of adaptation of genotypes across environments (Table 1). Presently we recommend the use of the restricted or residual maximum likelihood (REML) analysis of variance (Patterson and Thompson, 1971 and 1974; Harville, 1977) and prediction of genotype performance by use of best linear unbiased predictors (BLUPs) from these analyses. The second objective is to determine whether any of the GxE interactions are associated with environmental challenges considered to be repeatable. If this is found to be the case then the strategy of selection for broad adaptation is re-evaluated to determine whether it is necessary to add a component of selection for specific adaptation. Where research experiments are being analysed and more time is available a more comprehensive analysis of differences in broad and specific adaptation is conducted.

III. PARTITION OF THE GENOTYPE BY ENVIRONMENT INTERACTION

The traditional joint linear regression technique (JLR) (Yates and Cochran, 1938; Finlay and Wilkinson, 1963; Eberhart and Russell, 1966; Perkins and Jinks, 1988) partitions the GxE interaction into components due to differences among regression slopes and deviations from regressions. The evaluation of the techniques has proceeded either by investigation of model fit by R^2 methods or partition of variance components. The R^2 or model fit methods are those which investigate the proportion of the total sum of squares (TSS) attributable to predicted values from the fitted model. If either the genotypes or environments are assumed fixed or partially so (mixtures of repeatable and non-repeatable patterns) then the performance of individual genotypes or environments can be investigated by linear regression. The JLR method has not been successful, indeed it has been severely criticised (Mungomery et al., 1974; Shorter, 1981; Westcott, 1986), because the predictive part (that due to regression) has not explained enough of the GxE interaction to be considered useful. A more fruitful approach has been to partition the interaction by pattern analysis methods (DeLacy and Cooper, 1990), i.e. by grouping (classification) and ordination (including AMMI) methods.

Another partition of the GxE interaction that has been proposed, which has had little attention to date, is into a component due to scaling effects (heterogeneity of variances) and lack of correlation. The lack of correlation is the component which directly impacts on the efficacy of selection whereas that due to heterogeneity of variances has little influence. Muir et al. (1992) referred to the second component as that due to 'imperfect correlations'. Again these methods have been developed by partitioning of variance components (Falconer, 1952; Robertson, 1959; Dickerson, 1962; Yamada, 1962; Itoh and Yamada, 1990; Cooper and DeLacy, 1994) or by model fit methods (Muir et al., 1992).

Lack of correlation of genotype performance across environments would have substantial impact on selection if it led to change of ranking of performance in different environments. Hence we propose a third partition of the GxE interaction into that due to changes in ranking of genotype performance and that which is not due to such changes. We develop the theory using the model fit method and integrate this with the heterogeneity of variance technique proposed by Muir et al. (1992). We investigate this partition with the artificial data set used by Muir et al. (1992) (their Table 6) and the wheat data set used by Cooper and DeLacy (1994).

A. THEORETICAL DEVELOPMENT

The evaluation in a MET of n_g genotypes in n_e environments for any attribute results in a set of observations recorded as a matrix of data, Y, whose general term y_{ij} is the phenotypic value, usually the mean of several observations, of genotype i in environment j. Since similarities and differences in patterns associated with either genotypes or environments are reflected in the variations in the observations in Y, their extent may be quantified using a measure of variability, the total sum of squares (TSS), defined as

$$\text{TSS} = \sum_{i=1}^{n_g} \sum_{j=1}^{n_e} (y_{ij} - \bar{y}_{..})^2, \qquad i=1,...,n_g \quad j=1,...,n_e, \qquad (1)$$

where $\bar{y}_{..}$ is the mean of the $n_g n_e$ observations in Y. Also $\bar{y}_{i.}$ and $\bar{y}_{j}$ are the genotype (row) and environment (column) means, respectively. The values in Y can be represented by the linear model

$$y_{ij} = m + g_i + e_j + (ge)_{ij} \qquad (2)$$

where the following restrictions apply

$$\sum_i g_i = \sum_j e_j = \sum_i (ge)_{ij} = \sum_j (ge)_{ij} = 0 \ . \qquad (3)$$

Then

$$m = \bar{y}_{..} \ , \quad g_i = \bar{y}_{i.} - \bar{y}_{..} \ , \quad e_j = \bar{y}_{.j} - \bar{y}_{..} \text{ and}$$
$$(ge)_{ij} = y_{ij} - g_i - e_j - m \ . \qquad (4)$$

The TSS can be partitioned by the following identity

$$\text{TSS} = n_e \sum_i g_i^2 + n_g \sum_j e_j^2 + \sum_{ij} (ge)_{ij}^2 = SS_G + SS_E + SS_{GE} \quad (5)$$

into sources of variation due to differences among genotype and environment means and the failure of genotypes to perform consistently over environments, respectively. The latter is referred to as genotype by environment (GxE) interaction.

1. Partitioning GxE interaction by heterogeneity of variances

Muir et al. (1992) gave two methods (Table 2) for partitioning the GxE interaction, both of which separated the interaction into sources due to (1) heterogeneous variances and (2) lack of correlation. Their results follow from the principle of symmetric differences (Casella and Berger, 1990). This principle gives the relationship between the sum of squares among all pairwise differences among the n elements of a vector x and the sum of squares of deviations of the elements from their mean; i.e.,

$$\sum_{i=1}^{n-1} \sum_{i'(i'>i)}^{n} (x_i - x_{i'})^2 = \sum_{i'>i} (x_i - x_{i'})^2 = n \sum_{i=1}^{n} (x_i - \bar{x})^2 \qquad (6)$$

where $\bar{x}$ is the mean of the elements. Our results on re-ranking given in the next section use the same relationship. Method 1 (Muir et al., 1992) separates the GxE interaction into heterogeneity among environments in the scaling of differences among genotypes and the remainder arises from deviations from

Table 2

Partition of genotype by environment interaction into sources due to heterogeneity among variances and lack of correlation among performance values following the method outlined by Muir et al. (1992).

Type of interaction	Method 1[a]	Method 2[b]
Heterogeneous variances		
Individuals	$Z^2_{(HV)j} = \dfrac{1}{2n_e} \sum\limits_{j' \neq j} (Z_j - Z_{j'})^2$	$S^2_{(HV)i} = \dfrac{1}{2n_g} \sum\limits_{i' \neq i} (S_i - S_{i'})^2$
Overall	$Z^2_{(HV)} = \sum\limits_{j=1}^{n_g} (Z_j - \bar{Z}.)^2$	$S^2_{(HV)} = \sum\limits_{i=1}^{n_g} (S_i - \bar{S}.)^2$
Lack of correlation		
Individuals	$Z^2_{(LC)j} = \dfrac{1}{n_e} \sum\limits_{j' \neq j} (1 - r_{jj'}) Z_j Z_{j'}$	$S^2_{(LC)i} = \dfrac{1}{n_g} \sum\limits_{i' \neq i} (1 - r_{ii'}) S_i S_{i'}$
Overall	$Z^2_{(LC)} = \dfrac{2}{n_e} \sum\limits_{j,j'>j} (1 - r_{jj'}) Z_j Z_{j'}$	$S^2_{(LC)} = \dfrac{2}{n_g} \sum\limits_{i,i'>i} (1 - r_{ii'}) S_i S_{i'}$
Total SS$_{GE}$		
Individuals - heterogenous variance partition	$Z^2_{H(GE)j} = Z^2_{(HV)j} + Z^2_{(LC)j}$	$S^2_{H(GE)i} = S^2_{(HV)i} + S^2_{(LC)i}$
Individuals - traditional partition	$Z^2_{(GE)j} = \sum\limits_{i=1}^{n_g} (y_{ij} - \bar{y}_{i.} - \bar{y}_{.j} + \bar{y}_{..})^2$	$S^2_{(GE)i} = \sum\limits_{j=1}^{n_e} (y_{ij} - \bar{y}_{i.} - \bar{y}_{.j} + \bar{y}_{..})^2$
Overall	$SS_{GE} = \sum\limits_{i=1}^{n_g} \sum\limits_{j=1}^{n_e} (y_{ij} - \bar{y}_{i.} - \bar{y}_{.j} + \bar{y}_{..})^2$ for both models	

a: $Z^2_j = \sum\limits_i (y_{ij} - \bar{y}_{.j})^2$, $Z_{jj'} = \sum\limits_i (y_{ij} - \bar{y}_{.j})(y_{ij'} - \bar{y}_{.j'})$, $r_{jj'} = Z_{jj'}/[Z_j Z_{j'}]$

b: $S^2_i = \sum\limits_j (y_{ij} - \bar{y}_{i.})^2$, $S_{ii'} = \sum\limits_j (y_{ij} - \bar{y}_{i.})(y_{i'j} - \bar{y}_{i'.})$, $r_{ii'} = S_{ii'}/[S_i S_{i'}]$

perfect positive correlations of genotype performances among environments. Their method 2 separates the GxE interaction into heterogeneity among

64

genotypes in the scaling of differences among environments and the remainder arises from deviations from perfectpositive correlations among genotype performances across environments. It should be noted that the partitioning of the GxE interaction sums of squares into that attributable to individual environments or genotypes is not the same as that calculated by the traditional method (Table 2). For completeness we give the formulations for both methods in our notation (Table 2). Muir et al. (1992) did not give the results for the partitioning of the sum of squares for individual environments. They recommended method 1 as being suitable for random genotypes being tested in fixed or random environments and method 2 for evaluation of fixed genotypes.

2. Partitioning of the GxE interaction by reranking

The failure of genotypes to perform the same over environments depends on both the difference in mean performance and their failure to perform consistently. Hence a measure of this failure is given by the total genotype sum of squares, which is the sum of the genotype and the GxE interaction sum of squares, viz.,

$$SS_{(G-GE)} \cdot SS_G \cdot SS_{GE} . \tag{7}$$

Another measure of the failure of genotypes to perform the same across environments is the sum of squares of paired differences between the genotypes in each of the environments. This, the squared Euclidean distance between genotypes i and i' ($SED_{ii'}$), is

$$d_{ii'}^2 = \sum_j (y_{ij} - y_{i'j})^2 = n_e(g_i - g_{i'})^2 + \sum_j ((ge)_{ij} - (ge)_{i'j})^2 . \tag{8}$$

Thus, the $SED_{ii'}$ is a function of squared differences among genotype effects and GxE interaction effects.

There are direct relationships between the two measures of variability among genotypes given in equations (7) and (8). From equation (6) the sum of all SEDs among the n_g genotypes is

$$\sum_{i,i'>i} d_{ii'}^2 = \sum_{i,i'>i} \sum_j (y_{ij} - y_{i'j})^2 = \sum_j n_g \sum_i (y_{ij} - \bar{y}_j)^2 \tag{9}$$

and from equations (3) and (8) (with the relationship in equation (6))

$$\sum_{i,i'>i} d_{ii'}^2 = n_e \sum_{i,i'>i} (g_i - g_{i'})^2 + \sum_{i,i'>i} \sum_j ((ge)_{ij} - (ge)_{i'j})^2$$
$$= n_e n_g \sum_i g_i^2 + \sum_j n_g \sum_i (ge)_{ij}^2 . \tag{10}$$

From equations (5) and (10) and the definition of z_j^2 (Table 2), the corrected sum of squares for environment j, it follows that

$$\frac{1}{n_g} \sum_{i,i'>i} d_{ii'}^2 = \sum_j z_j^2 = SS_G + SS_{GE} \,. \tag{11}$$

Thus, the sum of the SEDs among genotypes is proportional to the genotype plus GxE sums of squares.

Genotype by environment interaction causes most problems (for selection) when it leads to reranking of genotype performance in different environments. The relationships above can be used to provide measures of this complicating factor. First reorder the rows of Y in order of decreasing genotype means, i.e. from largest to smallest g_i. Then the difference between the performance values of genotypes i and i', $i<i'$ at a location, will only be negative when there has been a reranking of performance in an environment compared to the ranking of the genotype mean over all environments. To formalise this, first define

$$d_{(ii')j} = (y_{ij} - y_{i'j}) = \begin{cases} d^+_{(ii')j} & \text{when } (y_{ij} - y_{i'j}) \geq 0 \text{ and} \\ d^-_{(ii')j} & \text{when } (y_{ij} - y_{i'j}) < 0 \end{cases} \quad , i<i' \,. \tag{12}$$

Now

$$d^-_{(ii')j} = (g_i - g_{i'}) + ((ge)_{ij} - (ge)_{i'j})^- = g_{(ii')} + (ge)^-_{(ii')j} \quad , i<i' \tag{13}$$

where the final terms in this equation are defined by the differences given previously in the equation. This difference, $d_{(ii')j}$, will only be negative, i.e. a rank change has occurred, when the difference between the GxE effects for the two genotypes in an environment is negative and greater in absolute terms than the difference between the genotype effects. The difference between the genotype effects will always be positive (or zero) as the genotypes have been ranked. Note that the difference between the GxE effects may be negative and not result in a rank change if it is smaller in absolute terms than the difference in genotype effects. Also the absolute value of $d^-_{(ii')j}$ is smaller than or equal to the absolute value of $(ge)^-_{(ii')j}$ as it is the difference in the GxE effects (which is negative) reduced by the difference between the genotype effects (which is zero or positive). The difference in genotype effects will be zero when their means are the same. Thus, we have two measures of the importance of reranking between the genotypes at an environment: *viz.* one calculated from the $d^-_{(ii')j}$ and one from those $(ge)^-_{(ii')j}$ which are large enough to make their associated $d^-_{(ii')j}$ negative.

Now we can partition z_j^2 into two parts (again using the relationship in equation (6)),

$$Z_j^2 = \frac{1}{n_g} \sum_{i,i'>i} d_{(\ddot{u}'\backslash j}^2$$

$$= \frac{1}{n_g} \left(\sum_{i,i'>i} (d_{(\ddot{u}'\backslash j}^{+})^2 + \sum_{i,i'>i} (d_{(\ddot{u}'\backslash j}^{-})^2 \right) \tag{14}$$

$$= Z_{(N)j}^2 + Z_{(R)j}^2$$

where we have separated the positive and negative terms in the summation. The second term in equation (14) $Z_{(R)j}^2$, referred to as the reranking sum of squares for environment j, is the sum of squares of all the pairwise differences among the genotypes which are negative, *i.e.* when there has been a change of rank of genotypes in environment j compared to the genotype mean over all environments. It provides a measure of the importance of changes in rank of genotype performance among environments. $Z_{(N)j}^2$ is the non-reranking sum of squares for environment j. From equations (13) and (14) we can write

$$Z_{(R)j}^2 = \frac{1}{n_g} \sum_{i,i'>i} (d_{(\ddot{u}'\backslash j}^{-})^2 = \frac{1}{n_g} \sum_{i,i'>i} \left(g_{(\ddot{u}'\backslash j} + (ge)_{(\ddot{u}'\backslash j}^{-} \right)^2 . \tag{15}$$

Define

$$Z_{R(GE)j}^2 = \frac{1}{n_g} \sum_{i,i'>i} \left((ge)_{(\ddot{u}'\backslash j}^{-} \right)^2 \tag{16}$$

as the sum of squares of the GxE effects which cause genotype reranking in an environment. As argued before,

$$\left| d_{(\ddot{u}'\backslash j}^{-} \right| \le \left| (ge)_{(\ddot{u}'\backslash j}^{-} \right| \tag{17}$$

so

$$\sum_{i,i'>i} (d_{(\ddot{u}'\backslash j}^{-})^2 \le \sum_{i,i'>i} \left((ge)_{(\ddot{u}'\backslash j}^{-} \right)^2 , \tag{18}$$

and hence

$$Z_{(R)j}^2 \le Z_{R(GE)j}^2 . \tag{19}$$

We know (Table 2 and the relationship in equation (6)) that

$$Z^2_{(GE)j} = \frac{1}{n_g} \sum_{i,i'>i} (ge)^2_{(ii')j} \tag{20}$$

is the traditional estimate of the GxE interaction sum of squares attributable to environment j and because it is the sum of all the positive and negative squared GxE interaction effects, we have

$$Z^2_{R(GE)j} \leq Z^2_{(GE)j} . \tag{21}$$

$Z^2_{R(GE)j}$ is another measure of the importance of change of genotype rank in different environments. Combining equations (19) and (21) we have

$$Z^2_{(R)j} \leq Z^2_{R(GE)j} \leq Z^2_{(GE)j} . \tag{22}$$

Hence both measures, $Z^2_{(R)j}$ and $Z^2_{R(GE)j}$, of the amount of change of rank in an environment compared to the overall ranking of the genotypes are partitions of the GxE interaction attributable to that environment.

Summing over all environments gives

$$Z^2_R = \sum_j Z^2_{(R)j} , \tag{23}$$

$$Z^2_{R(GE)} = \sum_j Z^2_{R(GE)j} \quad \text{and} \tag{24}$$

$$SS_{(GE)} = \sum_j Z^2_{(GE)j} \tag{25}$$

which provide a partition of the SS_{GE} into sources which measure the overall importance of change of rank. It follows that

$$Z^2_R \leq Z^2_{R(GE)} \leq SS_{GE} \tag{26}$$

and these two measures, Z^2_R and $Z^2_{R(GE)}$, of the importance of GxE interaction in causing changes in rank of genotype performance across all environment, are

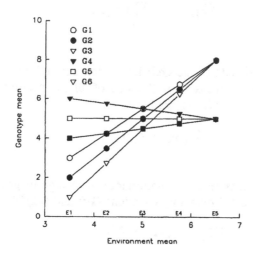

Figure 2: Performance plot of the yield of six genotypes in five environments (artificial data from Muir et al. (1992)).

partitions of the GxE interaction sum of squares. The first is the reranking sum of squares (z_R^2) which is $1/n_g$ times the sum of squares of all the pairwise differences that result in change of ranking of genotype performance in environments in comparison to the ranking of overall genotype means. This is less than or equal to the second measure, the GxE reranking sum of squares ($z_{R(GE)}^2$), which is again $1/n_g$ times the sum of squares of the differences of GxE effects which lead to change of rank of genotypes. Both partition the GxE sum of squares (SS_{GE}).

B. EXAMPLES
1. Numerical example of Muir et al. (1992)
Muir et al. (1992) used an artificial data set with six genotypes and five environments (Figure 2) to demonstrate the interpretation of the statistics for heterogeneity of variances (s_{HV}^2) and lack of correlation (s_{LC}^2) among genotypes. The example was devised such that genotypes 1, 2 and 3 (G1, G2, G3) were responsive but varied in their response to poor environments. Genotypes 4, 5 and 6 (G4, G5, G6) responded less, with G4 the only genotype better in the worst environments. In this example 28.33 or 78% of the SS_{GE} (36.25) was due to scaling differences and 7.92 or 22% due to lack of correlation among the

Table 3

Partition of the GxE sums of squares for the artificial data from Muir et al. for six genotypes in five environments into sources due to heterogeneous variances (HV) between lines and lack of correlation among line performance across environments.

Genotypes	HV	Lack of correlation	Partition of SS_{GE} by HV	Traditional partition of SS_{GE}	Percentages	
					HV[a]	Lack of correlation[b]
	$\hat{S}^2_{(HV)i}$	$\hat{S}^2_{LCi}$	$\hat{S}^2_{H(GE)i}$	S^2_{GEi}		
G1[c]	3.23	1.04	4.27	2.50	76	24
G4	4.06	3.96	8.02	10.00	51	49
G2	4.58	1.25	5.83	5.63	79	21
G5	5.83	0.00	5.83	5.63	100	0
G3	6.56	1.46	8.02	10.00	82	18
G6	4.06	0.21	4.27	2.50	95	5
Overall (sum)	28.33	7.92	36.25	36.25	78	22

(a) $S^2_{(HV)i}$ as a % of $\hat{S}^2_{H(GE)i}$; (b) S^2_{LCi} as a % of $S^2_{H(GE)i}$.
(c) Genotypes ordered by decreasing mean yield.

genotypes (Table 3). The partition of the SS_{GE} into parts attributable to single genotypes by the method of heterogeneous variances differed from that due to the partition due to the traditional method. In this example both methods gave a similar ranking of the genotypes for stability with G4 and G3 contributing the most to SS_{GE} and G1 and G6 the least. The lack of correlation parameter ($\hat{S}^2_{LCi}$) gave a useful summary of the genotype performance indicating that G4's performance was the least correlated with other genotypes followed by that of G3. Examination of the percentages of the GxE interaction sum of squares attributable to the genotype partitioned to lack of correlation shows that G4 is the highest followed by G1 and G2. This model fit or 'R²' parameter gives further information to that provided by the parameter. Both should be interpreted. Muir et al. (1992) give a more detailed discussion.

Muir et al. (1992) did not examine this example for heterogeneity among environments, but it is instructive to do so. In this case 19% of the SS_{GE} (Table 4) was due to scaling differences and 81% was due to lack of correlation; the reverse of the situation when examining genotypes. It is appropriate to examine individual environment differences when they are assumed fixed or a mixture, i.e. some repeatable differences among environments. In this example the GxE

Table 4

Partition of the GxE sums of squares for the artificial data from Muir et al. for six genotypes in five environments into sources due to heterogeneous variances (HV) between environments and lack of correlation among line performance in different environments.

Environment	HV	Lack of correlation	Partition of SS_{GE} by HV	Traditional partition of SS_{GE}	Percentage	
					HV[a]	Lack of correlation[b]
	$Z^2_{(HV)j}$	$Z^2_{(LC)j}$	$Z^2_{H(GE)j}$	$Z^2_{(GE)j}$		
E5[c]	1.22	9.65	10.88	14.50	11	89
E4	0.94	4.50	5.44	3.63	17	83
E3	2.00	1.63	3.63	0.00	55	45
E2	0.71	4.73	5.44	3.63	13	87
E1	1.89	8.99	10.88	14.50	17	83
Overall (sum)	6.75	29.50	36.25	36.25	19	81

(a) $Z^2_{(HV)j}$ as a % of $Z^2_{H(GE)j}$; (b) $Z^2_{(LC)j}$ as a % of $Z^2_{H(GE)j}$.
(c) Environments ordered by decreasing mean yield.

variability attributable to each environment was dominated by the lack of correlation of genotype performance in these environments with the others (Table 4). The lack of correlation statistics indicate that the performance of genotypes in environments 1 and 5 was the least correlated with their performance in the other environments. In contrast their R^2 (percentage) values indicated that a similar proportion of their GxE sum of squares for all environments (except 3) were due to this source.

Given that much of the SS_{GE} attributable to environments is caused by lack of correlation of performance from one environment to another, examination of the change in rank in genotype performance is indicated. From the reranking statistics (Table 5), differences in genotype performance leading to change of rank in environments were greater in environments 1 and 5 than in other environments and the rankings of the genotypes in environment 3 were the same as for the average over all environments.

2. Wheat example

Cooper and DeLacy (1994) used a wheat data set to study the relationships among analytical methods for the analysis of GxE interactions in plant breeding METs. Fifteen wheat lines were tested in ten environments in Queensland,

Table 5

Partition of the GxE sums of squares for the artificial data from Muir et al. for six genotypes in five environments into sources due to non-reranking (NRR) and reranking (RR) of line performance between environments.

Environment	NRR SS	RR SS	Usual partition of SS_{FE}	$SS_{(G+GE)}$	Percentage		ge NRR SS	ge RR SS	Percentage
					SS_{GE} due to RR[a]	$SS_{(G+GE)}$ due to RR[b]			SS_{GE} due to ge RR[c]
	$z^2_{(N)j}$	$z^2_{(R)j}$	$z^2_{(GE)j}$	z^2_j			$z^2_{N(GE)j}$	$z^2_{R(GE)j}$	
E5[d]	11.00	4.50	14.50	15.50	31	29	7.75	6.75	47
E4	3.94	0.69	3.63	4.63	19	15	1.94	1.69	47
E3	1.00	0.00	0.00	1.00	0	0	0	0	0
E2	3.31	1.31	3.63	4.63	36	28	1.98	1.65	45
E1	9.50	6.00	14.50	15.50	41	39	7.25	7.25	50
Overall (sum)	28.75	12.50	36.25	41.25	34	30	18.92	17.33	48

(a) $z^2_{(R)j}$ as a % of $z^2_{(GE)j}$; (b) z^2_{Rj} as a % of z^2_j; (c) $z^2_{R(GE)j}$ as a % of $z^2_{(GE)j}$.
(d) Environments ordered by decreasing mean yield.

Australia in 1988. The environments sampled the main regions of the Queensland wheat belt identified by Brennan et al. (1981). The lines included three local cultivars and 12 advanced lines from the 11th and 17th International Bread Wheat Screening Nurseries conducted by CIMMYT. Each experiment was laid out as a randomised complete block design with two replications. Line mean grain yield was rearranged into decreasing mean yield for lines as rows and decreasing mean yield of environments as columns (Table 6).

Cooper and DeLacy assumed a completely random model. This model implies that any individual pattern (described from or present in the data) occurs in low frequency. Under these conditions line selection devolves to selecting for broadly adapted lines and the selection strategy to determining the optimum number of test lines, test environments and replications. These decisions require a knowledge of the relative sizes of sources of variation and a description of the form of the GxE interaction. The form of the interaction is described by a number of different partitions of the interaction.

a. Quantitative genetic statistics from the pooled analysis of variance

When assuming a completely random model, the variance component for LxE interaction was significant and 0.8 times that for lines (Table 7, 0.09 *vs.*

Table 6

Line mean grain yield (t ha⁻¹) for 15 wheat lines tested in ten environments in Queensland in 1988; rows (lines) and columns (environments) arranged in decreasing mean yield.

Line	Environment										
	Fern[a]	Gatt	Emd-2	Emd-1	Gums	Jim	Pamp	Toob	King	Bil	Mean
17IB206	6.48	6.295	4.061	4.52	3.64	4.126	3.23	3.226	3.392	2.505	4.148
Genaro	5.801	6.21	4.481	4.324	3.491	4.12	3.356	3.817	3.25	2.167	4.102
17IB30	5.75	5.57	4.319	4.482	4.032	3.806	3.032	3.448	2.951	2.358	3.975
17IB173	4.834	4.956	4.006	4.687	3.792	3.305	3.063	3.124	3.357	2.098	3.722
17IB53	5.431	5.451	4.244	3.401	3.292	3.662	2.929	2.719	2.845	2.075	3.605
17IB64	5.433	4.694	4.088	3.639	3.497	3.515	3.411	3.169	2.704	1.899	3.605
Hartog	5.408	4.559	4.049	3.326	3.535	3.386	2.934	2.832	3.113	2.05	3.519
17IB92	5.172	4.379	3.751	3.839	3.473	3.154	3.315	2.86	2.267	2.142	3.435
17IB7	4.977	5.093	4.051	3.311	3.031	3.145	2.909	3.308	2.592	1.742	3.416
17IB38	5.285	4.606	4.399	3.042	3.471	3.284	2.919	2.447	2.693	1.942	3.409
17IB31	5.15	4.592	3.893	3.496	3.139	3.382	3.13	2.783	2.571	1.773	3.391
11IB50	5.039	4.596	3.558	3.511	3.438	3.293	2.923	2.99	2.883	1.592	3.382
Kite	4.306	5.409	3.363	3.089	3.024	2.985	2.456	2.719	2.173	1.91	3.143
17IB129	4.782	3.733	3.429	3.148	3.256	2.899	2.486	2.103	2.168	2.662	3.067
Banks	5.02	2.821	4.421	3.028	3.342	3.079	2.406	2.096	1.853	1.653	2.972
Mean	5.258	4.864	4.008	3.656	3.43	3.409	2.967	2.909	2.721	2.038	3.526

(a) Environment abbreviations: Fernlees (Fern), Gatton (Gatt), Emerald sowing date 2 (Emd-2), Emerald sowing date 1 (Emd-1), The Gums (Gums), Jimbour (Jim) , Pampas (Pamp), Toobeah (Toob), Kingsthorpe (King) and Biloela (Bil).

0.11). Line mean heritability was a relatively high at 0.90. However 0.06 or 69% of LxE interaction (0.09) was associated with lack of correlation resulting in a pooled genetic correlation of 0.65.

Even when a completely random model is assumed it is common to examine the patterns in the data, if for no other reason than to describe them in an attempt to understand their causes. Should some pattern among environments be recognised as persistent (repeatable) over years then the model assumptions for environments should be changed to either a mixture model, a mix of fixed and random effects or a fixed model. In any case repeatable patterns, when

Table 7

Environment parameters from analysis of variance of 15 wheat lines tested in ten environments in Queensland in 1988. Each trial was laid out as a randomised complete block design.

Environ-ment[a]	Mean grain yield (t ha^{-1})	Number of blocks	LSD 5%	Error variance component	Genotypic variance component	Phenotypic variance component	H^2 (line mean basis)
Fern	5.258	2	0.581	0.073	0.223	0.259	0.859
Gatt	4.864	2	1.270	0.351	0.616	0.792	0.778
Emd-2	4.008	2	0.575	0.072	0.089	0.125	0.711
Emd-1	3.656	2	0.600	0.078	0.294	0.333	0.883
Gums	3.430	2	0.469	0.048	0.049	0.073	0.673
Jim	3.409	2	0.430	0.040	0.121	0.141	0.858
Pamp	2.967	2	0.225	0.011	0.094	0.100	0.945
Toob	2.909	2	0.434	0.041	0.202	0.222	0.908
King	2.721	2	0.395	0.034	0.196	0.213	0.920
Bil	2.038	2	0.415	0.043	0.071	0.092	0.767
Overall from pooled analysis of variance				0.079	0.109	0.274	0.897

Line by environment interaction variance component - 0.086
Heterogeneity of genotypic variance - 0.027
Lack of genetic correlation - 0.059
Pooled genetic correlation - 0.651

(a) Environment abbreviations given in Table 6.

identified, have important consequences for line selection and selection strategy (Table 1).

b. Quantitative genetic statistics for individual locations

The two highest yielding environments, Fernlees in central Queensland and Gatton in south-east Queensland, produced differing performance characteristics among the lines grown in them (Table 7). Fernlees had a low phenotypic variance compared to Gatton but a higher heritability. Phenotypic correlations with other environments (Table 8) were more consistent for Fernlees (ranging from 0.40 to 0.90; 0.53 with Gatton) than those for Gatton (ranging from 0.16 to 0.79). These differences were also reflected in the genetic correlations.

The two lowest yielding environments also differed (Table 7) in that Biloela had the second lowest phenotypic variance and a moderate heritability while Kingsthorpe had the second highest heritability. The lowest yielding environment, Biloela, had the only negative phenotypic correlation, -0.07, with Emerald planting date two (Table 8), and consistently lower correlations with other environments than did Kingsthorpe.

Table 8

Phenotypic (lower triangle) and genotypic (upper triangle) correlations among ten environments for grain yield of 15 wheat lines tested in Queensland in 1988.

Environ- ment[a]	Fern	Gatt	Emd-2	Emd-1	Gums	Jim	Pamp	Toob	King	Bil
Fern		0.65	0.74	0.65	0.73	1.05	0.70	0.57	0.71	0.49
Gatt	0.53		0.21	0.74	0.34	0.93	0.62	0.94	0.85	0.42
Emd-2	0.58	0.16		0.32	0.59	0.73	0.40	0.36	0.41	-0.09
Emd-1	0.57	0.61	0.25		0.97	0.80	0.70	0.82	0.81	0.54
Gums	0.55	0.25	0.41	0.75		0.69	0.54	0.51	0.75	0.58
Jim	0.90	0.76	0.57	0.69	0.53		0.74	0.81	0.85	0.41
Pamp	0.63	0.53	0.33	0.64	0.43	0.66		0.79	0.68	0.13
Toob	0.50	0.79	0.29	0.73	0.40	0.71	0.73		0.78	0.10
King	0.63	0.72	0.33	0.73	0.59	0.76	0.63	0.71		0.32
Bil	0.40	0.32	-0.07	0.44	0.41	0.33	0.11	0.08	0.27	

(a) Environment abbreviations given in Table 6.

c. Heterogeneity statistics for environments (model 1, Table 2) Overall heterogeneity measures calculated from partitioning of the sum of squares (Table 9) confirmed those from variance components (Table 7). Overall 27% of the GxE interaction sum of squares was attributable to heterogeneity and 73% to lack of correlation. This compares to the 31% (0.027) to 69% (0.086) split (Table 7) when calculated by variance components. For individual environments the sum of squares attributable to each environment (Table 9) was closely correlated (r=0.88) with the phenotype variance (Table 7). There were considerable differences between the two highest yielding environments in the way they produced differences among the lines, with Gatton having a much higher total, GxE interactions and lack of correlation sum of squares associated

Table 9

Partition of the GxE sums of squares for each of the ten environments for gain yield of 15 wheat lines into sources due to heterogeneous variances between environments and lack of correlations among line performance in different environments.

Environment	Heterogeneous variances	Lack of correlation	Partition of SS_{GE} by heterogeneous variances	Traditional partition of SS_{GE}	Percentages	
					Heterogeneous variances[a]	Lack of correlation[b]
	$z^2_{(HV)j}$	$z^2_{(LC)j}$	$z^2_{H(GE)j}$	$z^2_{(GE)j}$		
Fern[c]	0.23	1.20	1.43	1.27	16	84
Gatt	1.55	2.04	3.59	5.59	43	57
Emd-2	0.28	1.43	1.71	1.84	16	84
Emd-1	0.32	1.25	1.57	1.55	20	80
Gums	0.44	0.86	1.30	1.02	34	66
Jim	0.25	0.68	0.93	0.27	27	73
Pamp	0.34	0.84	1.18	0.78	29	71
Toob	0.21	1.07	1.28	0.98	17	83
King	0.21	0.97	1.18	0.78	18	82
Bil	0.37	1.31	1.67	1.76	22	78
Overall (sum)	4.20	11.64	15.84	15.84	27	73

(a) $z^2_{(HV)j}$ as a % of $z^2_{H(GE)j}$; (b) $z^2_{(LC)j}$ as a % of $z^2_{H(GE)j}$.
(c) Environment abbreviations given in Table 6.

with it than did Fernlees (Table 9). Gatton, however, had a smaller proportion of its GxE sum of squares attributable to lack of correlation than did Fernlees (57% *vs.* 84%), albeit a smaller proportion of a much larger amount (Table 9). The two lowest yielding environments did not differ as much for these statistics. Biloela had a larger lack of correlation sum of squares but a similar proportion of its GxE interaction sum of squares attributable to this source as did Kingsthorpe.

d. Reranking statistics for environments

Given that a large proportion of the GxE interaction sum of squares was associated with lack of correlation, indicating complications for selection, inspection of the amount of change of rank in line performance is indicated

(Table 10). It is immediately obvious that while a large proportion of the GxE interaction was due to lack of correlation (73% overall, Table 9) a much smaller proportion is attributable to change of rank (12% overall, Table 10). A larger proportion (34% overall) of GxE effects was associated with reranking.

Table 10

Partition of the GxE interaction sums of squares for each of the ten environments for grain yield of 15 wheat lines into sources due to reranking and non-reranking of line performance in each environment.

Environ- ment	Non Reranking SS	reranking SS	traditional partition of SS_GE	SS_(G+GE)	percentage SS_GE due to reranking[a]	SS_(G+GE) due to reranking[b]	ge non reranking SS	ge reranking SS	percentage SS_GE due to ge reranking[c]
	$Z^2_{(N)j}$	$Z^2_{(R)j}$	$Z^2_{(GE)j}$	Z^2_j			$Z^2_{N(GE)j}$	$Z^2_{R(GE)j}$	
Fern[d]	2.81	0.17	1.27	2.98	13	16	0.87	0.40	31
Gatt	6.90	0.40	5.59	7.30	7	5	4.82	0.77	14
Emd-2	3.16	0.39	1.84	3.55	21	11	0.52	1.32	72
Emd-1	3.17	0.09	1.55	3.26	6	3	1.32	0.23	15
Gums	2.62	0.11	1.02	2.73	10	4	0.65	0.37	37
Jim	1.95	0.03	0.27	1.98	10	1	0.20	0.07	25
Pamp	2.41	0.08	0.78	2.49	10	3	0.44	0.34	43
Toob	2.54	0.14	0.98	2.68	14	5	0.67	0.30	31
King	2.40	0.10	0.78	2.49	12	4	0.60	0.19	24
Bil	3.04	0.43	1.76	3.47	24	12	0.41	1.35	77
Overall	30.99	1.92	15.84	32.92	12	6	10.50	5.34	34

(a) $Z^2_{(R)j}$ as a % of $Z^2_{(GE)j}$; (b) $Z^2_{(R)j}$ as a % of Z^2_j; (c) $Z^2_{R(GE)j}$ as a % of $Z^2_{(GE)j}$.
(d) Environment abbreviations given in Table 6.

Again Gatton had a much larger amount (0.40) of variation associated with change of rank, $Z^2_{(R)j}$, than did Fernlees (0.17) and Biloela's amount of variation associated with change of rank (0.43) was larger than Kingsthorpe's (0.10). Not only did these environments have 'larger lack of correlation' or lower correlation of line performance with other environments than did Fernlees and

Table 11

Partition of the GxE interaction sums of squares for grain yield for each of the 15 wheat lines tested in ten environments into sources due to heterogeneous variances between lines and lack of correlations among line performance across environments.

Genotypes	Heterogeneous variances	Lack of correlation	Partition of SS$_{GE}$ by heterogeneous variances	Traditional partition of SS$_{GE}$	Percentages Heterogeneous variances[a]	Lack of correlation[b]
	$\hat{s}^2_{HVi}$	$\hat{s}^2_{LCi}$	$s^2_{H(GE)i}$	$\hat{s}^2_{GEi}$		
17IB206	0.42	1.02	1.44	1.82	29	71
Genaro	0.20	0.95	1.15	1.24	17	83
17IB30	0.09	0.68	0.77	0.48	12	88
17IB173	0.13	1.14	1.26	1.47	10	90
17IB53	0.11	0.70	0.81	0.57	13	87
17IB64	0.08	0.61	0.68	0.30	11	89
Hartog	0.09	0.63	0.73	0.40	13	87
17IB92	0.12	0.74	0.86	0.67	14	86
17IB7	0.07	0.78	0.84	0.63	8	92
17IB38	0.07	0.81	0.88	0.70	8	92
17IB31	0.08	0.53	0.61	0.16	13	87
11IB50	0.10	0.62	0.72	0.38	13	87
Kite	0.07	1.18	1.25	1.44	5	95
17IB129	0.30	1.16	1.46	1.86	20	80
Banks	0.08	2.31	2.39	3.72	3	97
Overall (sum)	1.98	13.86	15.84	15.84	12	88

(a) $\hat{s}^2_{HVi}$ as a % of $s^2_{H(GE)i}$;   (b) $\hat{s}^2_{LCi}$ as a % of $s^2_{H(GE)i}$

Kingsthorpe, but they also had more differences in ranking of line performances. The second planting at Emerald also produced considerable change of ranking of line performance compared to the average. Because Gatton had a large GxE interaction sum of squares associated with it, the percentage that led to change in ranking was below average. Biloela and the second planting at Emerald had a relatively large proportion of their GxE interaction sum of squares associated with change in ranking of line performance. In this data set the information on change of rank of lines gained from the second statistic, $z^2_{R(GE)j}$, on reranking did not add significantly to that already gained from the first measure.

e. Heterogeneity statistics for lines (model 2, Table 2)

Overall 12% of the GxE interaction was associated with heterogeneity and 88% with lack of correlation among lines (Table 11). The percentage for lack of correlation was higher, 88%, compared with 73% for partitioning on an environment basis (Table 9). Interestingly, the higher partition for lack of correlation for model 2 is not due to lower correlations. The average of the phenotypic correlations among line performance across locations ($r_{ii'}$) is 0.89 while that for the correlations among line performance among locations ($r_{jj'}$) is 0.50, i.e. ($1-r_{ii'}$) is smaller on the average than ($1-r_{jj'}$); hence the larger proportion of lack of correlation for the line partition is due to the sums of squares among lines across environments (s_i^2). From an analogous argument to that in equations (10) and (11) the SS_E contributes to s_i^2. Hence the higher component due to lack of correlation for model 2 reflects the contribution of the sum of squares for environments, SS_E, to the sum of squares among lines across environments, the s_i^2. The sum of squares for environments is usually much larger than the sum of squares for lines, and, as such, dominates many statistics calculated to describe relationships among lines.

The three lowest yielding lines and the two highest were the highest contributors to GxE interaction sum of squares, with Banks the lowest yielding line contributing the most. To the extent that this is a measure of stability the lowest and highest yielding lines were the least stable in this data set. The lack of correlation parameter, a combination of lack of correlation of a genotype's performance with other line and the variability of the genotype's performance across locations have been suggested by Muir et al. (1992) as an alternative stability parameter. In this data set the lack of correlation only marginally altered the ranking of the lines for stability compared to the measure of GxE interaction sum of squares attributable to the lines.

IV. DISCUSSION

The objectives of this chapter were to: (1) demonstrate that the selection of a suite of analytical methods to analyse data from METs should be driven by the purpose of the analysis and (2) provide linkages among pooled statistics from the combined analysis of variance and the same statistics partitioned into averages (or sums) among individual environments and/or genotypes or partitioned to all pair-wise measures between individuals. Secondary objectives were to draw linkages between model fit and variance component investigations of the methods and to develop additional theory to link aspects of the work.

The main (though not the only) purpose of METs is to provide information for the selection among breeding lines or populations. In consequence, the models used in quantitative genetics to describe selection provide the appropriate framework for selection among analytical methods. It is the lack of correlation, or more specifically the component of lack of correlation which leads to change of rank of performance of genotypes from one environment to another that decreases efficiency in selection. This decrease in efficiency will occur when the composition of the selected group changes. This being the case, analytical methods designed to investigate reduction in correlations, change of rank or change in the selected group in different environments are appropriate. A completely random model implies that any individual pattern (described from or present in the data) occurs in low frequency. Under these conditions interpretation is focused on pooled statistics giving information for selecting broadly adapted lines and determining the optimum number of genotypes, test environments and replications. These decisions require a knowledge of the relative size of sources of variation and a description of the form of GxE interaction. The form of the interaction is described in a progressively detailed manner moving from the relative importance (size) of the interaction to its importance in causing lack of correlation and then change in ranking of performance across environments. Useful statistics in addition to the error, genetic, phenotypic and GxE variance components include heritability, heterogeneity and lack of correlation partitions of the interaction component and the pooled genetic correlation. To these can now be added reranking and non-reranking partitions of the GxE interactions.

Even when a completely random model is assumed it is common to examine the patterns in the data, if for no other reason than to describe them in an attempt to understand their causes. Should some pattern among environments be recognised as persistent (repeatable) over years then the model assumptions for environments should be changed to either a mixture model, a mix of fixed and random effects or a fixed model. In any case repeatable patterns, when identified, have important consequences for line selection and selection strategy (Table 1). To enable patterns to be described in the data the pooled statistics can be partitioned into components associated with either individual environments or genotypes. The partitions associated with the environments can be considered to be measures of discrimination power, and those associated with genotypes as stability measures. These include measures of size and form of the variation associated with individual genotypes or environments. Measures of size include the phenotypic, genotypic or GxE interaction variation and measures of form include linear, lack of correlation and now reranking partitions of the GxE interaction. There has been considerable concentration in the literature on devising and estimating stability parameters for genotypes but less for defining and estimating discrimination parameters for environments. This is unfortunate for two reasons. Explicit development and investigation of the parallel statistics associated with the partition of the variation into sources associated with individual genotypes or environments gives a better understanding of both their strengths and weaknesses. Also their combined use enables a better description of the patterns of performance and discrimination in the data.

These partitions of variation can be investigated by either model fit or partition of variance components methods. To illustrate the form of these relationships we give the connection between the sum of squares formulations, equations (10) and (11), and the analysis of variance. The sum of the squared Euclidean distances given in these equations can be rewritten in terms of mean squares

$$\frac{1}{n_g(n_g-1)} \sum_{i,i>i'} d_{ii'}^2 = \sum_j V_j^2 = MS_G + (n_e-1)MS_{GE} . \qquad (27)$$

where V_j^2 is the variance of phenotypic means in environment j and MS_G and MS_{GE} are genotype and GxE interaction mean squares amongst the y_{ij} phenotypic means, respectively. The LHS of equation (27) is half the average of all the SEDs among the genotypes. Also if the expected mean squares from the completely random analysis of variance model with n_r replications are substituted into equation (27) we have

$$\frac{1}{2} \overline{d_{ii'}^2} = n_e \left(\frac{1}{n_r} \sigma_e^2 + \sigma_{ge}^2 + \sigma_g^2 \right) \qquad (28)$$

where σ_e^2, σ_{ge}^2 and σ_g^2 are the error, GxE interaction and genotype variance components, respectively, and $\overline{d_{ii'}^2}$ is the average of all the pair-wise SEDs among the genotypes. Other relationships can be determined in a similar manner.

Many of the statistics can be partitioned into all pair-wise comparisons among individuals. Equation (6) enables the relationship between the pair-wise comparisons and the average or sum statistics for individuals or pooled statistics over all individuals to be determined. Pattern analysis using these matrices of paired comparisons or proximity matrices are powerful methods for exploring structure in data sets. Since it is averages or sums of these comparisons that form the discrimination statistics for environments or the stability statistics for genotypes a stability analysis is implicit in any pattern analysis. Pattern analysis performed on proximity measures derived from various transformations of the data parallel the suite of partitions of the GxE interactions described above. For instance when investigating discrimination among environments, proximity measures formed from untransformed data are equivalent to phenotypic variances of environments, those from environment centred data are equivalent to GxE variances of environments, those from environment standardised data are equivalent to phenotypic correlations among environments and those from ranked data are equivalent to changes of rank of genotype performance among environments. From the above it is clear that there is a continuum of analytical techniques based on pooled statistics from the analysis of variance through discrimination and stability parameters to proximity matrices of pair-wise measurements among individual environments and genotypes. How far along this progression is appropriate for any situation depends on the model assumptions and an understanding of the crop production system.

To illustrate the relationships among the statistics associated with individual environments (discrimination measures) or individual genotypes (stability measures) and investigation of the proximity matrices of all pair-wise comparisons among the individuals (pattern analysis) we use the same wheat data provided as an example in the previous section. The analysis was conducted on environment standardised values (Fox and Rosielle, 1982) as this procedure enables an investigation of the phenotypic correlations among environments and the associations among standardised yield values for lines. Hierarchical classifications were conducted by Ward's method (Ward, 1963; Burr, 1968, 1970; Wishart, 1969) and ordination by the singular value decomposition procedure (Eckart and Young, 1936) and a biplot (Gabriel, 1971) constructed. The ordination confirmed the relationship among the lines and environments depicted by the classifications (Cooper and DeLacy, 1994). The line and environment classifications were truncated at the five group level. The five groups for environments were: (1) Biloela, (2) Emerald second planting, (3) Emerald first planting and The Gums, (4) Fernlees and Jimbour and (5) Gatton, Kingsthorpe, Toobeah and Pampas. The five groups for lines were: (1) 17IB7, 17IB13, 17IB38, 17IB53, 17IB64, 17IB92, 11IB50, Hartog, (2) 17IB173, (3) 17IB129 and Kite, (4) Banks and (5) 17IB30, 17IB206, Genaro.

Biloela and Emerald second planting had high values for lack of correlation with other environments and grouped separately. But Gatton which had the highest value for lack of correlation and variable correlations with other environments grouped with three other environments. Fernlees formed a two member group with Jimbour even though it had a lower lack of correlation and higher and more consistent correlations with other environments than did Gatton which was a member of a group with four members. Examination of the phenotypic correlation matrix (Table 8) shows that Gatton had high correlations with Toobeah (0.79) and Kingsthorpe (0.72) and a moderate correlation with Pampas (0.53); the environments with which it grouped. Further, Pampas had high correlations with Toobeah (0.73) and Kingsthorpe (0.63), the other members of the group. While Gatton had a high correlation with Jimbour (0.76), the correlation of Jimbour with Fernlees (0.90) was higher. Hence Fernlees grouped with Jimbour and this grouping had a lower correlation with Gatton than did the environments (groups) which joined with Gatton. Similar relationships are present among line groups. Banks and 17IB173 occurred as single member groups which reflects their high lack of correlation parameter. Kite and 17IB129, which also had high lack of correlation parameters averaged over all lines, still grouped with each other. This reflected a relatively high correlation with each other compared to their correlation with other lines.

These relationships demonstrate both the power of the pattern analysis methods in examining the complex and subtle relationships among parameters

82

in proximity matrices and that a discrimination analysis for environments and a stability analysis among genotypes are implicit in such analyses. When a degree of repeatability has been detected in the GxE interactions we recommend a combined use of genotype and environment analyses.

V. CONCLUSIONS

There are clear relationships among the many analytical methods available to study the results from METs. These relations should be examined and understood to enable appropriate choices among them. An appropriate framework consists of pooled statistics from the analysis of variance, various forms of discrimination and stability parameters and two-way pattern analysis on appropriately transformed data. The suitability of methods should be judged in the context of their intended application. If this is selection, the methods should be chosen to reflect information pertinent to selection decisions, such as lack of correlation or change in rank of performance across or among locations. Also, and this has not been emphasised, the methods should be experimentally evaluated to determine their impact on realised response to selection (Cooper et al., 1993a).

GxE interactions should be investigated to identify those which are repeatable. Should repeatable patterns be identified, the random model should be re-evaluated and a mixture model may be more appropriate. Both a biological and statistical understanding of genotype adaptation in the target population of environments enables a strategy to be developed to exploit broad and/or specific adaptation. Developing such an understanding is multi-disciplinary and requires thinking beyond the analytical methodology (Eisemann et al., 1990). The biological and statistical models are complementary and knowledge of the crop production system is necessary to choose appropriate analytical methods. Detailed considerations of these relationships are discussed in Cooper and Hammer (in press).

REFERENCES

Baker, R.J. 1988. Tests for crossover genotype-environmental interactions. *Canadian Journal of Plant Science 68*: 405-410.

Baker, R.J. 1990. Crossover genotype-environmental interaction in spring wheat. p. 42-51. in M.S.Kang (ed.) *Genotype-by-Environment Interaction and Plant Breeding*. Louisiana State University, Baton Rouge, Louisiana 70803.

Bradu, D., and K.R. Gabriel. 1978. The biplot as a diagnostic tool for models of two-way tables. *Technometrics 20*: 47-68.

Brennan, P.S., D.E. Byth, D.W. Drake, I.H. DeLacy, and D.G. Butler. 1981. Determination of the location and number of test environments for a wheat cultivar evaluation program. *Australian Journal of Agricultural Research 32*: 189-201.

Burr, E.J. 1968. Cluster sorting with mixed character types. I. Standardization of character values. *Australian Computer Journal 1*: 97-99.

Burr, E.J. 1970. Cluster sorting with mixed character types. II. Fusion strategies. *Australian Computer Journal 2*: 98-103.

Byth, D.E., R.L. Eisemann, and I.H. DeLacy. 1976. Two-way pattern analysis of a large data set to evaluate genotypic adaptation. *Heredity 37*: 215-230.

Casella, G., and R.L. Berger. 1990. *Statistical inference*. Wadsworth and Brooks/Cole, California.

Cockerham, C.C. 1963. Estimation of genetic variances. in *Statistical Genetics and Plant Breeding*. Hanson, W.D. and Robinson, H.F., Eds., National Academy of Sciences - National Research Council, Publication 982, Washington, DC.

Comstock, R.E. 1977. Quantitative genetics and the design of breeding programmes. in E. Pollack, O. Kempthorne, and T.B.J. Bailey (eds.) *Proceedings of the International Conference on Quantitative Genetics*. The Iowa State University Press, Ames, Iowa, USA.

Cooper, M., and I.H. DeLacy. 1994. Relationships among analytical methods used to study genotypic variation and genotype-by-environment interaction in plant breeding multi-environment experiments. *Theoretical and Applied Genetics 88:* 561-572.

Cooper, M., and G.L. Hammer (eds.) In Press. *Plant adaptation and crop improvement*. CAB International, Wallingford, UK, ICRISAT (International Crops Research Institute for the Semi-Arid Tropics), Patancheu, 502 324, Andra Pradesh, India, IRRI (International Rice Research Institute), Manila, Philippines.

Cooper, M., and D.R. Woodruff. 1993. Predicting grain yield in Australian environments using data from CIMMYT international wheat performance trials 3. Testing predicted correlated response to selection. *Field Crops Research 35:* 191-204.

Cooper, M., D.E. Byth, and I.H. DeLacy. 1993a. A procedure to assess the relative merit of classification strategies for grouping environments to assist selection in plant breeding regional evaluation trials. *Field Crops Research 35:* 63-74.

Cooper, M., D.E. Byth, I.H. DeLacy, and D.R. Woodruff. 1993b. Predicting grain yield in Australian environments using data from CIMMYT international wheat performance trials. 1. Potential for exploiting correlated response to selection. *Field Crops Research 32:* 305-322.

Cooper, M., I.H. DeLacy, D.E. Byth, and D.R. Woodruff. 1993c. Predicting grain yield in Australian environments using data from CIMMYT international wheat performance trials. 2. The application of classification to identify environmental relationships which exploit correlated response to selection. *Field Crops Research 32:* 323-342.

Cooper, M., I.H. DeLacy, and R.L. Eisemann. 1993d. Recent advances in the study of genotype x environment interactions and their application to plant breeding. in B.C. Imrie, and J.B. Hacker (eds.) *Focused Plant Improvement: Towards Responsible and Sustainable Agriculture. Proceedings Tenth Australian Plant Breeding Conference. Vol. 1.*, Organising Committee, Australian Convention and Travel Service, Canberra.

Cooper, M., D.R. Woodruff, R.L. Eisemann, P.S. Brennan, and I.H. DeLacy. 1995. A selection strategy to accommodate genotype by environment interaction for grain yield of wheat: Managed environments for selection among genotypes. *Theoretical and Applied Genetics 90:* 492-502.

DeLacy, I.H., and M. Cooper. 1990. Pattern analysis for the analysis of regional variety trials. p. 301-334. in M.S. Kang (ed.) *Genotype-By-Environment Interaction and Plant Breeding*. Louisiana State University, Baton Rouge, Louisiana 70803 (12-13 February).

DeLacy, I.H., R.L. Eisemann, and M. Cooper. 1990. The importance of genotype by environment interaction in regional variety trials. p. 287-300. in M.S. Kang (ed.) *Genotype-By-Environment Interaction and Plant Breeding*. Louisiana State University, Baton Rouge, Louisiana 70803 (12-13 February).

DeLacy, I.H., P.N. Fox, J.D. Corbett, J. Crossa, S. Rajaram, R.A. Fischer, and M. van Ginkle. 1994. Long-term association of locations for testing spring bread wheat. *Euphytica 72,* 95-106.

Dickerson, G.E. 1962. Implications of genetic-environmental interaction in animal breeding. *Animal Production 4:* 47-64.

Eberhart, S.A., and W.A. Russell. 1966. Stability parameters for comparing varieties. *Crop Science 6:* 36-40.

Eckart, C., and G. Young. 1936. The approximation of one matrix by another of lower rank. *Psychometrika 1:* 211-218.

Eisemann, R.L., M. Cooper., and D.R. Woodruff. 1990. Beyond the analytical methodology - Better interpretation of genotype-by-environment interaction. p. 108-117. in M.S. Kang (ed.) *Genotype-By-Environment Interaction and Plant Breeding*. Louisiana State University, Baton Rouge, Louisiana 70803 (12-13 February).

Falconer, D.S. 1952. The problem of environment and selection. *The American Naturalist 86:* 293-298.

Finlay, K.W., and G.N. Wilkinson. 1963. The analysis of adaptation in a plant breeding programme. *Australian Journal of Agricultural Research 14:* 742-754.

Fox, P.N., and A.A. Rosielle. 1982. Reducing the influence of environmental main-effects on pattern analysis of plant breeding environments. *Euphytica 31:* 645-656.

Gabriel, K.R. 1971. The biplot-graphical display of matrices with applications to principal component analysis. *Biometrika 58:* 453-467.

Gauch, H.G. 1992. *Statistical analysis of regional yield trials: AMMI analysis of factorial designs.* Elsevier, Amsterdam.

Gilmour, R.F. 1988. *An analysis of the effectiveness of selection in early generation trials for grain yield of wheat.* PhD thesis, University of Queensland, Brisbane, Australia.

Gower, J.C. 1966. Some distance properties of latent root and vector methods used in multivariate analysis. *Biometrika 53:* 325-338.

Gower, J.C. 1967. Multivariate analysis and multidimensional geometry. *The Statistician 17:* 13-28.

Haldane, J.B.S. 1947. The interaction of nature and nurture. *Annals of Eugenics 13:* 197-205.

Harville, D.A. 1977. Maximum likelihood approaches to variance component estimation and related problems. *Journal of the American Statistical Association 72:* 320-340.

Itoh, Y., and Y. Yamada. 1990. Relationships between genotype x environment interaction and genetic correlation of the same trait measured in different environments. *Theoretical and Applied Genetics 80:* 11-16.

Kang, M.S. (ed.) 1990. *Genotype-By-Environment Interaction and Plant Breeding.* Louisiana State University, Baton Rouge, Louisiana 70803.

Kempton, R.A. 1984. The use of bi-plots in interpreting variety by environment interactions. *Journal of Agricultural Science 103:* 123-135.

McLachlan, G.J., and K.E. Basford. 1988. *Mixture Models: Inference and Applications to Clustering.* Marcel Dekker, New York.

Mirzawan, P.D.N., M. Cooper, I.H. DeLacy, and D.M. Hogarth. 1994. Retrospective analysis of the relationships among the test environments of the Southern Queensland sugarcane breeding program. *Theoretical and Applied Genetics 88:* 707-716.

Muir, W., W.E. Nyquist, and S. Xu. 1992. Alternative partitioning of the genotype-by-environment interaction. *Theoretical and Applied Genetics 84:* 193-200.

Mungomery, V.E., R. Shorter, and D.E. Byth. 1974. Genotype x environment interactions and environmental adaptation. I. Pattern analysis-application to soya bean populations. *Australian Journal of Agricultural Research 25:* 59-72.

Patterson, H.D., and R. Thompson. 1971. Recovery of inter-block information when block sizes are unequal. *Biometrika 58:* 545-554.

Patterson, H.D., and R. Thompson. 1974. Maximum likelihood estimation of components of variance. in *Proceedings of the Eighth International Biometrical Conference.* 197-207.

Pearson, K. 1894. Contributions to the mathematical theory of evolution. *Philosophical Transactions of the Royal Society, A 185:* 71-110.

Perkins, J.M., and J.L. Jinks. 1968. Environmental and genotype-environmental components of variability. III Multiple lines and crosses. *Heredity 23:* 339-356.

Robertson, A. 1959. The sampling variance of the genetic correlation coefficient. *Biometrics 15:* 469-485.

Shorter, R. 1981. Linear regression analyses of genotype x environment interactions. in D.E. Byth, and V.E. Mungomery (eds.) *Interpretation of plant response and adaptation to agricultural environments.* Queensland Branch, Australian Institute of Agricultural Science, Brisbane.

Ward, J.H. 1963. Hierarchical grouping to optimise an objective function. *Journal of the American Statistical Association 58:* 236-244.

Westcott, B. 1986. Some methods of analysing genotype-environment interaction. *Heredity 56:* 243-253.

Williams, W.T. (ed.) 1976. *Pattern Analysis in Agricultural Science.* Elsevier Scientific Publishing Company, Amsterdam.

Wishart, D. 1969. An algorithm for hierarchical classification. *Biometrics 25:* 165-170.

Yamada, Y. 1962. Genotype by environment interaction and genetic correlation of the same trait under different environments. *Japanese Journal of Genetics 37:* 498-509.

Yates, F., and W.G. Cochran. 1938. The analysis of groups of experiments. *Journal of Agricultural Science 28:* 556-580.

Chapter 4

AMMI ANALYSIS OF YIELD TRIALS

H. G. Gauch, Jr. and R. W. Zobel[1]

I. INTRODUCTION

The Additive Main effects and Multiplicative Interaction (AMMI) model combines analysis of variance for the genotype and environment main effects with principal components analysis of the genotype-environment interaction. It has proven useful for understanding complex genotype-environment interactions. The results can be graphed in a very informative biplot that shows both main and interaction effects for both genotypes and environments. Also, AMMI can partition the data into a pattern-rich model and discard noise-rich residual to gain accuracy.

The 1990 conference on genotype by environment interaction, producing 26 papers in Kang (1990), did much to summarize and assess the state of the art as of five years ago. Several papers mentioned AMMI, and two discussed it in detail. Zobel (1990) found AMMI powerful for analyzing numerous shoot and root traits of soybeans (*Glycine max* L.). He reported that interactions tend to be larger for traits, especially root traits, for which breeders have not imposed strong selection and hence reduced genetic variability. Gauch (1990b) found AMMI useful for understanding complex interactions, gaining accuracy, improving selections, and increasing experimental efficiency. Also, the expectation-maximization version, EM-AMMI, can impute missing data. The results presented in these and other papers published by 1990 now seem widely known and generally accepted.

The purpose of this paper is to review AMMI developments since the 1990 conference, emphasizing applications to yield trials, but also considering some additional agricultural or statistical developments. The text by Gauch (1992b) covers many of the developments until mid 1992, so special emphasis is placed here on the past two years.

More has been published about AMMI applications in agriculture in the past two years than during all previous years, so this review is reasonably comprehensive but cannot be complete. Fortunately, other contributions in this volume complement the material included here.

Two preliminary sections give the AMMI model equation and explain the challenges in yield-trial research. Then recent developments are reviewed under four headings: (1) understanding genotypes and environments, (2) evidence of

[1] Soil, Crop and Atmospheric Sciences, 1021 Bradfield Hall, Cornell University, Ithaca, NY 14853-1901 and USDA-ARS-NAA, 1017 Bradfield Hall, Ithaca, NY 14853-1901. This research is part of the USDA/Cornell program for Root-Soil Research, and was supported by the Rhizobotany Project of USDA-ARS. We appreciate helpful suggestions from R.E. Furnas, M.S. Kang, H.-P. Piepho, and A.F. Troyer.

accuracy gain, (3) experimental design, and (4) future directions. Finally, there are a few brief conclusions.

II. THE AMMI MODEL

AMMI combines analysis of variance (AOV) and principal component analysis (PCA) into a single model with additive and multiplicative parameters. The AMMI model equation is:

$$Y_{ger} = \mu + \alpha_g + \beta_e + \Sigma_n \lambda_n \gamma_{gn} \delta_{en} + \rho_{ge} + \epsilon_{ger}$$

where Y_{ger} is the observed yield of genotype g in environment e for replicate r. The additive parameters are: μ the grand mean, α_g the deviation of genotype g from the grand mean, and β_e the deviation of environment e. The multiplicative parameters are: λ_n the singular value for interaction principal component axis (IPCA) n, γ_{gn} the genotype eigenvector for axis n, and δ_{en} the environment eigenvector. The eigenvectors are scaled as unit vectors and are unitless, whereas λ has the units of yield. A convenient scaling for the multiplicative parameters is $\lambda^{0.5}\gamma_g$ and $\lambda^{0.5}\delta_e$, termed the "genotype IPCA scores" and "environment IPCA scores," because their product gives the expected interaction value directly without need of a further multiplication by the singular value. There are at most $\min(G-1, E-1)$ axes, but usually the number of axes N retained in the model is smaller, producing a reduced model denoted AMMI1 or AMMI2 if retaining 1 or 2 IPCAs, and so on. A reduced model leaves residuals, ρ_{ge}. Finally, if the experiment is replicated, there is also the error term ϵ_{ger}.

The least-squares solution is found for balanced data by AOV followed by PCA, as proved by Gabriel (1978). Gauch (1992b:85-92) gives a worked example. Missing data can be tolerated and imputed by an expectation-maximization version, EM-AMMI (Gauch and Zobel, 1990). The AMMI calculations are performed conveniently by the MATMODEL software (Gauch and Furnas, 1991). PCA was invented in 1901 (with precursors since 1870) and AOV in 1918, but their combination to form AMMI waited until Pike and Silverberg (1952) and Williams (1952). AMMI has gone under a host of names, including FANOVA, MI, doubly-centered PCA, and biplot analysis (although this last name is a misnomer since "biplot" refers to a type of graph, not to a particular model). Seminal work was done by Gilbert (1963), Gollob (1968), Mandel (1969, 1971, 1972), Corsten and van Eijnsbergen (1972), Johnson and Graybill (1972), Bradu and Gabriel (1978), Gabriel (1978, 1981), Cornelius (1980), Snee (1982), and others. Applications in agriculture were stimulated by Kempton (1984) and Zobel et al. (1988). The motivation in these papers was to model or understand interactions, and Gauch (1985, 1988) introduced another motivation, to gain accuracy and efficiency. Mandel (1995) explains AMMI and related statistical models.

III. RESEARCH CHALLENGES: INTERACTION AND NOISE

Before presenting AMMI as a helpful statistical tool for analyzing yield-trial data, one must first clearly grasp the practical research problems requiring solution. Yield-trial research, aimed at selecting superior genotypes in a breeding program or recommending varieties (or fertilizers or whatever) to farmers, faces two fundamental problems: interaction and noise. Were there no interaction, a single variety of wheat (*Triticum aestivum* L.) or corn (*Zea mays* L.) or any other crop would yield the most the world over, and furthermore the variety trial need be conducted at only one location to provide universal results. And were there no noise, experimental results would be exact, identifying the best variety without error, and there would be no need for replication. So, one replicate at one location would identify that one best wheat variety that flourishes worldwide. The purpose of this fiction is just to identify the two basic problems confronting agronomists and breeders. Returning now to reality, agricultural scientists must seek effective statistical tools for mitigating challenges caused by interaction and noise. AMMI addresses both challenges.

Regarding agricultural problems from genotype-by-environment interaction, there exist two basic options, one aimed at the genotypes, and the other at the environments (Ceccarelli, 1989; Simmonds, 1991; Zavala-Garcia et al., 1992). One option is to seek a high-yielding, widely-adapted genotype that wins throughout the growing region of interest. The other option, particularly relevant when the first fails, is to subdivide the growing region into several relatively homogeneous macro-environments (with little interaction within each macro-environment) and then breed and recommend varieties for each. As explained momentarily, AMMI can help with both of these options.

Regarding problems from noise, there exist four basic options: better experiments, more replications, analysis of the experimental design, and analysis of the treatment design (Gauch 1992b:15-18). First, accuracy can be increased by larger or better controlled yield plots. Second, greater accuracy results from a larger number of replications. Third, accuracy can be increased, relative to a completely randomized (CR) experiment, by a better experimental design, such as randomized complete blocks (RCB) or various incomplete block designs. For example, just 5 replications in a RCB design are often as accurate as 6 or 7 replications in a CR design, so the reward for a better statistical design and analysis is effectively 1 or 2 free replications. Complete block designs can reduce the error mean square and thereby increase statistical significance, whereas incomplete block designs can also adjust the yield estimates (so that an estimate is not simply an average over replications). Row and column designs (Williams and John, 1989) and α-lattice and related designs (Patterson and Williams, 1976; Patterson et al., 1978; Paterson et al., 1988) frequently outperform RCB designs. Fourth, accuracy can be gained, for a regional trial planted in several locations or years, by statistical analysis of the treatment design, namely, a two-way factorial design of genotypes and environments. The total sum of squares (SS) is partitioned into a pattern-rich model and a discarded noise-rich residual, thereby adjusting the yield estimates and gaining accuracy. For example, just 5 replications analyzed by AMMI frequently achieve equivalent accuracy as 10 to 25 replications without AMMI analysis, so the reward for AMMI analysis is effectively 5 to 20 free replications.

These four options for gaining accuracy are fundamentally different, so their advantages are complementary, and all can be implemented at once. The special attraction of the last two options is their remarkable cost effectiveness. Some training and work are required to implement an aggressive statistical analysis, but these costs are usually trivial relative to the cost of gaining comparable insight and accuracy by the brute-force means of conducting a far larger experiment (Gauch, 1993c). If ten seconds of microcomputer time for statistical analysis gains as much accuracy as 5000 additional yield plots (at a cost of, say, $20 each), then statistical analysis is tremendously cost effective for gaining accuracy, improving decision-making, and speeding progress. "Work smarter, not harder," as the saying goes. One may ask, "What are the costs for analyzing the data?," but one should also ask, "What are the costs for not analyzing the data?" Indeed, it can be very costly to spend thousands of dollars on yield-trial research and then extract only a fraction of the information contained in the data. Wasting data reduces agricultural progress and competitive advantages. Data worth collecting are also worth analyzing.

Finally, two markedly different statistical contexts must be distinguished: optimal theoretical analyses and routine agricultural applications. Because a data set is remarkably expensive or bears on vital decisions, or perhaps because of purely academic interests, statisticians often compare dozens of models to find the best (or very nearly the best) for a given data set and purpose. Such statistical luxury, however, is not always possible or warranted. For example, from the time a commercial corn breeder receives millions of dollars worth of data from this year's yield trials until he or she flies to Mexico or Hawaii to plant the winter nursery, there may be just three weeks. Enormous quantities of data must be analyzed and many decisions made quickly. In this context, what counts is a good statistical analysis (nearly optimal but not necessarily the very best) that can be applied in a relatively automatic, rapid fashion. The context emphasized here is the latter: good but easy statistical analyses for practical and often hurried agricultural applications.

IV. UNDERSTANDING GENOTYPES AND ENVIRONMENTS

In numerous studies, both basic and applied, AMMI has helped understand genotypes, environments, and their interactions. Crops analyzed include wheat (*Triticum aestivum* L. and *T. turgidum* L. var. *durum*) (Kempton, 1984; Crossa et al., 1991; Fernandez, 1991; Boggini et al., 1992; Calinski et al., 1992; Denis and Gower, 1992; Nachit et al., 1992a, 1992b, 1992c; Annicchiarico and Perenzin, 1994; Heenan et al., 1994; Eisenberg et al., 1995; Yau, 1995), corn (*Zea mays L.*) (Hirotsu, 1983; M'Benga, 1989; Crossa, 1990; Crossa et al., 1988, 1990; Goodman and Haberman, 1990; Ezuman et al., 1991; Cruz, 1992; Charcosset et al., 1993), rice (*Oryza sativa* L.) (Okuno et al., 1971; Sutjihno, 1993), oats (*Avena sativa* L.) (Peltonen-Sainio et al., 1993), barley (*Hordeum vulgare* L.) (Riggs, 1986; Romagosa et al., 1993; Van Oosterom and Ceccarelli, 1993; Van Oosterom et al., 1993), sorghum (*Sorghum bicolor* L. Moench) (Zavala-Garcia et al., 1992), triticale (× *Triticosecale* Wittmack) (Royo et al., 1993), ryegrass (*Lolium perene* L.) (Van Eeuwijk, 1992b; Van Eeuwijk and Elgersma, 1993), grass (Denis and Vincourt 1982), soybean (*Glycine max* L.) (Gauch, 1988, 1990b, 1992a, 1992b, 1993a; Gauch and Zobel, 1988, 1989a, 1989b, 1990; Smit and de Beer, 1989, 1991; Smit et al., 1992; Zobel, 1990, 1992a, 1992b, 1994; Zobel and Wallace, 1994; Zobel et al., 1988), dry bean (*Phaseolus vulgaris* L.) (Wallace and Masaya, 1988; Wallace et al., 1991, 1993, 1994a, 1994b; Saindon and Schaalje, 1993; Mmopi et al., 1994), field bean (*Vicia Faba* L.) (Piepho, 1994), alfalfa (*Medicago sativa* L.) (Annicchiarico, 1992; Smith and Gauch, 1992; Smith and Smith, 1992), potato (*Solanum tuberosum* L.) (Loiselle et al., 1990; Halseth et al., 1993; Steyn et al., 1993), tomato (*Lycopersicon Lycopersicum* (L.) Karst. ex Farw.) (Gilbert, 1963), cabbage (*Brassica oleracea* L. var. *capitata*) (Davik, 1989), rapeseed (*Brassica napus* L. subsp. *oleifera* (Metzg.) Sinsk. f. *biennis*) (Shafii et al., 1992), tea (*Camellia sinensis* (L.) O. Kuntze) (Eisenberg et al., 1995), and sour cherry (*Prunus Cerasus* L.) (Iezzoni and Pritts, 1991). Plant diseases analyzed include fungi on wheat (Kempton, 1984; Snijders and Van Eeuwijk, 1991; Van Eeuwijk, 1992b) and corn (Wilson et al., 1993), net blotch on barley (Sato and Takeda, 1993), nematodes on potato (Phillips and McNichol, 1986; Phillips et al., 1987, 1989; Dale and Phillips 1989), and virus on beet (*Beta vulgaris* L.) (Paul et al., 1993). The *Bradyrhizobium* symbiont of peanut (*Arachis hypogaea* L.) also has been analyzed (Alwi et al., 1989). AMMI analysis has also been applied to aphid distributions in England (Kempton, 1984), ant communities in citrus groves (Samways, 1983), and meteorological data (Tsianco and Gabriel, 1984; Gabriel and Mather, 1986; Glasbey, 1992).

Most of these papers apply additional statistical analyses besides AMMI and offer insightful comparisons. Also, many recent papers address important matters not covered before 1990. For a great diversity of crops and research objectives, sufficient experience with AMMI has accumulated to reach a balanced assessment of past applications and a realistic expectation of future applications. A comprehensive summary of these papers is not possible here, but some of the main findings may be reviewed briefly.

A. GENERAL PERSPECTIVE

The data structure typically emerging from a yield trial is a two-way factorial of genotypes and environments. The AMMI algorithm has two parts: AOV and PCA. The first part of AMMI analysis uses AOV to partition the total variation into three orthogonal sources: genotypes (G), environments (E), and genotype-environment interactions (GE). Two preliminary questions may be asked about this partition.

First, what fractions of the treatment variation are typically found in G, E, and GE? DeLacy, Eisemann, and Cooper (1990) present results for over 100 trials with many crops in numerous locations. Romagosa and Fox (1993) summarize that paper by saying, "In most yield trials, the proportion of sum of squares due to differences among sites ranged from 80 to 90% and variation due to G×E was usually larger than genotypic variation." One may propose a "70-20-10 Rule" saying that a median yield trial has about 70% of the variation in E, 20% in GE, and 10% in G, with, of course, wide differences in individual cases. Note that interaction is frequently larger than the genotype main effect. Furthermore, in AMMI analysis, just the first IPCA alone is usually larger than G. As the genotypes become more diverse (in contrast to extremely similar elite lines) and the environments become more diverse, GE tends to increase and easily reaches 40% to 60%.

Second, for which research questions is it fitting and helpful to impose this partition? For purposes of selection, only the rankings of the genotypes matter, which are determined by G and GE, with E irrelevant. Consequently, there is interest in reducing or eliminating the influence of environmental main effects for purposes of genotype selections (Fox and Rosielle, 1982; Gauch, 1992b:220-230).

Now, putting these two facts together, the awkward situation emerges that the environment main effect causes most of the variation in yields but is wholly irrelevant to selection. Consequently, analyses that provide a parsimonious summary of the treatment variation in the data (like PCA, SHMM, and most classification procedures) spend most of their effort on capturing an irrelevant E. They impose no fundamental distinction between variation due to G, E, and GE, so model parameters capture overall variation that combines and intermixes these three sources. Consequently, the results are diffuse relative to selection purposes, like a picture of the family pet in which the pet is just a small item, or like a double-exposed picture of a pet and a house. Why produce a multivariate summary of a yield trial for purposes of selection that devotes 70% of its picture to total irrelevancies? By contrast, AMMI can produce graphs that focus on data structure relevant to selection. This is like a clear picture of a pet, nicely filling the frame.

Moving on now to the second part of AMMI analysis, it uses PCA to partition the GE interaction into several orthogonal IPCA axes. Again, two preliminary questions can be asked about this partition.

First, are too many IPCA axes required to allow simple presentation of the results? Romagosa and Fox (1993) express concern that "When the best AMMI model includes more than one PCA axis, assessment and presentation of genetic stability are not as simple as the AMMI1 case." Two points may be offered in reply: (1) For AMMI2 also, a very satisfactory and simple graph can plot IPCA1 against IPCA2 (as contrasted with the usual main effects and IPCA1 shown for AMMI1; Kempton, 1984; Gauch, 1992b:85-96). Furthermore, polygons can be superimposed on the AMMI2 biplot to show which genotype is the winner in every region of the biplot (Gauch, 1992b:213-230). Routine scientific correspondence is limited to black-and-white printing on two-dimensional paper, but extension to AMMI3 is possible with computer graphics. No direct visualization is possible with four or more dimensions. (2) As Van Eeuwijk (1995) concludes, "experience has shown that only very infrequently there are sufficient grounds for including more than two axes." When AMMI3 and higher models are presented for agricultural data, usually a little scrutiny reveals that the third and higher IPCA axes are dominated by noise and have no predictive value and no biological interpretability.

Second, are significant IPCA axes interpretable in terms of known properties of the genotypes and environments? Researchers are concerned that "the chief difficulty with this approach is in the interpretation of the resulting principal components, which may not bear any obvious relationship to environmental conditions" (Westcott, 1986; also see Becker and Léon, 1988). Three points may be offered in reply. (1) From a statistical perspective, predictively significant model parameters indicate that physical or biological causes are at work, rather than random noise. IPCA axes "may be regarded as the first terms in the general Taylor expansion of multivariate nonlinear relationships," so models like AMMI can be expected to give "a simple, linear approximation to relationships in the data" (Aastveit and Martens, 1986). Or, to repeat this point in more concrete terms, if G captures 7% of the treatment SS and IPCA1 captures 21%, then it would be peculiar to interpret the genotype main effect but to be mystified by the IPCA1 effect that is three times as large. To say the least, if the interpretation of a large IPCA is not already evident, then it becomes a mighty good topic for future research! (2) From an agricultural perspective, extensive experience with AMMI is now available, so speculation is hardly necessary any more. As will be reviewed momentarily, the track record for easy interpretation of large IPCA is quite good. (3) Although interpretation of a given experiment's large IPCA axes is usually easy, a couple of related matters merit close scrutiny later in this section: repeatability and heritability. How many locations and years are required to quantify interaction patterns that are repeatable in other locations and future years? And how heritable are interaction traits, as quantified by AMMI or other statistical models?

A choice must be made between conceptualizing and analyzing a yield trial as either a two-way table of genotypes by environments, or as a three-way table of genotypes by locations by years. An "environment" is defined here as a location-year combination, for example Ithaca in 1994. As mentioned earlier, this conceptualization of a yield trial, with its genotypes and environments in a two-way factorial design, leads to the AOV partitions: G, E, and GE. But when locations (L) and years (Y) are considered separately, giving a three-way factorial design, this alternative conceptualization leads to the AOV partitions: G, L, Y, GL, GY, LY, and GLY. The simpler partitioning has one two-way interaction, whereas the more complex partitioning has three two-way interactions and one three-way interaction. Different yield trials can have quite different portions of the treatment SS in these various interactions, and different research questions and agricultural objectives are variously affected by these different kinds of interactions. A further complication is that changes in research locations over the years frequently generate a partial factorial design; that is, data are missing for some locations in some years. Similarly, the genotype roster may also change over years or locations or both.

First, consider the simpler two-way layout. Because the PCA portion of the AMMI algorithm must analyze a two-way matrix, and not a one-way design nor a three-way or higher design, this offers the natural conceptualization for AMMI analysis. This approach is particularly sensible when, as is often the case, the LY interaction is large, so that each location-year combination or "environment" merits individual consideration. Also, when not all locations have trials in all years, this conceptualization avoids generating large quantities of missing data. For a typical example, a soybean trial analyzed in Gauch and Zobel (1990) and Gauch (1992b:160-162) has 10 locations and 12 years, but instead of a full factorial design with 120 environments, data are actually available for a partial factorial design with only 55 environments. Hence, analysis as the full factorial would generate 54% missing data. Furthermore, in an AMMI biplot, points for environments can be labeled to show both the location and the year, such as "GN94" for Geneseo in 1994, so both location and year information is readily assimilated. For examples, see Riggs (1986), Zobel et al. (1988), and Gauch (1992b:85-96). Indeed, such a biplot offers an effective conceptualization of a location as a prediction region, with clear implications for variety recommendations (Gauch 1992b:217-219; also see pages 213-216 and 220-230).

Second, consider the more complicated three-way layout. When the three-way interaction, GLY, has a large SS, the situation is inherently tremendously complex. Unfortunately, genotype selections and variety recommendations possess little generality across locations or years. Anyway, in this case, statistical analyses for three-way data should be considered, such as Hohn (1979), Denis and Dhorne (1989), Gower (1990), Basford et al. (1991), and Génard et al. (1994). Bear in mind, however, that in general a GLY interaction with a large SS can originate from two quite different causes: large and real causal factors, or merely a large number of df that incorporates considerable noise. A simple, quick check is available for replicated experiments with a known EMS. This three-way GLY interaction has $(G-1)(L-1)(Y-1)$ df. Multiply that number by the EMS to estimate the amount of noise in this interaction. If the GLY interaction SS is nearly all noise SS, then it contains little if any real information, regardless of its size. For example, say GLY has a SS of 147200 with 2496 df and the EMS is 56.01. Then the noise SS is estimated to be about 139800, so this interaction is 95% noise. Such a terribly noisy interaction would be wisely ignored (apart, perhaps, from checking for a few large outliers), even if it happens to have a much larger SS than, say, G or GL or any other source in the AOV table. When the three-way interaction, GLY, is negligible, only two-way interactions remain and they can be tackled in the ordinary manner by AMMI. If GL, GY, and LY are all important, three AMMI analyses may elucidate the patterns in these three interaction matrices. In particularly fortunate instances, only one of the two-way interactions, say GL, may be large, leading to a simplified case easily handled by AMMI.

B. ENVIRONMENTAL INTERPRETATION

AMMI analysis of yield data has supplied to it no environmental data — just yield data. Nevertheless, significant AMMI parameters (for both main and interaction effects) ordinarily reflect identifiable causal factors (Gauch, 1992b:231-236). By various informal or formal means, the pattern in AMMI parameters or biplots can usually be interpreted clearly in terms of evident environmental or genetic causal factors.

One strategy for interpreting AMMI results is to analyze the yield data separately, followed by correlation of AMMI parameters with environmental data. Annicchiarico and Perenzin (1994) applied AMMI to a bread wheat trial in Italy. Besides yield data, four environmental factors were measured: rainfall, average daily maximum temperature, average daily minimum temperature, and number of frost days within a month before mean heading date. IPCA1 was strongly correlated with number of frost days, whereas IPCA2 related to terminal drought stress. The AMMI2 biplot revealed four clusters of environments, which related clearly to four geographical regions in Italy. They concluded, "Besides representing location differences more faithfully [than FW regression analysis], the use of AMMI analysis can also contribute to identification of major environmental and genotypic factors related to GL [genotype-location] interaction occurrence, thereby supporting breeding programme decisions regarding adaptation targets, selection environments, test sites, choice of parents and adaptive traits to breed for." Whereas AMMI2 captured 38.0% of the interaction (with higher axes nonsignificant noise), FW captured only 10.4%. Similarly, Annicchiarico (1992) analyzed alfalfa trials in Italy. IPCA1 reflected clay content and summer rainfall. The AMMI1 biplot revealed three subregions, so cultivar recommendations were given for each separate subregion. Also, Boggini et al. (1992) correlated IPCA1 to IPCA3 with environmental variables for a durum wheat trial in Italy. Nachit et al. (1992c) correlated AMMI main effects and IPCA1 scores with 11 environmental factors for a Mediterranean durum wheat trial. The environmental main effect reflected altitude, nitrogen fertilization, and a weather index, while IPCA1 reflected altitude and irrigation. They concluded, "These results show the usefulness of the AMMI model as a statistical tool for interpreting associations between environmental variables and components of GE interaction." Likewise, Van Eeuwijk and Elgersma (1993) readily interpreted AMMI results for a ryegrass trial. Many more citations could be added here, but it must suffice to remark that environmental interpretation of significant AMMI parameters is usually relatively straightforward.

Another strategy, requiring analyses other than AMMI, is joint analysis of the yield and environmental data together, such as by redundancy analysis. For an experiment concerning nitrate concentration in lettuce, Van Eeuwijk (1992a) applied both redundancy analysis and AMMI. Remarkably, he found that "AMMI and redundancy analysis gave comparable results, though the first extracts environmental scores as linear combinations of residuals from additivity [for the yield data], whereas the second forms environmental scores from linear combinations of measured environmental variables." That AMMI, which was not supplied the environmental data, and redundancy analysis, which was supplied the environmental data, should find corresponding patterns in the yield data is striking confirmation that significant AMMI parameters reflect underlying environmental and genetic causal factors. Likewise, Van Eeuwijk and Elgersma (1993) compared AMMI and various other analyses, concluding that "The different methods used in this paper to investigate relations between interaction and environmental factors all identified the same variables as important" (see also Van Eeuwijk, 1992b).

It becomes necessary to subdivide a growing region into several target environments or mega-environments when interaction is sizable. Otherwise, differences in yield rankings from environment to environment will imply suboptimal yields in many environments, were the entire region planted to a single variety. However, it is challenging to provide a clear conception of mega-environments and a rational procedure for defining mega-environments for a given crop and region. Breeders tend to develop a negative attitude toward interaction, seeing only an impediment to high heritability and selection gains. But it must be remembered, on balance, that positive interactions associated with reasonably predictable features of environments offer an opportunity for higher yields, although at the cost of subdividing a growing region (and extending a breeding program) into several mega-environments. Interaction is not merely a problem, it is also an opportunity (Simmonds, 1991). Specific adaptations can make the difference between a good variety and a superb variety.

To understand mega-environments, first of all it must be understood that GL and GY interactions have very different implications. Plant breeders and farmers have, for better or for worse, quite different interests in the GL and GY interactions. A breeder wants a small GL interaction so that the new variety will become a big success, grown over a wide region (which is also an asset to seed producers and sellers). But a farmer wants a small GY interaction: yield predictability and dependability from year to year at a given location. The farmer cares little about GL interactions, that is, about how the variety performs somewhere else in the world. This difference between breeders' and farmers' interests raises a difficult question. Just how widely adapted can a variety be and still give optimal yields at a given location? That is, should breeders focus their resources on a few varieties with wide adaptation, or should they subdivide a crop's region into numerous particular target environments and run separate breeding programs to optimize yields for each target environment? At one extreme, a single variety for a whole region gives many farmers suboptimal yields when GE interactions could have been exploited; at the opposite extreme, an ideal variety for each individual farm imposes impossible demands on plant breeders; so a compromise must be struck somewhere between these extremes. The larger the GL is relative to GY, the larger the number of mega-environments that should be defined (Annicchiarico and Perenzin, 1994).

A special kind of AMMI biplot graph can help researchers comprehend and define mega-environments (Gauch, 1992b:213-230). To indicate a location's probable position in an AMMI biplot in future years, circle the points for past results for that location (or give each location a distinctive symbol, or whatever). Various locations will occupy different prediction regions, with a small region indicating high predictability and a large region low predictability. Then superimpose the regions in which various genotypes are winners, which turn out to be horizontal bands in an AMMI1 biplot or polygons in an AMMI2 biplot. Locations situated within a single genotype's winning turf, or which rarely and slightly cross over into another genotype's turf, have an obvious variety recommendation. But locations that straddle genotype boundaries are inherently less predictable and more problematic. In such cases, income stability may be promoted best by planting at least two different varieties to hedge one's bets, rather than by attempting to breed a single genotype with awesome stability. Those genotype turfs that contain numerous locations (or that represent significant crop regions) are candidates for mega-environments.

Note that when new genotypes are developed and introduced, the map of genotype winners can change. For example, a remarkably superior new variety might have a winning turf that encompasses two or three predecessors' turfs, thereby reducing the number of mega-environments required. That is, useful mega-environments depend not only on the environments, but also on the current genotype roster. Although this outcome may seem paradoxical at first, in fact, it is just right. The name "mega-environment" may suggest that this concept involves only environments, but it also involves genotypes because interaction is what makes various mega-environments necessary, and GE interaction is inherently a concept that involves both genotypes and environments.

This specialized biplot, showing locations as prediction regions and showing turfs of winning genotypes, is the most powerful tool yet developed for visualizing mega-environments. Numerous agricultural questions can be addressed effectively with this tool, such as whether various mega-environments currently have an appropriate sampling intensity, whether certain locations (or mega-environments) are inherently less predictable than others as regards genotype rankings, whether breeding effort should increase or decrease for the various mega-environments, and which crosses might produce new genotypes performing best in a specified region of the biplot. This specialized biplot shows all predictively relevant information for the most common case that AMMI1 or AMMI2 is the most predictively accurate member of the AMMI family. For rare cases with higher-order AMMI models, these general concepts can be extended to more dimensions, because a straightforward two-dimensional graph no longer suffices.

AMMI has also been used to guide and modify the selection of test sites to increase research efficiency. M'Benga (1989) found that fewer sites could suffice for the corn trials in the Gambia. Likewise, Saindon and Schaalje (1993) found that fewer sites would be adequate for testing dry bean cultivars in western Canada. Davik (1989) identified redundant sites in the Norwegian cabbage trials. Shafii et al. (1992) applied AMMI to rapeseed, a relatively new crop in the United States to define specific areas that could consistently produce high-quality canola or industrial rapeseed. More efficient sampling of customary growing regions can free up needed resources to better sample potential new growing regions.

C. PLANT MORPHOLOGY AND PHYSIOLOGY

AMMI results can illuminate plant physiological processes that cause genotypes to interact with environments. They can also reveal the relative importance of various environmental factors or stresses. Most agricultural papers using AMMI provide a biological interpretation of the AMMI genotype parameters, so it will suffice here to cite just a few examples. Zobel et al. (1988)

found genotype IPCA1 scores for a New York soybean trial to reflect maturity groups (and correspondingly environment IPCA1 scores reflected growing degree days). Smit and de Beer (1989, 1991) and Smit et al. (1992) found the same pattern in the national soybean trials in South Africa. Crossa et al. (1991) found three major groups of genotypes in an international bread wheat yield trial that reflected pedigree information and identified potential mega-environments. Nachit et al. (1992b) associated AMMI genotype parameters with morpho-physiological traits in durum wheat. Height and tillering explained 59% of the genotype main effect, and height and post-meridian leaf rolling explained 51% of the genotype IPCA1 scores. The analysis helped identify morphological and physiological traits related to stress tolerance. In an AMMI analysis of alfalfa trials in Italy, Annicchiarico (1992) found a clear association between the origin of parental materials and the adaptation pattern of progeny. Annicchiarico and Perenzin (1994) report that genotype IPCA1 scores for a bread wheat trial reflect lateness of heading and frost tolerance (corresponding to environment IPCA1 scores that reflect level of late frosts), and IPCA2 reflected lodging and terminal drought stress. Van Eeuwijk and Elgersma (1993) compared FW and AMMI for a ryegrass yield trial. The genotype IPCA1 scores contrast early and late cultivars, and IPCA2 scores reflect responsiveness to favorable environments. FW analysis picked up only this second, smaller interaction pattern.

Zobel (1992b) applied AMMI to numerous shoot and root traits of commercial soybean cultivars. A clear pattern emerges that root traits typically have a considerably larger GE interaction than shoot traits. For example, the percentage of the treatment SS in interaction was 66.1% for total root number and 64.3% for taproot diameter, but only 24.4% for seed yield and 16.1% for plant height. The probable explanation is that plant breeders have eliminated much of the genetic diversity causing interaction for those traits that have experienced intensive selection. Such results have important implications for breeding programs. They also imply that researchers dealing with traits (or crops) without a prior history of intensive selection are likely to need effective statistical procedures for handling large interactions. Zobel (1992b, 1994) found AMMI helpful in comprehending large interactions for root traits and in relating root systems to stress resistance.

Wallace and Masaya (1988), Wallace et al. (1991, 1993, 1994a, 1994b), Mmopi et al. (1994), and Zobel and Wallace (1994) used AMMI to explicate the physiological genetics of yield accumulation in beans, particularly photoperiod and temperature interactions affecting days to flowering, biomass partitioning, and yield. Wallace et al. (1993) recommended a combination of yield system analysis and AMMI analysis to increase efficiency of breeding for higher crop yield. Annicchiarico and Perenzin (1994) reported that earliness of heading was the main wheat trait related to cultivar adaptation, in agreement with the general conclusions of Wallace et al. (1993).

D. BREEDING, SELECTION, AND HERITABILITY

Wallace et al. (1993) concluded that: "An additive main effects and multiplicative interaction (AMMI) statistical analysis can separate and quantify the genotype × environment interaction (G×E) effect on yield and on each physiological component that is caused by each genotype and by the different environment of each yield trial. The use of yield trials to select parents that have the highest rates of accumulation of both biomass and yield, in addition to selecting for the G×E that is specifically adapted to the site, can accelerate advance toward the highest potential yield at each geographical site." Yield system analysis examines rate of biomass accumulation, harvest index, and time to harvest maturity. It seems especially relevant for crops, like rice and wheat, for which harvest index has already been nearly maximized, so further gains must come from other components of yield.

Charcosset et al. (1993) applied several statistical models to a top-cross mating design with 58 corn inbreds. AMMI was found most efficient in predicting hybrid performance. With the AMMI model, a manageable amount of data from line × tester crosses can identify promising hybrids, which is helpful when direct field evaluation of all line × line hybrids is not feasible because of the large number of all possible crosses, $(N(N-1))/2$ or 1653 in this case.

Annicchiarico and Perenzin (1994) applied AMMI to a bread wheat trial in Italy and thereby defined four macro-environments. Heritability was considerably higher in these mega-environments than in the entire region.

Romagosa et al. (1993) studied several mutations in near-isogenic barley lines. The GE interaction SS was large, three times the size of the genotype main effect, and even IPCA1 alone had twice the SS of the genotype main effect. IPCA1 showed close association with a mutation causing a smaller leaf area at anthesis and earlier maturity, and IPCA2 with a different mutation causing denser spikes and more leaves at maturity. A third mutation had no detectable interaction effects. They concluded, "The interpretation of the empirical or statistical approach to G×E with the physiological measurements has proven fruitful in this study."

Zavala-Garcia et al. (1992) expected stability to improve performance for grain sorghum grown in highly unpredictable environments. Accordingly, they examined selection gains from indirect selection based on three stability indices: the FW regression coefficient, deviations from regression, and genotype IPCA scores from AMMI analysis. Regarding AMMI, for two sorghum populations, they selected about 50 families having above average genotype main effects and good stability as indicated by small (near zero) IPCA1 scores. They found indirect selection for stability alone to be ineffective, but a combined index using a stability index and genotype means did increase selection efficiency dramatically. Perhaps this indicates that the original criterion of being in the top 50% for the main effect would have been better replaced by a more exclusive criterion, such as the top 20%.

Several additional studies are interesting, although they do not concern AMMI. Helms (1993) examined selection for yield and stability (using several indices) in oats and found that selection solely for yield gave low stability and selection solely for stability gave low yield. Lin and Binns (1991) used a diallel cross to compare heritabilities for four stability indices and found considerable differences. Van Sanford et al. (1993) found that weighting data from primary locations based on correlations between primary and target locations increases predictive accuracy. Pham and Kang (1988) and Jalaluddin and Harrison (1993) examined the repeatability of several stability indices across various subsets of environments. Berke et al. (1992) identified quantitative trait loci (QTLs) in wheat for stability of grain yield as assessed by FW regression. More work needs to be done to assess the heritability of AMMI interaction parameters and to associate them with specific QTLs in various crops. However, in comparison to other interaction or stability indices including FW, AMMI interaction parameters may be expected to achieve exceptionally high heritability simply because they are efficient in capturing interaction SS and therefore achieve a high signal-to-noise ratio.

Gauch and Zobel (1989a) and Gauch (1992b:171-204, 1993c) show that a serious impediment to high heritability and breeding progress is poor selections caused by noise. Poor accuracy means poor selections, which means reduced progress. Order statistics show that the problems are far more severe than agronomists and breeders tend to realize. Great effort and expense produce numerous truly superior lines which, in an inaccurate trial, are summarily discarded. And accidental discarding of superior lines makes superior lines appear rarer than they actually are (Simmonds, 1989). Thereby, inaccurate yield trials tempt breeders to test excessive numbers of genotypes in desperate hopes of finding a great rarity, with the inevitable tradeoff of reducing the number of replications, which then exacerbates the problem of inaccuracy leading to accidental discarding of superior lines. For these reasons, an important means for increasing heritability and selection gains is to use the AMMI model to increase accuracy, as discussed next.

V. EVIDENCE OF ACCURACY GAIN

Historically, the first motivation for using AMMI, clearly reflected in the influential papers by Bradu and Gabriel (1978) and Kempton (1984), was dimensionality reduction: to produce a parsimonious, low-dimensional summary of complex, high-dimensional data. But Gauch (1982) and Huber (1985) show that PCA-related analyses can effectively separate pattern from noise. Accordingly, a second motivation for using AMMI, first reported in Gauch (1985, 1988) and Gauch and Zobel (1988), was noise reduction: to separate a pattern-rich model from a discarded noise-rich residual, thereby gaining accuracy and improving decisions. Routinely, AMMI analysis gains accuracy as much as would collecting two to four times as much data. But a few seconds of microcomputer time costs far less than collecting hundreds or thousands of additional yield observations, so AMMI analysis offers a remarkably cost-effective means for improving research efficiency and increasing the returns on research investments (Gauch, 1993c).

Furthermore, AMMI partitions the treatment SS into a model and residual, whereas traditional error-control strategies based on blocking partition the error SS into, say, pure error and blocks. These two strategies analyze different and orthogonal df, so they are fundamentally separate strategies, and for exactly that reason, both can be used if desired (Gauch, 1992b:15-18). However, accuracy gains from modeling are typically several times as large as those from blocking, so it is regrettable that the larger opportunity has been so severely neglected. In consequence, crop yields are now lower than they could have become, had yield trial data been analyzed more aggressively. This section reviews accuracy gain under three headings: validation evidence, additional evidence, and practical implications.

A. VALIDATION EVIDENCE
Numerous statistical procedures exist for quantifying a model's predictive accuracy, and hence for choosing the most accurate among several models (Hamilton, 1991; Fourdrinier and Wells, 1994). A simple method is data splitting or cross validation, for which part of the data are used for model construction and the remainder for model validation (Gauch, 1985, 1988; Gauch and Zobel, 1988). This validation procedure has been implemented for AMMI in the MATMODEL program (Gauch and Furnas, 1991). The resultant statistic is the RMS prediction difference (RMSPD), which is the root mean square difference between a model's predicted value and a validation observation's actual value. A low RMSPD is good, meaning that the model's predictions are close to the validation data. A table (or graph) showing RMSPD values for the AMMI family can be used to select the most predictively accurate member of the AMMI family for a given data set. The typical outcome is "Ockham's hill," with an intermediate model (often AMMI1 or AMMI2) most accurate, with simpler models underfitting real patterns, and with more complex models overfitting spurious noise (Gauch, 1993c). Although data splitting with RMSPD calculations is ordinarily applied to replicated experiments, with EM-AMMI, validation can also be applied to unreplicated experiments, even if there are missing data (Gauch and Zobel, 1990; Gauch, 1992b:157-162).

Accuracy gain can be expressed as statistical efficiency, which equals the number of replications merely averaged that would be required to equal the predictive accuracy of an AMMI model supplied fewer replications. For example, if an AMMI model supplied 2 replications produces yield estimates as accurate as averages of 5 replicates, then AMMI's statistical efficiency is 2.5. Incidentally, the statistical efficiency also equals the error MS of the full model (averages over replications) divided by the error MS of the reduced AMMI model.

Accuracy gain from AMMI analysis has been reported for yield trials with wheat (Crossa, et al. 1991; Nachit et al., 1992a; Eisenberg et al., 1995), corn (Crossa et al., 1988, 1990; Crossa and Cornelius, 1993), rice (Sutjihno, 1993), barley (Romagosa et al., 1993; Van Oosterom et al., 1993), soybean (Gauch, 1985, 1988, 1990b, 1992a, 1992b:139 and 160, 1993a, 1993c; Gauch and Zobel, 1988, 1989a, 1989b, 1990; Smit and de Beer, 1989, 1991; Smit et al., 1992; Crossa, 1990; Gauch and Furnas, 1991), alfalfa (Smith, 1992; Smith and Gauch, 1992; Smith and Smith, 1992), field bean (Piepho, 1994), potato (Steyn et al., 1993), and winter rapeseed (Shafii et al., 1992). For these 31 reports, the most predictively accurate model was AMMI1 or AMMI2 except for two reports in which AMMI0 won occasionally (Crossa et al., 1990; Eisenberg et al., 1995; also see Weber and Westermann, 1994). Future applications are expected occasionally to recommend a higher model, but also to continue the pattern that ordinarily AMMI1 or AMMI2 wins.

A RMSPD table (or graph), indicating the most predictively accurate member of the AMMI family, is given by Gauch (1985, 1988, 1992b:139), Crossa et al. (1990, 1991), Nachit et al. (1992a), Romagosa et al. (1993), Steyn et al. (1993), Van Oosterom et al. (1993), Eisenberg et al. (1995), and Piepho (1994). Gauch and Zobel (1988) report a statistical efficiency of 2.5 for the AMMI1 model of a soybean trial, Gauch (1992b:139) reports 1.45 for AMMI2 for a different and small soybean trial, Nachit et al. (1992a) report 2.9 for AMMI1 for a wheat trial, and Crossa et al. (1990) report 4.3 for AMMI1 for a corn trial. Statistical efficiency is expected to be highest for large and noisy data sets with relatively simple interaction structure fit well by a low-order AMMI model. Experience indicates that efficiencies of 2 to 4 will prove to be common. This means that with AMMI analysis, 5 replications become as accurate as 10 to 20, or that the analysis is equivalent to receiving 5 to 15 "free" replications.

For perspective, Snedecor and Cochran (1989:263-264) indicate that typical statistical efficiencies for the randomized complete block (RCB) design are 1.2 to 1.4, meaning that 5 replications in the RCB design are as accurate as 6 to 7 in the completely randomized design, giving 1 or 2 free replications. Note that the typical benefit from AMMI is about an order of magnitude greater than that from blocking. But again, addressing the treatment design and the experimental design are two fundamentally different strategies so both can be done. Furthermore, although AMMI and RCB both decrease the estimate of the (predictive or postdictive) error MS and hence both increase the ability to achieve statistical significance, only AMMI actually adjusts and changes the yield estimates to make them closer to the true values. Complete block designs, like RCB, do not adjust yields, although incomplete block designs do.

Estimated yields from the AMMI1 model were published for 18 wheat varieties in 25 environments by Crossa et al. (1991). The RMS yield adjustment from AMMI was considerable, 10% of the grand mean (but quite reasonable relative to the standard error of the mean of their 3 replications, which was 8% of the grand mean). These adjustments caused AMMI to pick a different winner than the raw data in 18 of the 25 environments. Steyn et al. (1993) gave AMMI2 estimates for 11 potato varieties in 24 environments, finding that AMMI2 selected a different winner than did the raw data in 13 of the 24 environments, or 54% different. Sutjihno (1993) summarized AMMI2 results for 15 fertilizers applied to rice grown in 3 provinces. Smit and de Beer (1989, 1991) and Smit et al. (1992) summarized AMMI1 results for a national soybean trial for 5 mega-environments. Smith and Smith (1992) gave AMMI1 yield rankings for an alfalfa trial.

How does AMMI gain accuracy? Fundamentally, the pattern in the data due to real treatment effects tends to be simpler than the noise in the data due to complicated chance effects and numerous uncontrolled factors. Consequently, early model df can capture large amounts of pattern, whereas late df capture most of the noise, so a parsimonious model can gain accuracy by discarding a noise-rich residual. In other terms, a parsimonious model accepts a little bias but avoids a lot of variance, thereby gaining accuracy overall — the statisticians' "Stein effect" (Gauch 1990a, 1992b:146-147, 1993c).

Gauch (1992b:142-143) shows that RMSPD values reflect two imperfections: the model is not perfect in estimating the true yield values, and the validation data are also not perfect. By subtracting the validation noise from RMSPD, the model's own RMS noise can be estimated. Also, the model noise can be partitioned further into model variance and model misfit or bias. For a sequence of higher-order AMMI models, the variance becomes larger and the bias becomes smaller.

Accuracy gain is a widespread phenomenon in multivariate analysis, not restricted just to AMMI. Crossa and Cornelius (1993) provided the first comparative results, applying 27 statistical models to a wheat trial. AMMI1 achieved a RMSPD of 949 kg/ha, while SHMM2 with nearly the same number of df achieved 953, and some other models were also comparable. Although AMMI was marginally superior to SHMM in this case, this minute difference is probably not significant relative to the accuracy of the validation calculations. Furthermore, it would be hasty to base general expectations on a single case. Although empirical comparisons are presently too few to support general conclusions, statistical theory strongly suggests, in agreement with this example, that similar model families with comparably parsimonious models will achieve comparable statistical efficiencies.

However, some other multivariate models, especially agglomerative classification procedures, are unlike AMMI, PCA, and SHMM in that they do not effectively concentrate pattern in parsimonious models (Smith, 1992; Smith and Gauch, 1992). Consequently, when a classification of genotypes (or environments or both) is required, one may consider first analyzing the data by AMMI to remove much of the noise. Crossa et al. (1991) reported that a classification based on AMMI-cleaned data was much clearer than a classification based on the original noisy data. Hence, pre-analysis by AMMI can help other analyses that lack the ability to discriminate between pattern and noise.

B. ADDITIONAL EVIDENCE

As just reviewed above, validation evidence for accuracy gain uses only yield data, ordinarily from a replicated experiment. In brief, the data are split, with some used for model construction and the remainder used for model validation. Then a parsimonious AMMI model is shown to be closer to the independent validation data than is the model's raw data. The model is more accurate than its data because the model discards a noise-rich residual. By contrast, the evidence for accuracy gain reviewed in this section concerns yield data *and* something else, such as pedigrees, environmental information, or performance in locations or years beyond those included in a given yield-trial data set. That an AMMI model, selected for predictive accuracy with its own yield data, should *also* show clearer relationships with other genetic or environmental factors external to the AMMI

102

analysis itself offers striking additional evidence that the AMMI model really is selectively recovering pattern and thereby gaining accuracy.

Crossa et al. (1991) compared wheat rankings over locations, for both the raw data and the predictively accurate AMMI1 estimates, and found the latter more consistent. The explanation for this result is that spurious noise added to genuine interactions makes the interactions (and rankings) appear to be more complicated than they really are. Eliminating much of this spurious complexity is helpful to breeders. Similarly, speaking of multivariate analyses in general, including AMMI, Romagosa and Fox (1993) claim that multivariate association of new lines with known lines based on a wide test in just one year provides results with high repeatability in future years. Combining these two observations, by reducing noise and clarifying interaction patterns, AMMI analysis can help limited test results to achieve greater predictability and reliability across other locations and future years.

Another indication that AMMI gains accuracy is that routinely the parameters of the predictively accurate AMMI model do have biological interpretability *and* the parameters of the model's residual do not have biological interpretability. Romagosa et al. (1993) found AMMI2 most predictively accurate for a barley trial. The model parameters were biologically interpretable. IPCA3 was declared significant at the 5% level by a postdictive F-test, but more importantly, it was not predictively significant and was not biologically interpretable. This example fits the usual pattern that postdictive tests are more generous than predictive tests in declaring parameters significant. As another example, Snijders and Van Eeuwijk (1991) found AMMI2 most predictively accurate for a study of head blight on wheat. Predictively significant parameters were also biologically interpretable, whereas higher IPCA axes deemed mostly noise were not interpretable. Again following the usual pattern, an F-test found one more axis, IPCA3, postdictively significant. As a third example, Nachit et al. (1992a, 1992c) found numerous large correlations between predictively significant AMMI parameters and measured environmental variables. Although not reported in these brief articles, it was also the case that significant correlations for higher IPCA axes were only as numerous as would be expected by chance.

The general conclusion, justified by statistical theory and agreeing with these three examples, is that predictive assessment of significance is a better indicator of identifiable biological causes than is postdictive assessment. A further advantage is that a given IPCA axis either is or is not predictively helpful and significant using RMSPD, whereas postdictive assessment by F-tests requires the specification of an essentially arbitrary significance level, so an axis may be significant judged at the 5% level, but not at the 1% level. Incidentally, F-tests at 1% generally mimic predictive tests better than F-tests at 5%.

Three comments add perspective to this general conclusion. First, differences between replicates and the consequent noise in the data do have causes, and these causes might even be known, such as an obvious soil gradient across an experimental field. The term "noise" does not imply uncaused or unknown, but rather that some variation is not associated with the imposed treatments. Variation between replicates is caused by uncontrolled factors associated with the experimental design rather than controlled factors associated with the treatment design. The residuals from an AMMI analysis may be inspected for any obvious patterns, such as a clear trend in the field plot map.

Second, were an experiment repeated with more replications or greater accuracy, additional IPCA axes might become predictively significant and biologically interpretable. So, "significant" means significant relative to a given experiment with its particular accuracy and limitations.

Third, a good policy is to carefully inspect the predictively significant IPCA axes plus one more. The separation between pattern and noise is strong but imperfect, so all IPCA axes have some mixture of pattern and noise. Early axes are mostly pattern, and late axes mostly noise, but one to a few intermediate or transitional axes can have a substantial mix of both pattern and noise. In some cases, a transitional IPCA axis might fail to increase predictive accuracy and yet have a hint of known biological pattern. Such an IPCA axis might suggest a simpler model, such as a classification of the genotypes (or environments) into two or three groups, that has far fewer df than the entire IPCA axis and might even contribute to predictive accuracy. Just as the whole interaction may not be predictively significant, whereas a simpler AMMI model captures the predictively significant structure within the interaction, just so a still simpler model might capture real structure within a rather noisy IPCA axis.

Emphasis on predictive significance amounts to saying that a claimed pattern in the data is confirmed and solid only if it emerges repeatedly from numerous random subsets of the data. Additional supposed "pattern" in a less parsimonious model chosen for postdictive accuracy is likely to be unique to a given single data set, and is unlikely to generalize well. Hence, in such cases, "correlation" with known genetic or environmental properties is probably due to chance and is unlikely to be found again were it possible to repeat the experiment. A pattern that is postdictively significant but not predictively significant should ordinarily be regarded as weak evidence. On balance, weak evidence sometimes helps scientists in formulating worthwhile hypotheses for further research, but more research is still required to reach strong conclusions. Nevertheless, in cultivar advancement to commercial status, prediction is everything.

This discussion of AMMI accuracy gain has used one model as the standard of comparison, the raw data or averages over replications, which is the full model, AMMIF. However, one additional standard of comparison merits brief mention, the additive model without interaction, AMMI0, which stands at the opposite end of the spectrum from AMMIF. Surprisingly frequently, researchers conduct geographically extensive trials but then base selection decisions largely or entirely on genotypic means, ignoring interactions. The most accurate model, say AMMI1 in a given case, is more accurate than AMMIF because AMMI1 does not overfit spurious noise; but AMMI1 is also more accurate than AMMI0 for the different reason that AMMI1 does not underfit real pattern. Underfitting pattern causes a model to lose biological realism and interpretability. For example, Smith and Smith (1992) concluded that "Expected trends in results from the [alfalfa] yield trials were not obvious using the additive main effects model" denoted AMMI0 here, "while the application of the AMMI model resulted in rankings of cultivars in different environments which could readily be explained by the breeding history and dormancy of the cultivars."

C. PRACTICAL IMPLICATIONS

The claim that AMMI and some other multivariate analyses can gain accuracy has received a mixed review. Romagosa and Fox (1993) say, "Despite the evidence, use of AMMI adjustments in genotypic trials is controversial." On the other hand, Cornelius et al. (1993) say that adjusted means from multivariate models "should replace empirical cell means routinely reported in experiment station crop variety trial bulletins." So, do unadjusted raw means or adjusted AMMI estimates merit researchers' confidence? And, what are the stakes in this debate?

Before tackling this debate directly, first consider a simpler question about accuracy. Imagine a yield trial with four replications. To estimate a yield, should a researcher prefer the mean of the four replications, or just pick one of the replicates at random? If you answer the former, exactly why is the mean of four observations more accurate than one observation? Is it always more accurate, or just most of the time?

This exceedingly common practice, of averaging several replicates, is justified because it usually gives an estimate closer to the true mean than would one replicate. A simple simulation quantifies the outcome: generate and average four standard Gaussian random deviates (with mean 0 and variance 1) and check whether this average or a single value (picked at random) is closer to the true value, which is zero in this case. Repeating this procedure many times, it turns out that by increasing an experiment from one replicate to four replicates, the estimates move closer to the truth 72.9% of the time and farther from the truth 27.1% of the time. Not only are good adjustments from replicating 2.69 times as likely as bad adjustments, they are also larger on average by a factor of 2.31, as expressed in terms of RMS. Bad adjustments are relatively infrequent and small as compared with good adjustments, and these tendencies increase with the number of replications. So, why replicate? Because it usually, but not always, gets researchers closer to the truth. Expressed in terms of RMS errors around the true means, the standard error of a mean of N observations equals the standard deviation of a single observation divided by the square root of N, so 4 replicates halve the error.

Incidentally, the best of 4 replicates beats their average 65.1% of the time, so it would usually be better to use the one best replicate than the average of four replicates. Furthermore, this outcome obtains for any number of replicates from 2 up. However, this comment is merely academic since in a real-world experiment, unlike the present mathematical simulation, one cannot identify the best replicate. Returning to the main point here, averaging replicates is justified because it usually, but not always, improves results. As common-sense experience and statistical theory indicate, more data produce better results.

Similarly, moving on to a second preliminary example, why use some fancy experimental design with incomplete blocks to adjust the yield estimates, rather than merely the simpler completely randomized design with unadjusted means?

Just like the preceding case, the justification is that the RMS error is reduced and this usually, but not always, moves estimates closer to the truth.

Having justified these two simple practices of replicating and blocking, let us now turn to the present issue of AMMI. Why use AMMI to adjust estimates? Because AMMI reduces the RMS predictive error, and this usually, but not always, moves estimates closer to the truth. For example, if the statistical efficiency of AMMI is 2 for a given experiment, then on average 64.8% of the adjusted estimates are closer to the truth; and if the statistical efficiency is 4, then 72.9% are better. So, the three practices — replicating, blocking, and AMMI — all have the same justification: these practices usually, but not always, help. For real-world experiments, that is the best we can do. Any cost-effective means for gaining accuracy deserves serious consideration.

Let us then turn the question around: Why not use AMMI-adjusted means? When AMMI achieves a typical statistical efficiency of 2 to 4, the raw averages have a *larger* RMS predictive error and they are *worse* than the AMMI estimates from 64.8% to 72.9% of the time. What is the justification for using these worse estimates when better ones are readily available? That is the real question! Indeed, the raw averages over replications equal the full model, AMMIF, which is just one particular member of the AMMI family. It has no privileged, special, or default status. For a given data set, why use AMMIF rather than a more parsimonious member of the AMMI family? Some reason must be given for this choice, based on optimizing some particular criterion, such as predictive accuracy.

Having examined the average or statistical justification for AMMI, what about atypical yields with specific biological causes? A parsimonious AMMI model captures general patterns in the data, but what about just one variety that has, say, an extreme disease susceptibility in the particular conditions found at just one test location? Won't AMMI focus on the general pattern and not pick up this atypical but real result? Indeed, such is the case, that AMMI would largely ignore rare and atypical responses. Citing Crossa et al. (1991), Romagosa and Fox (1993) express legitimate concern that AMMI might be "over-riding a message that is asynchronous with the [general] pattern." This concern invites two further questions. Is there some method for detecting cases in which this problem is present? And, in actual practice, how frequent is this problem?

Regarding the first question, an atypical or asynchronous result generates a large residual (which is exactly what it means to say that AMMI does not pick up a pattern in its model). It is always good statistical council to examine a model's residuals, and AMMI offers no exception. Hence, a sizable atypical result would ordinarily *not* go undetected, but rather would generate exceptionally large residuals that cluster around the particular genotypes and locations exhibiting some unusual biological response. For such cases, the raw data are preferable to AMMI adjusted means, especially when the cause of the atypical response is entirely obvious, such as a devastating insect attack.

Regarding the second question, however, rather extensive practical experience so far has indicated residuals consistent with general noise rather than with special biological responses. Doubtless, as AMMI is applied more widely, occasional exceptions will occur. But because this complication is both apparently rare and ordinarily detectable, it subtracts little from the pervasive and substantial accuracy gains available from AMMI analysis.

Just how does AMMI gain accuracy? The mechanism of its accuracy gain has been explained in earlier publications (Gauch, 1988, 1990a, 1992b, 1993c; Gauch and Zobel, 1988), so a brief summary suffices here. Consider a yield trial with G genotypes, E environments, and R replications, and ask, "What is the yield of a particular genotype g in a particular environment e?" The full model, AMMIF, responds that the relevant data are the R replicates of this particular trial, and offers their average as its yield estimate (perhaps with adjustments for an incomplete block design). But a reduced model, say AMMI1, responds that the relevant data are the entire GER observations, and fits a multivariate model to the entire data set to calculate its yield estimate. The AMMI1 model is more accurate than AMMIF because it uses more data: GER observations instead of only R observations. There is no magic, no mystery; just more data. The superficial difference between AMMIF and AMMI1 is that the former uses trivial arithmetic (averaging R numbers) whereas the latter requires millions of arithmetic steps (which is easy with merely an ordinary microcomputer). But the substantial difference is that the latter uses more data. Obviously the R direct data for a given yield estimate are relevant, but whether the remaining (GE-1)R indirect data are also truly informative and worth analyzing is not a question for philosophical speculation, but rather is a question receiving a firm empirical answer from RMSPD or similar calculations. As reviewed earlier in this section, at this stage, AMMI analysis has been applied already to numerous and diverse yield trials, and statistical efficiencies of 2 to 4 appear commonplace. Again, why does AMMI provide more accurate estimates? Because it uses more data.

Lastly, what are the stakes in using or not using AMMI? Frequently, the greatest benefit from AMMI analysis is a better understanding of the genotypes and environments and their complex interactions, which can be applied to numerous decisions including reducing redundant test sites and defining mega-environments. The reply given here, however, is limited to benefits from gaining accuracy. As regards accuracy, the stakes in using AMMI are equivalent to the stakes in providing researchers with resources to increase replications by a factor of two to four, except that unlike the enormous costs of literally collecting several times as much data, the cost of running a microcomputer for a few seconds is trivial. Better accuracy means better selections which mean more food. Gauch (1992b:261-262, 1993c) estimates conservatively that AMMI analysis applied to the major breeding programs would increase the rate of yield gains by about 0.4% per year. After several years, the additional yield attributable to AMMI analysis would amount to enough food to feed hundreds of millions of people. Agricultural research is important, and analyzing the data aggressively is critical for efficiency and progress. The stakes are high.

VI. EXPERIMENTAL DESIGN

AMMI has implications for experimental design of multilocation trials, particularly for tradeoffs among the numbers of genotypes, environments, and replications included in a yield trial (Gauch 1992b:241-258). Researchers would like to have many genotypes G to increase the chances of including markedly superior lines; many environments E to characterize interactions thoroughly and thereby avoid ugly and costly surprises; and numerous replications R to increase accuracy. But the cost of a trial rises roughly with the number of yield plots, GER, so tradeoffs are inevitable.

An efficient yield trial must balance these three requirements, resulting in three moderate problems, rather than solving one problem excessively while neglecting another that gets out of control. For example, there is no reward for entering excessive numbers of genotypes in a trial if the consequent tradeoff with fewer replications lets noise get out of control, causing inaccuracy to mix up the good and bad entries to such an extent that most selections are poor.

Experiment station research with replicated small plots shows a strong historical trend toward more environments and fewer replications (Bradley et al., 1988), and likewise on-farm trials with large plots have little or no replication (Johnson et al., 1992; Johnstone et al., 1993), so in either case, there is a trend toward less replication. However, this tradeoff between G, E, and R should be considered carefully because it strongly impacts efficiency and rate of progress (Gauch and Zobel, 1989a; Gauch, 1992b:171-204). Unfortunately, because of experimental errors, the genotype winning the trial with the highest empirical, measured yield may not be the same as the genotype possessing the highest true yield. Noise causes an imperfect correlation between phenotype and genotype (Federer, 1951). Frequently, a suboptimal tradeoff causes even half or more of the data to be wasted effort in that a wiser experiment with half as many plots could perform equally well.

Much could be said about optimal tradeoffs among G, E, and R — about the selection triathlon. But in the present context of this review of AMMI analysis, just one matter is explored: the interaction between design and analysis of experiments. The best design can depend on the intended analysis, particularly whether or not AMMI analysis is planned. The main interaction is intuitively obvious, that AMMI analysis can support designs with fewer replications. For example, if for yield trials of a particular kind, AMMI routinely has achieved a statistical efficiency of 2 to 3 in the past so that in future trials at least 2 can be expected, then an experiment that had 4 replications without AMMI could switch to 2 replications with AMMI and still perform just as well. AMMI analysis reduces the pressure to design yield experiments with large numbers of replications. Reducing replications then frees resources to examine more genotypes or more environments or both.

To illustrate this interaction between design and analysis, consider a hypothetical yield trial that is representative of actual trials. The true genotype means are modeled as a normal distribution with mean 2000 (in whatever units) and standard deviation of 150, and errors for individual yield plots as a normal distribution with mean 0 and standard deviation 200. Further assume that the yield trial is allocated resources for 400 yield plots at each location, so the choices of design are 400 genotypes with 1 replication, 200 with 2, 133 with 3, 100 with 4, and so on. The design objective is to select the experiment that maximizes the average true yield of that genotype selected on the basis of having the highest empirical yield.

For this example without AMMI analysis, the choice of 100 genotypes with 4 replications strikes the optimal tradeoff between genotypes and replications, achieving an average top selection with a yield of 2307. For comparison, an experiment having 200 genotypes with 2 replications has the same number of plots and same cost, but it achieves only 2290. However, with an AMMI analysis achieving a statistical efficiency of 2, it is best to switch from 4 to 2 replications, increasing the top selection from 2333 to 2403. So, the same design change that is good with AMMI analysis is bad without it. This is what it means for design and analysis to interact. Consequently, optimal design of experiments requires foresight, with data analysis planned in advance of conducting the experiment. Better design and better analysis both improve selections.

Bad designs waste data. For the above setup, an experiment with 400 genotypes and 1 replication achieves an average top selection of only 2267. However, the smallest experiment achieving this same yield has 53 genotypes and 3 replications, or only 159 yield plots instead of 400 (Gauch, 1992b:194). Hence, the original, second-rate experiment has a data efficiency of only 159/400 or 40%. For every research dollar, it effectively uses 40 cents and wastes 60 cents.

Early-generation testing faces the challenge of characterizing interactions as well as main effects. Accordingly, it has changed considerably in recent decades. Previously 1 or 2 locations with 3 or 4 replications was common, but 7 to 12 locations with 1 replication are becoming more common now. When the interaction SS is comparable or even larger than the genotype SS, as is often the case, this increased interest in interactions is well motivated indeed. In this context, AMMI can help to describe and understand interaction patterns and to maintain accuracy despite fewer replications.

Late-generation testing of elite materials faces the challenge of achieving tremendous accuracy. For farmers achieving a typical profit margin of 10% to 20%, a yield increase of merely 1% translates to a 10% to 5% profit increase, respectively, so slight yield differences are economically important. Similarly, for breeders, even slight improvements in selection systems of 1% or less accumulate over breeding seasons to cause sizable yield improvements. Hence, to meet agricultural objectives, researchers would like an accuracy of 1% or better, or two significant digits. Yet a typical yield experiment's coefficient of variation of about 10% implies results with only one significant digit. Unfortunately, noisy data compromise agricultural objectives. AMMI can help to reduce the accuracy shortfall and can improve the targeting of particular genotypes to particular growing regions.

VII. FUTURE DIRECTIONS

Significant application of AMMI to agricultural research began only a decade ago with Kempton (1984), although this analysis was invented three decades earlier. Rather extensive results at the present time support the claims that AMMI routinely helps researchers to understand complex interactions and AMMI frequently increases the accuracy of yield estimates substantially. However, more research is still needed on a number of topics. This section sketches some particularly interesting and important directions for future research.

A. ALGORITHM VARIATIONS AND REFINEMENTS

A method for choosing the most predictively accurate member of the AMMI family, by data splitting with RMSPD calculations, was popularized by Gauch (1988) and Gauch and Zobel (1988). It constitutes a truncation model in that each IPCA axis is given an extreme weight of either 0 or 1. For example, the AMMI2 model includes IPCA1 and IPCA2 with weights of 1, and gives IPCA3 and all higher axes weights of 0. But a wider class of models results from allowing continuous weights from 0 to 1, such as a model placing weights of 0.987, 0.642, and 0.293 on IPCA1, IPCA2, and IPCA3 (and negligible or zero weights on all higher axes). The particular weights that optimize predictive accuracy for a given data set could be determined by empirical Bayes procedures (Efron and Morris, 1972; Chuang, 1982) or by a BLUP criterion (Cornelius et al., 1993; Piepho, 1994). The best AMMI model from this wider class with continuous weights may achieve greater predictive accuracy than the best model from the narrower class of truncated models. In cases where the data are quite costly or the derived decisions are especially important, even a marginal increase in accuracy may justify the additional computational effort.

One may suspect that this additional effort would normally yield a small increment in accuracy, but no actual comparisons are yet available, let alone enough comparisons for a balanced and general assessment. Perhaps the increment would be larger than many researchers would expect. On the other hand, perhaps the expected improvement would not be forthcoming. In a general statistical context involving simultaneous inference on model choice and model parameters, Venter and Steel (1993) favor truncation, thereby completely eliminating high-order model parameters too noisy to help prediction. Some general considerations in Fourdrinier and Wells (1994) also merit reflection. Anyway, the needed comparative research has not been done, so the verdict is not yet in. The future verdict should balance theoretical and practical considerations. For example, if a hundred-fold increase in computational effort typically increases accuracy by a tiny amount but results in a model with numerous axes that cannot be visualized or graphed in two or three dimensions, then practical needs would ordinarily be served better by the present simple truncation approach, especially when understanding interactions is an objective as well as estimating yields.

Furthermore, it is not clear that this particular class of models with continuous weights would offer *the* most predictively accurate model anyway. To the contrary, one may strongly suspect that still more accurate yield estimates could result from replacing the regular PCA portion of AMMI with a projection pursuit algorithm that optimizes the signal-to-noise ratio (see Huber, 1985; Jones and Sibson, 1987). But again, this research has not been done, so comparisons in terms of predictive accuracy, model complexity, and computational load are not yet available.

Incidentally, Fakorede (1986) replaced the PCA portion of AMMI with factor analysis followed by varimax rotation. This is an interesting variation on the AMMI theme, but is not expected to be of general value to agricultural researchers because the resultant model of the interaction does not have orthogonal axes.

Gower and Harding (1988) generalized the classical biplot of Bradu and Gabriel (1978). AMMI uses a Euclidean metric and produces linear biplots, but other metric and nonmetric scalings give rise to nonlinear biplots. Nonlinear biplots are considerably more complicated to interpret, containing both points and trajectories instead of just points.

A generalized version of AMMI, denoted GAMMI, has been developed to accommodate an error structure other than the usual assumption of additive normal errors (Van Eeuwijk, 1995). The natural application of GAMMI has been to disease incidence on plants, which frequently has an error distribution that is not even roughly normal. However, AMMI is considerably simpler than GAMMI and more suitable when the errors are approximately additive and normal, as is frequently the case with yield-trial data.

B. DEGREES OF FREEDOM

Two systems for assigning df in the AMMI model were popularized by Gollob (1968) and Mandel (1971). Recently, a third has been introduced by Gauch (1992b:112-129, 1993b) and still others by Chadoeuf and Denis (1991), Cornelius (1993), and Cornelius et al. (1993). Assigning df is important because these values enter into F-tests, so different accountings easily lead to different assessments regarding which IPCA axes are significant. Hence, an IPCA axis may be significant (at, say, the 5% level) using one df accounting, and yet be nonsignificant using another accounting. In comparing these several systems, both theoretical and practical considerations are relevant. Unfortunately, no general agreement can be reported yet. Perhaps this controversy persists in part because of subtle differences in definitions of df implicit in the various systems, so foundational issues would also require clarification as well as technical issues. Although a completely satisfying evaluation is not yet possible, a few helpful remarks seem justified already.

Although the literature has generally been fairly clear on this technical point, greater recognition is being given currently to the difference in noise recovery patterns for a pure-noise matrix and a pattern-plus-noise matrix. The simulation results reported in Mandel (1971) were derived from pure-noise matrices containing random normal deviates, but were then applied to a chemistry experiment with about five significant digits and hence an enormous signal-to-noise ratio, that is, to a pattern-plus-noise matrix. Such an extension is not justified because pure-noise and pattern-plus-noise matrices behave differently (Gauch, 1992b:112-129, 1993b).

On a practical level, without dispute, the system from Gollob (1968) is tremendously easier to use than any other. It requires trivial arithmetic. For a matrix with R rows and C columns, IPCA axis n receives $R+C-1-2n$ df. Most other systems require results from extensive simulations or from published tables.

Most important of all, debates regarding the various methods for conducting F-tests (based on different df assignments) should not be allowed to distract attention from the far more significant choice between any postdictive F-test and any predictive test. The larger point is that only predictive accuracy is helpful and relevant when extending a conclusion from the sample data to a larger population of inference, which is nearly always precisely what agronomists and breeders are doing in practice. Consequently, the primary motivation for a simple F-test, specifically that test due to Gollob (1968), would be computational ease rather than theoretical exactness. This easy test can be made to mimic predictive tests more closely by the simple expedient of increasing the significance level from the customary 5% to the more demanding 1%, or perhaps 0.1%. However, the choice of a computationally difficult F-test requiring extensive simulation is difficult to appreciate when the same computational effort could apply a predictive test using data splitting and RMSPD, and is doubly difficult to appreciate when manifestly the experiment's results are being applied to some larger population of inference.

Although the data-splitting predictive test with RMSPD requires considerable computation, an extremely simple rough approximation performs remarkably similarly (Gauch 1992b:147-148). The existence of this trivial predictive test makes the option of a predictive test more competitive in situations where computational ease is a weighty factor that might otherwise recommend the easy postdictive F-test of Gollob. For an estimate of the noise SS within the interaction, merely multiply the interaction df by the error MS. Then discard as many of the highest IPCA axes as are required to discard this much SS. If this procedure happens to land near the middle between two successive AMMI models, generally prefer the simpler model on grounds of parsimony, but instead retain that next axis when available genetic or environmental knowledge leads to a clear biological interpretation of that axis.

Finally, the above studies of various F-tests have assumed that the errors are distributed normally with a common variance. Piepho (1995) examined the robustness of several F-tests to departures from these assumptions and found some other tests more robust than Gollob's test.

C. CONFIDENCE INTERVALS

Besides testing whether an IPCA axis is statistically significant, a researcher may want to determine the confidence intervals around IPCA scores for individual genotypes and environments and to test whether, say, two genotypes have significantly different IPCA1 scores. Recently, very interesting work on significance and confidence intervals has been done by Goodman and Haberman (1990), Chadoeuf and Denis (1991), Denis and Gower (1992, 1994), Van Eeuwijk and Elgersma (1993), and Williams and Wood (1993). Azzalini and Diggle

(1994) present a PCA graph with 95% confidence ellipses. More work is needed, however, to clarify how dependent these results are on the choice of df assignments and on whether a pure-noise or pattern-plus-noise matrix is envisioned. If the dependency is strong, then the preliminary issue of df assignments requires clarification before one can have confidence in these confidence intervals. Also, it would be good to check the results for robustness against departure from normal errors. Further work along these lines is needed to make the required calculations readily accessible to agricultural researchers and to provide worked examples of the correct application and interpretation of these statistical methods.

D. IMPUTING MISSING DATA

Because of intentional changes in genotypes and locations over the years and unintentional losses of data, yield trials frequently have missing data. The expectation-maximization version, EM-AMMI, which is implemented in the MATMODEL software, imputes maximum likelihood values for the missing data (Gauch and Zobel, 1990; Gauch and Furnas, 1991). Under the assumption of normal errors, the maximum likelihood values are also the least-squares values. EM-AMMI has been applied with good results to matrices as large as 40,000 cells with 30,000 missing cells. It is possible, however, for an EM algorithm to get stuck in a local minimum, so some caution is warranted. Different starting configurations can check for local-minimum problems. Another algorithm for imputing missing data uses alternating least squares (Denis, 1991; Calinski et al., 1992; Denis and Baril, 1992). More work could be done to compare alternative approaches and to optimize the robustness and computing speed of the generally best approach.

E. RELATED ANALYSES

Much more could be done to compare AMMI with other multivariate analyses for various kinds of yield trials and various research purposes, and also to find effective combinations of analyses. So many old and new analyses are available that it is difficult to get a clear overview. Comparisons can be based on statistical theory, empirical results, or both. When empirical comparisons are necessary, a large number of diverse data sets must be compared to reach a balanced conclusion.

Until several years ago, FW regression provided the most common analysis of interaction. FW and AMMI1 have the same mathematical form: additive main effects plus the product of a genotype and environment parameter to model each interaction. The only difference is the fitting procedure: FW constrains the environment interaction parameter to equal the environment main effect, whereas AMMI uses PCA to capture the greatest SS possible by a multiplicative term. By mathematical theory, AMMI1 always captures as much or more of the treatment SS as does FW, so it is not possible for FW to capture more SS than AMMI1. Because this outcome is already known from theory, there is no need laboriously to scan dozens or hundreds of empirical results to determine the outcome. For those occasional cases with the SS captured by AMMI1 and FW nearly equal, the

AMMI biplot diagnoses FW as the applicable sub-model (Bradu and Gabriel, 1978), and the simpler FW model is preferable. For the more frequent case that AMMI1 captures far more SS than FW, the AMMI model is preferable, and furthermore additional IPCA axes can be included as needed to describe the interaction (Cruz, 1992; Yau, 1995). Genotypes (or environments) with small scores on all significant IPCA axes are "stable" in the same sense as genotypes (or environments) with nearly average slopes in FW analysis, so stable genotypes are near the center of an IPCA1-IPCA2 biplot (Shafii et al., 1992).

Since the paper by Finlay and Wilkinson (1963) appeared, numerous additional stability indices and analyses have been proposed (Westcott, 1986, 1987; Lin and Binns, 1988; Eskridge, 1991; Eskridge and Johnson, 1991; Kang and Pham, 1991; Souza and Sunderman, 1992). Important recent work has examined the genetic basis of interaction and the repeatability of interaction patterns across locations and years (Becker and Léon, 1988; Lin and Binns, 1991; Wallace et al., 1991, 1994a, 1994b; Berke et al., 1992; Braun et al., 1992; Helms, 1993; Jalaluddin and Harrison, 1993; Van Sanford et al., 1993). Stability has been important in plant breeding because it is often among the considerations used in making selections, thereby affecting yield gains, released varieties, and even germplasm collections. More work is needed to compare these approaches with AMMI in terms of capturing real interaction patterns, avoiding most of the spurious interaction noise, and producing readily comprehended parameters and graphs.

AMMI can be combined with classifications by circling groups of points in a biplot, or equivalently, by using different symbols for different groups. The classification may come from a clustering algorithm applied to the yield data to classify genotypes or environments or both, or it may come from intrinsic, known factors, such as the pedigrees of the genotypes or the geographical locations of the environments. AMMI biplots with superimposed classifications are presented by Okuno et al. (1971), Crossa et al. (1991), Gauch (1992b:218-222), Zobel (1992b), Romagosa and Fox (1993), Sutjihno (1993), and Annicchiarico and Perenzin (1994). Some interesting and relatively new clustering procedures include Corsten and Denis (1990), Cornelius et al. (1992, 1993), Crossa et al. (1992, 1994), Crossa and Cornelius (1993), and Baril et al. (1994).

AOV, FW regression, and PCA can all be combined in a single model called the "vacuum cleaner" (VC) by Tukey (1962). The VC model applies PCA to the residual from FW regression (extended to include row, column, and joint regressions), rather than to the interaction (which is the residual from AOV) as in ordinary AMMI. Freeman (1985) mentions the VC model under the name CLINCO, referring to an unpublished manuscript by Gauch.

Sometimes the data structure is a two-way genotypes-by-environments matrix of yields *and* a two-way factors-by-environments matrix of environmental factors measured in each environment, and a primary research objective is to relate yield and environmental patterns, largely to suggest or determine environmental causes of yield differences. This data structure is outside AMMI's domain because AMMI addresses only a single two-way matrix, not two matrices. For the two-matrix problem, one may consider redundancy analysis (Van Eeuwijk, 1992a, 1992b) or partial least squares (PLS) regression (Aastveit and Martens, 1986).

F. SPATIAL ANALYSIS

With the one exception of Eisenberg et al. (1995), agricultural researchers have applied AMMI to the treatment design, rather than the experimental design. However, when yield plots are laid out in a rectangular arrangement (such as 10 rows by 15 columns), AMMI can be used as a method for spatial analysis that can discard a noise-rich residual and thereby gain accuracy. Much work remains to test this AMMI error-control strategy in a number of yield trials and to reach a balanced assessment of its effectiveness in comparison with several of the leading methods for blocking and for spatial analysis. If a yield trial is repeated in several locations or years, as was the case in Eisenberg et al. (1995), then AMMI can be applied to both the experimental and treatment designs for two rounds of accuracy gain.

G. SELECTION TRIATHLON AND EXPERIMENTAL DESIGN

To make good selections, breeders like to increase three quantities: numerous genotypes G to play the "numbers game" and increase the likelihood of including markedly superior lines, numerous environments E to characterize interactions thoroughly and avoid ugly surprises, and numerous replications R to increase accuracy and thereby decrease the probabilities of accidentally discarding good material or accidentally retaining poor material. Since the cost of a yield trial is roughly proportional to the total number of plots, GER, an inevitable tradeoff ensues between G, E, and R. The "selection triathlon" refers to finding a nearly optimal tradeoff, balancing these three requirements. The stakes in this tradeoff are research efficiency and selection gains. When the tradeoff is suboptimal, data are wasted, meaning that a better experimental design could achieve the same selection gains with only a half or quarter as many plots. Research inefficiency is expensive. As an informal estimate of the worldwide average, based on moderately extensive experience, about half of yield trial plots are wasted by inefficient designs. For example, there may be too many genotypes and too few replications, so the numbers game is played excessively well but extreme neglect of noise problems causes numerous, superb lines to be discarded. The selection triathlon requires solving three problems, and there can be remarkably little reward for solving two while being clobbered by a third.

In one sense, experimental design is a completely separate topic from AMMI, and, indeed, the error and treatment df are orthogonal. But in another sense, these two topics are related because AMMI accuracy gain has crucial implications for experimental design. The most obvious implication is that an AMMI statistical efficiency of 2 or 3 may justify reducing replications from 4 to only 2, thereby releasing half of the research plots to plant more genotypes or to test more locations, or even to conduct additional other experiments. Preliminary results on the interaction between experimental design and AMMI are given by Gauch and Zobel (1989a) and Gauch (1992b:241-258). An important conclusion is that the best experimental design can change depending on whether or not AMMI analysis is planned for the resultant data. For example, switching from 4 to 2 replications can be both a good idea if AMMI is applied to the data and a bad idea if AMMI is not used. So far, results on experimental design and AMMI are available for a number of typical yield-trial scenarios. It remains, however, to work out the implications in greater generality so that breeders could specify a few simple parameters about a given yield trial and then easily determine a nearly optimal tradeoff between G, E, and R. It is plausible to claim that clearer insight into the selection triathlon could double the efficiency of yield-trial research worldwide.

VIII. CONCLUSIONS

The claim by Kempton (1984) that AMMI with biplots offers a helpful statistical tool for understanding interactions is now well supported by the numerous applications to date. Likewise, the claim by Gauch (1985, 1988) and Gauch and Zobel (1988) that AMMI with RMSPD or similar validation calculations can routinely gain accuracy is now well supported. More research is required, however, to clarify df assignments and confidence intervals, to compare numerous suggested variations, refinements, and related analyses, and to optimize experimental designs that integrate AMMI analysis to gain accuracy and thereby improve selections.

Three statistical tools can increase the efficiency of most yield-trial research. First, if not already optimal, adjust the numbers of genotypes and replications to balance the tradeoff between including rare superior genotypes and successfully selecting them. Second, consider fancier experimental designs than randomized complete blocks, such as row and column designs and α-lattice designs, especially for trials with over 30 entries. Third, for trials repeated over locations (or years or both), use AMMI to understand genotype-environment interactions and to gain accuracy and improve selections. For the great majority of current yield trials, this prescription would increase progress as much as would getting 2 to 5 times as much funds for planting and harvesting more yield plots. Improved statistical design and analysis offer cost-effective means for increasing returns on research investments, whereas inefficient procedures make it necessary to spend several times as much on data collection. Effective design and analysis makes a difference.

REFERENCES

Aastveit, A.H., and H. Martens. 1986. ANOVA interactions interpreted by partial least squares regression. *Biometrics* 42:829-844.

Alwi, N., J.C. Wynne, J.O. Rawlings, T.J. Schneeweis, and G.H. Elkan. 1989. Symbiotic relationship between *Bradyrhizobium* strains and peanut. *Crop Science* 29:50-54.

Annicchairico, P. 1992. Cultivar adaptation and recommendation from alfalfa trials in northern Italy. *Journal of Genetics and Breeding* 46:269-277.

Annicchiarico, P., and M. Perenzin. 1994. Adaptation patterns and definition of macro-environments for selection and recommendation of common wheat genotypes in Italy. *Plant Breeding* 113:197-205.

Azzalini, A., and P.J. Diggle. 1994. Prediction of soil respiration rates from temperature, moisture content and soil type. *Applied Statistics* 43:505-526.

Baril, C.P., J.-B. Denis, and P. Brabant. 1994. Selection of environments using simultaneous clustering based on genotype × environment interaction. *Canadian Journal of Plant Science* 74:311-317.

Basford, K.E., P.M. Kroonenberg, and I.H. DeLacy. 1991. Three-way methods for multiattribute genotype x environment data: an illustrated partial survey. *Field Crops Research* 27:131-157.

Becker, H.C., and J. Léon. 1988. Stability analysis in plant breeding. *Plant Breeding* 101:1-23.

Berke, T.G., P.S. Baenziger, and R. Morris. 1992. Chromosomal location of wheat quantitative trait loci affecting stability of six traits, using reciprocal chromosome substitution. *Crop Science* 32:628-633.

Boggini, G., P. Annicchiarico, A. Longo, and L. Pecetti. 1992. Produttività e addattamento di nuove costituzioni di frumento duro (*Triticum durum* Desf.). *Rivista di Agronomia* 26:482-488.

Bradley, J.P., K.H. Knittle, and A.F. Troyer. 1988. Statistical methods in seed corn product selection. *Journal of Production Agriculture* 1:34-38.

Bradu, D., and K.R. Gabriel. 1978. The biplot as a diagnostic tool for models of two-way tables. *Technometrics* 20:47-68.

Braun, H.-J., W.H. Pfeiffer, and W.G. Pollmer. 1992. Environments for selecting widely adapted spring wheat. *Crop Science* 32:1420-1427.

Calinski, T., S. Czajka, J.-B. Denis, and Z. Kaczmarek. 1992. EM and ALS algorithms applied to estimation of missing data in series of variety trials. *Biuletyn Oceny Odmian* 24-25:7-31.

Ceccarelli, S. 1989. Wide adaptation: How wide? *Euphytica* 40:197-205.

Chadoeuf, J., and J.-B. Denis. 1991. Asymptotic variances for the multiplicative interaction model. *Journal of Applied Statistics* 18:331-353.

Charcosset, A., J.-B. Denis, M. Lefort-Buson, and A. Gallais. 1993. Modelling interaction from top-cross design data and prediction of F1 hybrid value. *Agronomie* 13:597-608.

Chuang, C. 1982. Empirical Bayes methods for a two-way multiplicative-interaction model. *Communications in Statistics, Theory and Methods* 11(25):2977-2989.

Cornelius, P.L. 1980. Functions approximating Mandel's tables for the means and standard deviations of the first three roots of a Wishart matrix. *Technometrics* 22:613-616.

Cornelius, P.L. 1993. Statistical tests and retention of terms in the additive main effects and multiplicative interaction model for cultivar trials. *Crop Science* 33:1186-1193.

Cornelius, P.L., J. Crossa, and M.S. Seyedsadr. 1993. Tests and estimators of multiplicative models for variety trials. *In* Proceedings of the 1993 KSU Conference on Applied Statistics in Agriculture, W. Noble (Ed.), pages 156-169. Department of Statistics, Kansas State University, Manhattan, Kansas.

Cornelius, P.L., M. Seyedsadr, and J. Crossa. 1992. Using the shifted multiplicative model to search for "separability" in crop cultivar trials. *Theoretical and Applied Genetics* 84:161-172.

Cornelius, P.L., D.A. Van Sanford, and M.S. Seyedsadr. 1993. Clustering cultivars into groups without rank-change interactions. *Crop Science* 33:1193-1200.

Corsten, L.C.A., and J.-B. Denis. 1990. Structuring interaction in two-way tables by clustering. *Biometrics* 46:207-215.

Corsten, L.C.A., and A.C. Van Eijnsbergen. 1972. Multiplicative effects in two-way analysis of variance. *Statistica Neerlandica* 26:61-68.

Crossa, J. 1990. Statistical analyses of multilocation trials. *Advances in Agronomy* 44:55-85.

Crossa, J., and P.L. Cornelius. 1993. Recent developments in multiplicative models for cultivar trials. *In* International Crop Science I, D.R. Buxton, R. Shibles, R.A. Forfberg, B.L. Blad, K.H. Asay, G.M. Paulsen, and R.F. Wilson (Eds.), pages 571-577. Crop Science Society of America, Madison, Wisconsin.

Crossa, J., P.L. Cornelius, K. Sayre, and J.I. Ortiz-Monasterio. 1994. A shifted multiplicative model fusion method for grouping environments without cultivar rank change. *Crop Science* 35:54-62.

Crossa, J., P.L. Cornelius, M. Seyedsadr, and P. Byrne. 1992. A shifted multiplicative model cluster analysis for grouping environments without genotypic rank change. *Theoretical and Applied Genetics* 85:577-586.

Crossa, J., P.N. Fox, W.H. Pfeiffer, S. Rajaram, and H.G. Gauch. 1991. AMMI adjustment for statistical analysis of an international wheat yield trial. *Theoretical and Applied Genetics* 81:27-37.

Crossa, J., H.G. Gauch, and R.W. Zobel. 1988. Estimacion estidastica predictiva de rendimento en ensayos de variedades. *In* Investigacion en el Manejo de Cultivos, L. Alvarado (Ed.). PCCMCA, San Jose, Costa Rica.

Crossa, J., H.G. Gauch, and R.W. Zobel. 1990. Additive main effects and multiplicative interaction analysis of two international maize cultivar trials. *Crop Science* 30:493-500.

Cruz, R.M. 1992. More about the multiplicative model for the analysis of genotype-environment interaction. *Heredity* 68:135-140.

Dale, M.F.B., and M.S. Phillips. 1989. Genotype by environment interaction in the assessment of tolerance of partially resistant potato clones to potato cyst nematodes. *Annals of Applied Biology* 119:69-76.

Davik, J. 1989. Assessing three methods for identification of desirable genotypes in white cabbage (*Brassica oleracea* L. var. *capitata*). *Theoretical and Applied Genetics* 77:777-785.

DeLacy, I.H., R.L. Eisemann, and M. Cooper. 1990. The importance of genotype-by-environment interaction in regional variety trials. *In* Genotype-by-Environment Interaction and Plant Breeding, M.S. Kang (Ed.), pages 287-300. Department of Agronomy, Louisiana State University, Baton Rouge, Louisiana.

Denis, J.-B. 1991. Ajustements de modèles linéaires et bilinéaires sous contraintes linéaires avec données manquantes. *Revue de Statistique Appliquée* 39:5-24.

Denis, J.-B., and C.P. Baril. 1992. Sophisticated models with numerous missing values: The multiplicative interaction model as an example. *Biuletyn Oceny Odmian* 24-25:33-35.

Denis, J.-B., and T. Dhorne. 1989. Orthogonal tensor decomposition of 3-way tables. *In* Multiway Data Analysis, R. Coppi and S. Bolasco (Eds.), pages 31-37. North-Holland, Amsterdam, The Netherlands.

Denis, J.-B., and J.C. Gower. 1992. Biadditive Models. Laboratoire de Biométrie, INRA, Versailles, France.

Denis, J.-B., and J.C. Gower. 1994. Asymptotic covariances for the parameters of biadditive models. *Utilitas Mathematica* 46:193-205.

Denis, J.-B., and P. Vicourt. 1982. Panorama des méthodes statistiques d'analyse des interactions génotype x milieu. *Agronomie* 2:219-230.

Efron, B., and C. Morris. 1972. Empirical Bayes on vector observations: An extension of Stein's method. *Biometrika* 59:335-347.

Eisenberg, B.E., H.G. Gauch, R.W. Zobel, and K. Killian. 1995. Spatial analysis of field experiments: Fertilizer experiments with wheat (*Triticum aestivum*) and tea (*Camellia sinensis*). *In* Genotype-by-Environment Interaction, M.S. Kang and H.G. Gauch (Eds.). CRC Press, Boca Raton, Florida.

Eskridge, K.M. 1991. Screening cultivars for yield stability to limit the probability of disaster. *Maydica* 36:275-282.

Eskridge, K.M., and B.E. Johnson. 1991. Expected utility maximization and selection of stable plant cultivars. *Theoretical and Applied Genetics* 81:825-832.

Ezumah, H.C., W.T. Federer, H.G. Gauch, and R.W. Zobel. 1991. Analysis and Interpretation of the Maize Component of Maize and Cassava System with Nitrogen in a Multilocation Experiment. Technical Report BU-1129-MA, Biometrics Unit, Cornell University, Ithaca, New York.

Fakorede, M.A.B. 1986. Factor analysis of genotype x environment interaction in maize. *Maydica* 31:315-324.

118

Federer, W.T. 1951. Evaluation of Variance Components from a Group of Experiments with Multiple Classifications. Research Bulletin 380, Agricultural Experiment Station, Iowa State College, Ames, Iowa.

Finlay, K.W., and G.N. Wilkinson. 1963. The analysis of adaptation in a plant-breeding programme. *Australian Journal of Agricultural Research* 14:742-754.

Fourdrinier, D., and M.T. Wells. 1994. Risk Comparisons of Variable Selection Rules. Statistics Center Technical Report, Statistics Center, Cornell University, Ithaca, New York.

Fox, P.N., and A.A. Rosielle. 1982. Reducing the influence of environmental main-effects on pattern analysis of plant breeding environments. *Euphytica* 31:645-656.

Freeman, G.H. 1985. The analysis and interpretation of interaction. *Journal of Applied Statistics* 12:3-10.

Gabriel, K.R. 1978. The biplot graphic display of matrices with application to principal component analysis. *Biometrika* 58:453-467.

Gabriel, K.R. 1981. Biplot display of multivariate matrices for inspection of data and diagnosis. *In* Interpreting Multivariate Data, V. Barnett (Ed.), pages 147-173. John Wiley, New York, New York.

Gabriel, K.R., and G.K. Mather. 1986. Exploratory data analysis of 1951-82 summer rainfall around Nelspruit, Transvaal, and possible effects of 1972-81 cloud seeding. *Journal of Climate and Applied Meteorology* 25:1077-1087.

Gauch, H.G. 1982. Noise reduction by eigenvector ordinations. *Ecology* 63:1643-1649.

Gauch, H.G. 1985. Integrating Additive and Multiplicative Models for Analysis of Yield Trials with Assessment of Predictive Success, Mimeo 85-7. Soil, Crop, and Atmospheric Sciences, Cornell University, Ithaca, New York.

Gauch, H.G. 1988. Model selection and validation for yield trials with interaction. *Biometrics* 44:705-715.

Gauch, H.G. 1990a. Full and reduced models for yield trials. *Theoretical and Applied Genetics* 80:153-160.

Gauch, H.G. 1990b. Using interaction to improve yield estimates. *In* Genotype-by-Environment Interaction and Plant Breeding, M.S. Kang (Ed.), pages 141-150. Department of Agronomy, Louisiana State University, Baton Rouge, Louisiana.

Gauch, H.G. 1992a. AMMI analysis of yield trials. *In* Wheat Special Report No. 8: Management and Use of International Trial Data for Improving Breeding Efficiency, P.N. Fox and G.P. Hettel (Eds.), pages 9-12. CIMMYT, Mexico City, Mexico.

Gauch, H.G. 1992b. Statistical Analysis of Regional Yield Trials: AMMI Analysis of Factorial Designs. Elsevier, Amsterdam, The Netherlands.

Gauch, H.G. 1993a. MATMODEL Version 2.0: AMMI and Related Analyses for Two-way Data Matrices. Microcomputer Power, Ithaca, New York.

Gauch, H.G. 1993b. Non-additivity and biplots. *Biometrics* 49:952-954.

Gauch, H.G. 1993c. Prediction, parsimony, and noise. *American Scientist* 81:468-478.

Gauch, H.G., and R.E. Furnas. 1991. Statistical analysis of yield trials with MATMODEL. *Agronomy Journal* 83:916-920.

Gauch, H.G., and R.W. Zobel. 1988. Predictive and postdictive success of statistical analyses of yield trials. *Theoretical and Applied Genetics* 76:1-10.

Gauch, H.G., and R.W. Zobel. 1989a. Accuracy and selection success in yield trial analyses. *Theoretical and Applied Genetics* 77:473-481.

Gauch, H.G., and R.W. Zobel. 1989b. Using interaction in two-way data tables. *In* Proceedings of the 1989 Kansas State University Conference on Applied Statistics in Agriculture, G.A. Milliken and J.R. Schwenke (Eds.), pages 205-213. Department of Statistics, Kansas State University, Manhattan, Kansas.

Gauch, H.G., and R.W. Zobel. 1990. Imputing missing yield trial data. *Theoretical and Applied Genetics* 79:753-761.

Génard, M., M. Souty, S. Holmes, M. Reich, and L. Breuils. 1994. Correlations among qualilty parameters of peach fruit. *Journal of the Science of Food and Agriculture* 66:241-245.

Gilbert, N. 1963. Non-additive combining abilities. *Genetical Research* 4:65-73.

Glasbey, C.A. 1992. A reduced rank regression model for local variation in solar radiation. *Applied Statistics* 41:381-387.

Gollob, H.F. 1968. A statistical model which combines features of factor analytic and analysis of variance techniques. *Psychometrika* 33:73-115.

Goodman, L.A., and S.J. Haberman. 1990. The analysis of nonadditivity in two-way analysis of variance. *Journal of the American Statistical Association, Theory and Methods* 85:139-145.

Gower, J.C. 1990. Three-dimensional biplots. *Biometrika* 77:773-785.

Gower, J.C., and S.A. Harding. 1988. Nonlinear biplots. *Biometrika* 75:445-456.

Halseth, D.E., K.S. Yourstone, and R.W. Zobel. 1993. Analysis of yield trial data with the AMMI model. *American Potato Journal* 70:814.

Hamilton, M.A. 1991. Model validation: An annotated bibliography. *Communications in Statistics, Theory and Methods* 20(7):2207-2266.

Heenan, D.P., A.C. Taylor, B.R. Cullis, and W.J. Lill. 1994. Long term effects of rotation, tillage and stubble management on wheat production in southern N.S.W. *Australian Journal of Agricultural Research* 45:93-117.

Helms, T.C. 1993. Selection for yield and stability among oat lines. *Crop Science* 33:423-426.

Hirotsu, C. 1983. An approach to defining the pattern of interaction effects in a two-way layout. *Annals of the Institute of Statistical Mathematics* 35:77-90.

Hohn, M.E. 1979. Principal components analysis of three-way tables. *Journal of the International Association for Mathematical Geology* 11:611-626.

Huber, P.J. 1985. Projection pursuit. *Annals of Statistics* 13:435-525.

Iezzoni, A.F., and M.P. Pritts. 1991. Applications of principal component analysis to horticultural research. *HortScience* 26:334-338.

Jalaluddin, M., and S.A. Harrison. 1993. Repeatability of stability estimators for grain yield in wheat. *Crop Science* 33:720-725.

Johnson, D.E., and F.A. Graybill. 1972. Estimation of σ^2 in a two-way classification model with interaction. *Journal of the American Statistical Association, Theory and Methods* 67:388-394.

Johnson, J.J., J.R. Alldredge, S.E. Ullrich, and O. Dangi. 1992. Replacement of replications with additional locations for grain sorghum cultivar evaluation. *Crop Science* 32:43-46.

Johnstone, P.D., W.L. Lowther, and J.M. Keoghan. 1993. Design and analysis of multi-site agronomic evaluation trials. *New Zealand Journal of Agricultural Research* 36:323-326.

Jones, M.C., and R. Sibson. 1987. What is projection pursuit? *Journal of the Royal Statistical Society, Series A* 150:1-36.

Kang, M.S. (Ed). 1990. Genotype-by-Environment Interaction and Plant Breeding, Department of Agronomy, Louisiana State University, Baton Rouge, Louisiana.

Kang, M.S., and H.N. Pham. 1991. Simultaneous selection for high yielding and stable crop genotypes. *Agronomy Journal* 83:161-165.

Kempton, R.A. 1984. The use of biplots in interpreting variety by environment interactions. *Journal of Agricultural Science, Cambridge* 103:123-135.

Lin, C.S., and M.R. Binns. 1988. A superiority measure of cultivar performance for cultivar x location data. *Canadian Journal of Plant Science* 68:193-198.

Lin, C.S., and M.R. Binns. 1991. Genetic properties of four types of stability parameter. *Theoretical and Applied Genetics* 82:505-509.

Loiselle, F., G.C.C. Tai, and B.R. Christie. 1990. Genetic components of chip color evaluated after harvest, cold storage and reconditioning. *American Potato Journal* 67:633-646.

Mandel, J. 1969. The partitioning of interaction in analysis of variance. *Journal of Research of the National Bureau of Standards, B. Mathematical Sciences* 73:309-328.

Mandel, J. 1971. A new analysis of variance model for non-additive data. *Technometrics* 13:1-18.

Mandel, J. 1972. Principal components, analysis of variance and data structure. *Statistica Neerlandica* 26:119-129.

Mandel, J. 1995. Analysis of Two-Way Layouts. Chapman & Hall, New York, New York.

M'Benga, M. 1989. The Use of Genotype x Environment Interactions to Enhance Maize (*Zea mays* L.) Cultivar and Test Site Selection in the Eastern Part of the Gambia. M.S. Thesis, Cornell University, Ithaca, New York.

Mmopi, S., D.H. Wallace, P.N. Masaya, R. Rodriguez, and R.W. Zobel. 1994. Control of days to flowering of bean (*Phaseolus vulgaris* L.) by interaction of a photoperiod gene and a non-photoperiod gene. *In* Handbook of Plant and Crop Physiology, M. Pessarakli (Ed.), pages 917-939. Marcel Dekker, New York, New York.

Nachit, M.M., G. Nachit, H. Ketata, H.G. Gauch, and R.W. Zobel. 1992a. Use of AMMI and linear regression models to analyze genotype-environment interaction in durum wheat. *Theoretical and Applied Genetics* 83:597-601.

Nachit, M.M., M.E. Sorrells, R.W. Zobel, H.G. Gauch, R.A. Fischer, and W.R. Coffman. 1992b. Association of morpho-physiological traits with grain yield and components of genotype-environment interaction in durum wheat I. *Journal of Genetics and Breeding* 46:363-368.

Nachit, M.M., M.E. Sorrells, R.W. Zobel, H.G. Gauch, R.A. Fischer, and W.R. Coffman. 1992c. Association of environment variables with sites' mean grain yield and components of genotype-environment interaction in durum wheat II. *Journal of Genetics and Breeding* 46:369-372.

Okuno, T., F. Kikuchi, K. Kumagai, C. Okuno, M. Shiyomi, and H. Tabuchi. 1971. Evaluation of varietal performance in several environments. *Bulletin of the National Institute of Agricultural Sciences, Tokyo, Series A* 18:93-147.

Paterson, L.J., P. Wild, and E.R. Williams. 1988. An algorithm to generate designs for variety trials. *Journal of Agricultural Science, Cambridge* 111:133-136.

Patterson, H.D., and E.R. Williams. 1976. A new class of resolvable incomplete block designs. *Biometrika* 63:83-92.

Patterson, H.D., E.R. Williams, and E.A. Hunter. 1978. Block designs for variety trials. *Journal of Agricultural Science, Cambridge* 90:395-400.

Paul, H., F.A. Van Eeuwijk, and W. Heijbroek. 1993. Multiplicative models for cultivar by location interaction in testing sugar beets for resistance to beet necrotic yellow vein virus. *Euphytica* 71:63-74.

Peltonen-Sainio, P., K. Moore, and E. Pehu. 1993. Phenotypic stability of oats measured by different stability analyses. *Journal of Agricultural Science, Cambridge* 121:13-19.

Pham, H.N., and M.S. Kang. 1988. Interrelationships among and repeatability of several stability statistics estimated from international maize trials. *Crop Science* 28:925-928.

Phillips, M.S., and J.W. McNicol. 1986. The use of biplots as an aid to interpreting interactions between potato clones and populations of potato cyst nematodes. *Plant Pathology* 35:185-195.

Phillips, M.S., H.J. Rumpenhorst, D.L. Trudgill, K. Evans, G. Gurr, D. Heinicke, M. Mackenzie, and S.J. Turner. 1989. Envoronmental interactions in the assessment of partial resistance to potato cyst nematodes. I. Interactions with centres. *Nematologica* 35:187-196.

Phillips, M.S., D.L. Trudgill, K. Evans, C.N.D. Lacey, M. Mackenzie, and S.J. Turner. 1987. The assessment of partial resistance of potato clones to cyst nematodes at six test centres. *Potato Research* 30:507-515.

Piepho, H.-P. 1994. Best linear unbiased prediction (BLUP) for regional yield trials: A comparison to additive main effects and multiplicative interaction (AMMI) analysis. *Theoretical and Applied Genetics* 89:647-654.

Piepho, H.-P. 1995. Robustness of statistical tests for multiplicative terms in the AMMI model for cultivar trials. *Theoretical and Applied Genetics* 90:438-443.

Pike, E.W., and T.R. Silberberg. 1952. Designing mechanical computers. *Machine Design* 24:131-137, 159-163.

Riggs, T.J. 1986. Collaborative spring barley trials in Europe 1980-1982 analysis of grain yield. *Plant Breeding* 96:289-303.

Romagosa, I., and P.N. Fox. 1993. Genotype x environment interaction and adaptation. *In* Plant Breeding: Principles and Prospects, M.D. Hayward, N.O. Bosemark, and I. Romagosa (Eds.), pages 373-390. Chapman and Hall, London.

Romagosa, I., P.N. Fox, L.F. Garcia del Moral, J.M. Ramos, B. Garcia del Moral, F. Roca de Togores, and J.L. Molina Cano. 1993. Integration of statistical and physiological analyses of adaptation of near-isogenic barley lines. *Theoretical and Applied Genetics* 86:822-826.

Royo, C., A. Rodriguez, and I. Romagosa. 1993. Differential adaptation of complete and substituted triticale. *Plant Breeding* 111:113-119.

Saindon, G., and G.B. Schaalje. 1993. Evaluation of locations for testing dry bean cultivars in western Canada using statistical procedures, biological interpretation and multiple traits. *Canadian Journal of Plant Science* 73:985-994.

Samways, M.J. 1983. Community structure of ants (Hymenoptera: Formicidae) in a series of habitats associated with citrus. *Journal of Applied Ecology* 20:833-847.

Sato, K., and K. Takeda. 1993. Pathogenic variation of *Pyrenophora teres* isolates collected from Japanese and Canadian spring barley. *Bulletin of the Research Institute for Bioresources Okayama University* 1:147-158.

Shafii, B., K.A. Mahler, W.J. Price, and D.L. Auld. 1992. Genotype x environment interaction effects on winter rapeseed yield and oil content. *Crop Science* 32:922-927.

Simmonds, N.W. 1989. How frequent are superior genotypes in plant breeding populations? *Biological Reviews* 64:341-365.

Simmonds, N.W. 1991. Selection for local adaptation in a plant breeding programme. *Theoretical and Applied Genetics* 82:363-367.

Smit, M.A., and G.P. de Beer. 1989. Report on the National Soybean Cultivar Trials 1988/89. Oil and Protein Seed Centre, Department of Agriculture, Potchefstroom, South Africa.

Smit, M.A., and G.P. de Beer. 1991. Report on the National Soybean Cultivar Trials 1990/91. Oil and Protein Seed Centre, Department of Agriculture, Potchefstroom, South Africa.

Smit, M.A., G.P. de Beer, H.S.J. Grobler, and S.C. Sullivan. 1992. Report on the National Soybean Cultivar Trials 1991/92. Oil and Protein Seed Centre, Department of Agriculture, Potchefstroom, South Africa.

Smith, M.F. 1992. The AMMI Model in Agriculture: A Comparison with Standard Procedures. M.S. Thesis, University of South Africa, Pretoria, South Africa.

Smith, M.F., and H.G. Gauch. 1992. Effects of noise on AMMI and hierarchical classification analyses. *South African Statistical Journal* 26:121-142.

Smith, M.F., and A. Smith. 1992. The success of the AMMI model in predicting lucerne yields for cultivars with differing dormancy characteristics. *South African Journal of Plant and Soil* 9:180-185.

Snedecor, G.W., and W.G. Cochran. 1989. Statistical Methods, Eighth edition. Iowa State University, Ames, Iowa.

Snee, R.D. 1982. Nonadditivity in a two-way classification: is it interaction or nonhomogeneous variance? *Journal of the American Statistical Association, Applications* 77:515-519.

Snijders, C.H.A., and F.A. Van Eeuwijk. 1991. Genotype x strain interactions for resistance to *Fusarium* head blight caused by *Fusarium culmorum* in winter wheat. *Theoretical and Applied Genetics* 81:239-244.

Souza, E., and D.W. Sunderman. 1992. Pairwise rank superiority of winter wheat genotype for spring stand. *Crop Science* 32:938-942.

Steyn, P.J., A.F. Visser, M.F. Smith, and J.L. Schoeman. 1993. AMMI analysis of potato cultivar yield trials. *South African Journal of Plant and Soil* 10:28-34.

Sutjihno. 1993. Statistical analysis of N-fertilizer trial on rice using "AMMI" model. *Contributions from the Central Research Institute for Food Crops Bogor* 81:1-15.

Tsianco, M.C., and K.R. Gabriel. 1984. Modeling temperature data: An illustration of the use of biplots in nonlinear modeling. *Journal of Climate and Applied Meteorology* 23:787-799.

Tukey, J.W. 1962. The future of data analysis. *Annals of Mathematical Statistics* 33:1-67.

Van Eeuwijk, F.A. 1992a. Interpreting genotype-by-environment interaction using redundancy analysis. *Theoretical and Applied Genetics* 85:89-100.

Van Eeuwijk, F.A. 1992b. Multiplicative models for genotype-environment interaction in plant breeding. *Statistica Applicata* 4:393-406.

Van Eeuwijk, F.A. 1995. Multiplicative interaction in generalized linear models. *Biometrics* (in press).

Van Eeuwijk, F.A., and A. Elgersma. 1993. Incorporating environmental information in an analysis of genotype by environment interaction for seed yield in perennial ryegrass. *Heredity* 70:447-457.

Van Oosterom, E.J., and S. Ceccarelli. 1993. Indirect selection for grain yield of barley in harsh Mediterranean environments. *Crop Science* 33:1127-1131.

Van Oosterom, E.J., D. Kleijn, S. Ceccarelli, and M.M. Nachit. 1993. Genotype-by-environment interactions of barley in the Mediterranean region. *Crop Science* 33:669-674.

Van Sanford, D.A., T.W. Pfeiffer, and P.L. Cornelius. 1993. Selection index based on genetic correlations among environments. *Crop Science* 33:1244-1248.

Venter, J.H., and S.J. Steel. 1993. Simultaneous selection and estimation for the some zeros family of normal models. *Journal of Statistical Computation and Simulation* 45:129-146.

122

Wallace, D.H., J.P. Baudoin, J. Beaver, D.P. Coyne, D.E. Halseth, P.N. Masaya, H.M. Munger, J.R. Myers, M. Silbernagel, K.S. Yourstone, and R.W. Zobel. 1993. Improving efficiency of breeding for higher crop yield. *Theoretical and Applied Genetics* 86:27-40.

Wallace, D.H., P.A. Gniffke, P.N. Masaya, and R.W. Zobel. 1991. Photoperiod, temperature, and genotype interaction effects on days and nodes required for flowering of bean. *Journal of the American Society for Horticultural Science* 116:534-543.

Wallace, D.H., and P.N. Masaya. 1988. Using yield trial data to analyze the physiological genetics of yield accumulation and the genotype x environment interaction effects on yield. *Annual Report of the Bean Improvement Cooperative* 31:vii-xxiv.

Wallace, D.H., P.N. Masaya, R. Rodriguez, and R.W. Zobel. 1994a. Genotype, temperature and genotype x temperature interaction effects on yield of bean (*Phaseolus vulgaris* L.). *In* Handbook of Plant and Crop Physiology, M. Pessarakli (Ed.), pages 893-915. Marcel Dekker, New York, New York.

Wallace, D.H., K.S. Yourstone, J.P. Baudoin, J. Beaver, D.P. Coyne, J.W. White, and R.W. Zobel. 1994b. Photoperiod x temperature interaction effects on the days to flowering of bean (*Phaseolus vulgaris* L.). *In* Handbook of Plant and Crop Physiology, M. Pessarakli (Ed.), pages 863-891. Marcel Dekker, New York, New York.

Weber, W.E., and T. Westermann. 1994. Prediction of yield for specific locations in German winter-wheat trials. *Plant Breeding* 113:99-105.

Westcott, B. 1986. Some methods of analysing genotype-environment interaction. *Heredity* 56:243-253.

Westcott, B. 1987. A method of assessing the yield stability of crop genotypes. *Journal of Agricultural Science, Cambridge* 108:267-274.

Williams, E.J. 1952. The interpretation of interactions in factorial experiments. *Biometrika* 39:65-81.

Williams, E.R., and J.A. John. 1989. Construction of row and column designs with contiguous replicates. *Applied Statistics* 38:149-154.

Williams, E.R., and J.T. Wood. 1993. Testing the significance of genotype-environment interaction. *Australian Journal of Statistics* 35:359-362.

Wilson, D.O., S.K. Mohan, and E.A. Knott. 1993. Evaluation of fungicide seed treatments for *Shrunken-2* ("Supersweet") sweet corn. *Plant Disease* 77:348-351.

Yau, S.K. 1995. Regression and AMMI analyses of genotype × environment interactions: An empirical comparison. *Agronomy Journal* 87:121-126.

Zavala-Garcia, F., P.J. Bramel-Cox, and J.D. Eastin. 1992. Potential gain from selection for yield stability in two grain sorghum populations. *Theoretical and Applied Genetics* 85:112-119.

Zobel, R.W. 1990. A powerful statistical model for understanding genotype-by-environment interaction. *In* Genotype-by-Environment Interaction and Plant Breeding, M.S. Kang (Ed.), pages 126-140. Department of Agronomy, Louisiana State University, Baton Rouge, Louisiana.

Zobel, R.W. 1992a. Root morphology and development. *Journal of Plant Nutrition* 15:677-684.

Zobel, R.W. 1992b. Soil environment constraints to root growth. *Advances in Soil Science* 19:27-51.

Zobel, R.W. 1994. Stress resistance and root systems. *In* Proceedings of the Workshop on Adaptation of Plants to Soil Stresses, August 1-4, 1993, INTSORMIL Publication Number 94-2, pages 80-99. Institute of Agriculture and Natural Resources, University of Nebraska, Lincoln, Nebraska.

Zobel, R.W., and D.H. Wallace. 1994. The AMMI statistical model and interaction analysis. *In* Handbook of Plant and Crop Physiology, M. Pessarakli (Ed.), pages 849-862. Marcel Dekker, New York, New York.

Zobel, R.W., M.J. Wright, and H.G. Gauch. 1988. Statistical analysis of a yield trial. *Agronomy Journal* 80:388-393.

Chapter 5

IDENTIFICATION OF QUANTITATIVE TRAIT LOCI THAT ARE AFFECTED BY ENVIRONMENT

W. D. Beavis and P. Keim[1]

I. INTRODUCTION

An integral part of plant breeding is the well known but little understood phenomenon of genotype by environment interactions (GxE). GxE are manifested as phenotypes that exhibit inconsistent values among lines when tested in multiple environments. Because it is important for breeders to identify lines that perform consistently well in a set of test environments a number of statistical methods have been developed to characterize GxE, many of which were compiled and reviewed by Kang (1990). These methods have been useful for breeding purposes, but have done little to identify the underlying genetic factors or mechanisms responsible for the phenomenon.

Quantitative genetic theory has been developed on the premise that the expression of a quantitative trait is the net result of multiple genetic loci interacting with the environment (Fisher, 1918). Recent developments of molecular and statistical techniques have made it possible to identify segregating genomic regions (QTL) associated with expression of quantitative traits. It is useful to the breeder to identify lines that are consistent across a set of test environments, assuming the test environments represent the target environments of the breeding program. In the emerging breeding techniques that utilize genetic marker information, it will be useful to identify QTL with consistent effects across a set of test environments. Several QTL studies have investigated consistency of QTL across environments (Guffy et al., 1989; Paterson et al., 1991; Stuber et al., 1992; Zehr et al., 1992; Bubeck et al., 1993; Hayes et al., 1993; Schon et al., 1994). The motivation for these investigations has been to enhance the development of marker-aided selection (MAS). However, because it is likely that QTL with inconsistent effects across environments (QxE) represents the genetic factors responsible for GxE, the motivation for the identification of QxE does not need to rely solely on development of MAS. The study of QxE is justified as a source of information about the genetic component of GxE. Identification of QxE coupled with research on gene expression and signal transduction pathways will ultimately provide a genetic understanding and possible regulation of the phenomenon.

[1]Department of Agronomic Traits, Trait and Technology Group of Pioneer Hi-Bred International, Inc., 7300 NW 62nd, PO Box 1004, Johnston, Iowa 50131-1004 and Department of Biology, Northen Arizona University, Flagstaff, Arizona 86011-5640.

124

To illustrate the potential of QxE information to enhance our understanding of GxE, consider a simply inherited trait in an inbred population where the expression of the trait is under genetic control at a single segregating locus with two alleles (A and a). Plant breeders tend to associate GxE with quantitative traits, but there also are examples of simply inherited phenotypic traits that are affected by environment (Langridge and Griffing, 1959; Hoisington et al., 1982). From recent advances in molecular biology and studies on metabolic pathways, it is not difficult to imagine that the proteins encoded by the alleles at the locus could be involved in a single step of a metabolic pathway such as nutrient transport or DNA binding and the subsequent phenotype. For example, Schactman and Schroeder (1994) have identified and cloned the $HKT1$ gene from a cDNA library of barley roots. $HKT1$ is responsible for encoding a membrane protein that is a high affinity K^+ uptake transporter. When $HKT1$ is active, the ability of the plant to take up K^+ under dilute concentrations is enhanced. It is not difficult to imagine that environment could have an impact on the activity of such an enzyme. Allele A could have a low substrate Km and a temperature optimum of 20° C, while allele a could have a high Km with a temperature optimum of 30° C. If the segregating population were evaluated under four environmental conditions (low substrate, low temperature; low substrate, high temperature; high substrate, low temperature; high substrate, high temperature) the phenotypes might express an extreme case of cross-over GxE, (Figure 1). There would be no superior lines when averaged across environments but within environments there would be differences among lines that would change ranks in different environments.

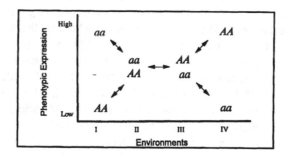

Figure 1. A single locus model for GxE in an inbred population. Genotypes, AA and aa are at a hypothetical locus that encodes for an enzyme with activities, Km, that are affected by substrate and temperature. Genotypes are arranged by rank order of phenotypic expression in each of four environments: I low substrate, low temperature; II low substrate, high temperature; III high substrate, low temperature; IV, high substrate, high temperature.

The genome could then be searched using molecular markers to identify the region of the genome that co-segregates with the trait in individual environments. This would reveal the genomic location of the underlying genetic locus responsible for the responses, and is referred to as Co-segregation analysis. Co-segregation analysis is simply a matter of looking for genetic markers that co-segregate with the phenotype.

It is easy to extend the single locus model to two loci. Assume there are two independently segregating loci (a and b) in an inbred population that are involved in the expression of the phenotype. Furthermore, assume that the enzymes encoded at each locus are sensitive to independent environmental components, such as nutrients and temperature. Depending upon the actual environment, different alleles at these loci could effect a positive additive response to the trait. An inbred population segregating at both the a and b loci would have four different genotypes having different adaptive features for different environments. As in the first example, phenotypic expression of the trait in four environments representing the four possible optimal conditions would result in an extreme case of cross-over GxE, i.e., there would be no superior lines across environments but within environments there would be differences among lines that would change ranks in different environments (Figure 2). The underlying genetic basis for this result also could be revealed through identification of regions of the genome that co-segregate with the trait. Co-segregation analyses would identify loci a and b as being associated with performance in each individual environment, but not across environments. The analysis also would provide estimates of the genetic effects of each allele in each environment, and, thus, identify allelic interactions with environments.

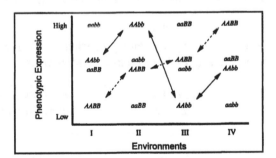

Figure 2. A two locus model for GxE in an inbred population. Four inbred genotypes at two independent loci, a and b, that encode for enzymes with activities, Km that are affected by either nutrient concentration or temperature are arranged by rank order of phenotypic expression in each of four environments: I low nutrients, low temperature; II low nutrients, high temperature; III high nutrients, low temperatures; IV high nutrients, high temperatures.

Co-segregation analyses will not provide an understanding of the underlying mechanisms of genetic expression or of the physiological bases for the phenotypic expression of the trait, but they will identify loci and alleles that respond differently to the different environments. Such knowledge could then be used to design controlled experiments to elucidate the underlying genetic mechanisms and physiology of the trait.

Most agronomic traits of economic importance exhibit not only GxE, but also are affected by multiple genetic loci. Because QTL analyses have been developed to identify regions of the genome associated with the expression of such traits, they also have been applied to the task of identifying QxE. However, the statistical issues involved with applying QTL analyses to investigate QxE have not been studied. Indeed, the statistical issues associated with models that do not include environmental effects are still being resolved (Doerge et al., 1994). In this paper, we will assume that most of the readers are familiar with GxE analyses, but will introduce the emerging field of QTL data analyses techniques and discuss some of the issues in the use of QTL analyses for QxE data analyses. We also will analyze and discuss the results of these analysis techniques on topcross (TC) and $F_{2:4}$ progeny from the cross of two maize inbred lines, B73 and Mo17.

II. DATA ANALYSIS TECHNIQUES

A. QTL ANALYSES

Numerous data analysis techniques for the identification and mapping of QTL have been developed (Thoday et al., 1961; Soller et al., 1976; Lander and Botstein, 1989; Knapp et al., 1990; Jansen, 1992, 1993; Zeng, 1993). These techniques are based on detecting co-segregation between a marker locus and the phenotypic expression of a quantitatively inherited trait (Table 1). The obvious distinction between the concept and its application is that quantitative traits such

Table 1

The basis for identification of QTL: Co-segregation between marker locus 2 and the expression of a quantitative trait.

Segregating line	Marker loci			Phenotype of a quantitative trait
	1	2	3	
1	B	A	B	High
2	A	H	A	Medium
3	H	H	H	Medium
4	A	B	A	Low
5	A	B	A	

as yield cannot be classified into distinct phenotypes (Figure 3). Nonetheless, proof of principle was shown a number of years ago (Sax, 1923; Rasmussen, 1935). The inherent statistical issues associated with these types of data analyses were not recognized until the development of molecular markers. Weller (1993) reviewed many of the proposed statistical methodologies and Doerge et al. (1994)

reviewed the statistical issues involved in QTL analyses. To date, none of the proposed techniques has been developed specifically for detecting QxE, but the statistical models can be adapted for this purpose. Before describing how QTL analyses can be extended to QxE analyses, we will introduce the QTL analysis techniques that have been proposed and discuss some of the statistical issues that are relevant to QxE analyses.

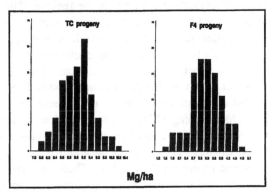

Figure 3. Frequency distribution of averaged grain yield (Mg/ha) in 112 topcross and F_{2-4} progeny from the cross B73xMo17 grown at several central U.S. cornbelt environments.

The first challenge of QTL analyses is to detect significant associations between segregating marker loci and phenotypes, given there is an underlying mixture of genotype and trait distributions (Weller, 1986). It is well known that the t-test is a robust test statistic for detecting differences between two groups and was proposed for the case with two genetic marker classes nearly 20 years ago (Soller et al., 1976). Likewise the F statistic is useful in the situation where the number of genotypic classes is greater than two (Soller et al., 1976; Edwards et al., 1987). Analyses where statistical associations between segregating marker genotypes and phenotypic traits are detected can be referred to as marker-trait (MT) analyses.

The development of ubiquitous sets of molecular markers that span the genome provided a means to investigate the entire genome for MT associations with quantitative traits. However, a statistical issue also emerged because multiple tests for MT associations are conducted using a single data set. The usual inferences about significance of the calculated statistic can no longer be applied because multiple simultaneous inferences are considered in the experiment. Furthermore, tests are not independent because many of the markers are genetically linked. Lander and Botstein (1989) addressed this by showing that an appropriate threshold can be determined analytically for a genome with an infinite number of markers and can be estimated for less saturated genomes. For a genome with less than infinite saturation but with linked markers, the appropriate threshold depends upon the genome size and density of genetic markers. Lander and Botstein (1989) sampled data from a normal distribution and assigned these values to a simulated genome to illustrate how to estimate an appropriate threshold. It is also possible to obtain empirical estimates suitable for each experiment by randomly assigning phenotypic values obtained in the experiment to the individuals or lines from which the genome map is derived (Churchill and Doerge, 1994). A random assignment will be associated with the null hypothesis of no QTL effects at any point in the genome. The randomization process is then iterated to obtain an estimate of the appropriate threshold for the QTL data analysis technique that is being employed. Sampling phenotypic data with replacement is essentially a "bootstrap" approach whereas sampling without replacement is essentially a "permutation test" (Lehman, 1986; Good, 1994). Our experience indicates that both provide similar genome-wide threshold values for QTL experiments, but a definitive comparison of the techniques is needed.

In addition to detecting QTL in a manner that avoids an excessive frequency of false positives, QTL analyses need to provide estimates of the underlying genetic parameters. These include the additive and dominance effects as well as the genomic position of the QTL. Thus, the statistical issues of accuracy and precision need to be addressed. Estimates of genetic effects and position of the QTL are confounded when MT analyses are used (Soller et al., 1976; Edwards et al., 1987).

Although the development of molecular markers that span the genome has meant that the researcher needs to empirically determine an appropriate threshold for each genome, the advantage of a genome saturated with genetic markers is that it also provides information that can be used to increase the estimated precision and accuracy of genetic effects and QTL placement. Lander and Botstein (1989) showed how to utilize this additional information to place the QTL within intervals of linked markers and referred to the method as interval mapping, (IM). IM utilizes genetic information from flanking markers and maximizes the likelihood function to find the most likely position and estimated genetic effects to explain the distribution of the quantitative trait. Maximum likelihood IM has become a widely used technique primarily because it was made available through the release of MAPMAKER/QTL (Lincoln and Lander, 1990). Interval mapping also can be accomplished through the use of least squares techniques (Knapp et al., 1990; Haley and Knott, 1992). Although least squares methods are not as computationally intensive as maximum likelihood methods, they have not been widely used because they have not been developed and supported in publicly available software.

IM represents a significant conceptual contribution to QTL analyses, but it does not address the fact that the expression of most QT is due to multiple QTL. Unless all QTL are included in the genetic model and the effects are estimated simultaneously, the estimates of genetic effects as well as inferences about significance of QTL will be biased (Knapp et al., 1992). These are two well known principles associated with general linear models (Searle, 1971) and reflect the axiom that an incorrectly specified model can produce biased and misleading results. The challenge is to correctly specify the underlying genetic model using data that are derived from exploratory research where the underlying genetics are unknown. One approach has been to build the genetic model using one of the various algorithms associated with multiple regression techniques (Cowen, 1989; Stam, 1991; Moreno-Gonzales, 1992). These algorithms can be applied in the context of maximum likelihood IM (Paterson et al., 1991; Stuber et al., 1992; Beavis et al., 1994) or least squares IM (Jansen, 1992; Knapp et al., 1992), but as Lander and Botstein (1989) pointed out, choosing an appropriate threshold for the test statistic needs to be resolved. This is another manifestation of the challenge associated with applying multiple regression algorithms to data obtained from exploratory research (Miller, 1990).

The inaccuracy of estimating QTL position and genetic effects is acute in cases where QTL are linked (Haley and Knott, 1992; Martinez and Curnow, 1992). Recently, these issues have been addressed independently by Jansen (1993), Rodolphe and Lefort (1993) and Zeng (1993). These methods, referred to as composite interval mapping by Zeng (1994), are essentially a synthesis of multiple regression and IM. The methods proposed by Zeng (1993) and Rodolphe and Lefort (1993) are similar and appear to provide more accurate and precise estimates of the QTL positions and genetic effects than the method proposed by Jansen (1993) (Zeng, 1994). Composite IM is not yet generally available, but there is considerable effort being devoted to development of software (Zeng, personal communication; Jansen, personal communication).

A third statistical issue for QTL analyses is the ability or power of the technique to identify real QTL. For a given set of experimental data, decreasing the frequency of false positives detected with an analysis technique will be associated with decreased power to identify real QTL. Soller et al. (1976) showed how the power of MT analyses depends not only on the magnitude of the genetic effects at the QTL, but also is a function of the amount of recombination between marker and QTL. Asins and Carbonell (1988) elaborated by showing that the power of MT analyses could also be affected by unequal variances of the marker classes resulting from either dominance at the QTL or recombination between QTL and the marker. Lander and Botstein (1989) argued that IM is more powerful than MT, but Darvasi et al. (1993) showed that the increased power was negligible for genomes with markers distributed at densities greater than one per 20 cM. Stuber et al. (1992) found little evidence for a difference between IM and MT analyses using experimental data from maize. Empirical evidence suggests that the use of multiple QTL models may be more powerful than single QTL models at detecting QTL in some cases (Knapp et al., 1992; Beavis et al., 1994; Utz and Melchinger, 1994), but the issue of an appropriate threshold for inclusion of QTL when using the algorithm associated with multiple QTL models where multiple hypothesis tests are conducted has not been addressed. There is compelling evidence that composite IM is less powerful

(Zeng, 1994), but Jansen (1994) recently proposed a two-step approach for improving the power of composite IM, although statistical issues concerning significance tests for multiple hypotheses still need to be addressed. Also, the power of composite IM may be significantly improved when data from replicated progeny grown in multiple environments are analyzed in a combined analysis. However, the extension of the method to multiple environments is still being developed and involves a number of difficult statistical issues (see below).

Given the tremendous intellectual effort that has been devoted to developing QTL analysis techniques, it may seem surprising that there have not been large differences in the results generated by the various data analysis techniques. However, it is good to keep in mind that the primary limitation on information about QTL in segregating populations is from the limitations imposed by the design of the experiments (Soller et al., 1976; Darvasi et al., 1993; Beavis, 1994). In other words, the number of progeny, the type of progeny, and the resources used to assay the genotypes of the markers and evaluate the quantitative traits will have the greatest influence on the amount of QTL information that is available.

B. QxE ANALYSES

The issues involved in extending QTL analyses to QxE analyses can be shown by using MT analyses in a multiple environment context. The extension of these concepts to IM using multiple QTL models is currently being pursued by Zeng and co-workers at North Carolina State University.

Because most GxE analyses are based on linear models, we first consider the expansion of the linear model to include QTL and QxE effects in a combined analysis of replicated progeny. Most QTL analyses were developed for evaluation of individual progeny *per se* from bi-parental crosses. However, it is well known among plant breeders that the phenotype measured on a single plant basis has low heritability and, therefore, quantitative traits are typically evaluated on a line mean basis. The improved heritability on a line mean basis can be shown in the context of a linear model. The phenotype, y_{ijk}, can be modeled as

$$y_{ijk} = \mu + e_j + r(e)_{k/j} + l_i + el_{ij} + \epsilon_{ijk}, \qquad \{1\}$$

where i=1,2,3...I, j=1,2,3...J, k=1,2,3...K, μ is the population mean, e_j is the effect of jth environment, $r(e)_{k/j}$ is the effect of the kth rep within environment j, l_i is the effect of the ith line, el_{ij} is the effect associated with the interaction between the ith line and the jth environment, and e_{ijk} is the error associated with measuring the phenotype, y, of line i grown in the kth rep of environment j. The averaged phenotypic value for the line is then:

$$y_{i.} = \mu + l_i + el/J + \epsilon/JK, \qquad \{2\}$$

because by definition the averaged effects, e and r(e), are 0. Thus, the impact of el and ϵ on the expression of the phenotype is decreased as the number of environments, J, increases. Because the average phenotypic value of the line more closely estimates the genotypic value of the line, QTL analyses, often, have been conducted by searching the genome for associations with the mean phenotypic value of the lines, averaged over reps and environments, in both marker-trait analyses and IM analyses (Cowen, 1988; Soller and Beckman, 1990; Stuber et al., 1992; Beavis et al., 1994). The linear model based upon phenotypic means that incorporates QTL effects that can be detected with the marker genotypic classes is:

$$y_{i/m} = \mu + q_m + l(q)_{i/m} + al/J + \epsilon/JK \qquad \{3\}$$

where μ and i are as in $\{1\}$, m designates the genotypic class at the marker locus, q_m is the effect of the mth genotypic class of the QTL, and $l(q)_{i/m}$ is the averaged effect of the ith line across reps and environments nested in the mth genotypic class. Note that $\{3\}$ is a re-expression of $\{2\}$ in which line effects are apportioned to fixed effects associated with the marker classes and random effects of lines nested within each marker class. The influence of el and ϵ on the estimated value of the line within the marker classes is decreased as the number of environments and reps increases. However, such analyses lose information on QxE and may lose power relative to a combined analysis. Thus, a combined analysis, rather than an analysis of the means, will be more informative. The linear model for a combined analysis that incorporates QTL and QxE effects detected by marker genotypic classes is:

$$y_{ijkm} = \mu + e_j + r(e)_{k/j} + q_m + l(q)_{i/m} + qe_{mj} + l(qe)_{i/mj} + \epsilon_{ijkm} \qquad \{4\}$$

where m designates the genotypic class at the marker locus , q_m is the effect of the mth genotypic class of the QTL, $l(q)_{i/m}$ is the effect of the ith line nested in the mth genotypic class, qe_{mj} is the interaction effect of the mth QTL with environment j, and $l(qe)_{i/mj}$ is the effect of the ith line nested in the mth QTL grown in environment j. The remaining terms were defined in $\{1\}$. Inferences about QTL effects and QxE can then be made from markers that are linked to QTL using Least Squares techniques (Knapp, 1994). Keep in mind that when MT analyses are used, estimates of q_m and qe_{mj} are biased by the amount of recombination between the marker locus and the QTL (Soller et al., 1976). In the case where lines are evaluated in multiple environments, Knapp (1994) advocated the calculation of a complex F statistic from a linear function of mean squares obtained from the analysis of variance.

Notice that $\{4\}$ is an expansion of $\{1\}$ in which the effects of lines and lines x environments are defined as a function of QTL effects. QTL effects are estimated by progeny that fall within fixed marker classes, whereas the other effects in the model are usually considered random effects because the environments in which the lines are evaluated represent a random sample of all possible environments to which inferences will be made. Occasionally, environmental effects, such as those regulated in a growth chamber or by management practices, can be treated as fixed effects because the inferences from the experiment will apply only to a limited set of conditions.

If environments represent random effects, then in order to avoid estimation problems, one needs to recognize that {4} is a Generalized Linear Model with Mixed effects (GLMM). To obtain the best linear unbiased estimates (BLUE) of the QTL effects and the best linear unbiased predictions (BLUP) of the QxE effects, optimal estimates of the variance components are needed (Searle, 1988). Most linear model procedures found in standard statistical packages estimate variance components by equating the mean squares of an analysis of variance to their expected values, which Searle (1988) refers to as the ANOVA method. For unbalanced data in a random effects model, use of the ANOVA method is appropriate if the sums of squares are calculated using Henderson's Method III; however, for a GLMM, the ANOVA method does not provide optimal estimates of the variance components (Searle, 1988). Often the ANOVA method results in negative estimates of the variance components, which can adversely affect the hypothesis tests as well as the estimates of effects in the model. Another disadvantage of using the ANOVA method for estimating variance components from {4} is that the F tests are complex because the linear combination of mean squares consists of variance components in which the coefficients of the variance components are "hopelessly intractable" (Searle, 1988). Thus, the use of statistical procedures in standard statistical packages such as SAS' GLM will not provide the best linear unbiased estimates (BLUE) of QTL effects nor the best linear unbiased predictions (BLUP) of QxE effects because the variance components will not be estimated correctly (Wolfinger et al., 1991).

To illustrate the importance of using GLMM for QxE studies, consider the problem of determining an appropriate threshold for the test statistics associated with QTL and QxE effects. We applied the permutation test to a maize genome consisting of 96 markers distributed on 10 linkage groups covering ~20 M of recombination in a set of 112 F_2 derived lines from the cross B73xMo17 (described in Beavis et al., 1994). Each line was evaluated for grain yield in TC progeny at five central corn belt environments. We randomly assigned the realized grain yield values from each rep of the environments to the lines 100 times. Thus, there were environmental effects and random error in each set of permuted data, but no QTL or QxE effects. We then obtained F statistics at all 96 markers for each permutation based on the complex linear combinations of mean squares, suggested by Knapp et al. (1994), and calculated by the type III sums of squares estimates from SAS' GLM procedure (SAS, 1990). From each set of permuted data, we found at least one F statistic for QTL effects in the genome that exceeded a value of 100 and for ten of the permutations, we obtained F values in excess of 650,000. The primary reason for these unusual values is that the estimates of σ^2_{qe} were negative. Using the ANOVA method of estimating variance components, it is possible, especially under the null hypothesis, to obtain negative estimates because the type III sums of squares calculations are done as though all effects are fixed (Wolfinger et al., 1991).

Fortunately, GLMM theory automatically incorporates the appropriate estimates of variance components into the calculation of the inferential statistics, so that complex linear combinations of estimated mean squares need not be specified (Wolfinger et al., 1991). Also, the implementation of GLMM in procedures such as SAS' MIXED procedure (Wolfinger, 1992) produces more reliable estimates based on maximum likelihood methods in which the expected estimates can be restricted to non-negative values (REML) (Patterson and Thompson, 1971). Obviously, caution should be exercised in the application of REML. There are valid reasons for obtaining negative estimates of variance components. These include data that are excessively variable, incorrectly recorded observations, negative correlations among lines grown in the same environment, and data that do not exhibit the same covariance among lines from one environment to the next (Hocking, 1985). All of these violations of the assumptions underlying the use of linear models certainly need to be investigated prior to data analyses, but none of these is applicable when using permuted data for threshold determinations. By using a GLMM and REML as implemented by SAS Mixed Procedure (Wolfinger, 1992) to estimate an appropriate threshold from the permuted data, we generated F statistics for each permutation that were more reasonable (Figure 4). Based upon this plot, it was possible to establish that an appropriate threshold for declaring existence of QTL effects in these experimental data is about 9.7 (P<.1).

These same issues will need to be considered when extending IM or composite IM to the situation of combined analyses across multiple environments, especially if the inference space is beyond the specific set of environments in which the lines were grown. In other words, if the test environments represent a sample of the target environments in which the progeny could be grown, then the effects due to environments need to be considered random and the resulting linear model {4} will be a GLMM.

III. AN APPLICATION OF THE ANALYSES

A. BACKGROUND

Beavis et al. (1994) evaluated 96 RFLP loci and 27 agronomic traits for 112 F_2 derived lines as both $_2\bar{F}$ and TC progeny from the maize cross B73xMo17 in nine central U.S. corn belt environments during 1987 and 1988

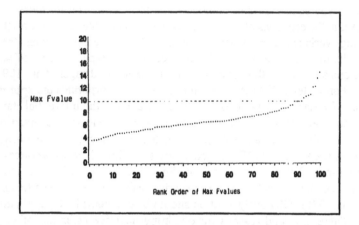

Figure 4. Rank order of Maximum 'F' values generated for each of 100 permutations of data from six central U.S. cornbelt environments. Each permutation consisted of randomly assigning two replicates of grain yield (Mg/ha) evaluated at each environment on 112 topcross (TC) progeny from the cross B73xMo17 to each of the 112 TC progeny. F values were calculated for each population at each of 96 marker loci that were arranged on 10 linkage groups and spanned 20M of recombination in the maize genome using SAS' Mixed Procedure. (Wolfinger, 1992).

(Table 2). The tester used in the TC progeny was a proprietary inbred V78 that was unrelated to either B73 or Mo17. We utilized grain yield (Mg/ha) data from both types of progeny in these experiments to illustrate the application and interpretation of the QxE analyses that we have proposed. The development of the plant materials and methods used to evaluate the genotypic and phenotypic data is described elsewhere (Beavis et al., 1994).

B. CHARACTERIZATION OF LINES AND ENVIRONMENTS

The progeny represent a random sample of $F_{2.4}$ and V78-TC progeny from the cross B73xMo17. Each location-year combination was intended to represent a typical central U.S. cornbelt environment. In other words, the initial intent of the experiment was to apply inferences about QTL and QxE effects from this experiment to all of these types of progeny from B73xMo17 grown in the central U.S. cornbelt.

The summer of 1987 was a "typical" growing season throughout most of the central U.S. cornbelt, whereas 1988 was characterized by extreme heat and drought. The environmental conditions in 1988 were so severe at Windfall, Indiana, Kellog, Iowa, and Winterset, Iowa that many plots failed to produce harvestable grain. The difference in growing conditions between 1987 and 1988 is reflected by the BLUP estimates (Table 2) of grain yield. Yields were greater for both $F_{2.4}$ and TC progeny in 1987 than in 1988. An exception to this trend

Table 2

BLUP estimates of grain yield (Mg/ha adjusted to 15% moisture) for ten cornbelt environments evaluated with 112 random $F_{2:4}$ and TC progeny from the maize cross B73xMo17.

Environment	Progeny	
	$F_{2:4}$	TC
Ankeny, Iowa, 1987 (An 87)	3.452	9.259
Johnston, Iowa, 1987 (Jh 87)	3.775	9.814
Kellog, Iowa, 1987 (Kl 87)	Not Planted	9.369
Windfall, Indiana, 1987 (Wn 87)	3.490	Not Planted
Winterset, Iowa, 1987 (Ws 87)	Not Planted	10.310
Ankeny, Iowa, 1988 (An 88)	2.006	8.282
Johnston, Iowa, 1988 (Jh 88)	3.551	8.814
Kellog, Iowa, 1988 (Kl 88)	Not Planted	Not Harvested
Windfall, Indiana, 1988 (Wn 88)	Not Harvested	Not Planted
Winterset, Iowa, 1988 (Ws 88)	Not Planted	Not Harvested

was that grain yield of the $F_{2:4}$ progeny at Johnston in 1988 was about equivalent to grain yields of $F_{2:4}$ progeny at Ankeny and Windfall in 1987.

Despite the variable growing conditions between 1987 and 1988, the broad sense heritability, on a progeny mean basis, of the lines in both types of progeny was fairly high (Table 3). The estimated heritability was greater among the $F_{2:4}$ than the TC progeny. The estimated plot to plot variability (experimental error) was greater in the TC progeny, which probably reflects the greater care that was taken in handling and growing the inbred progeny. The variance component describing variability among lines was less in the TC progeny. This was expected because half of the genome in the TC progeny is from V78 (Beavis et al., 1994). Also, the estimated GxE variance component was less in the TC progeny. This also was expected because maize inbreds are more susceptible to environmental stress than hybrids.

In addition to characterizing the environments by the average grain yield of the progeny, we estimated GxE deviations across environments using BLUP effects from SAS Mixed Procedure (Wolfinger, 1992). As expected from the estimated variance components, the BLUP values for GxE $F_{2:4}$ progeny were larger than the BLUP values for GxE from the TC progeny. The sum of the squared GxE estimates from the TC progeny was about 8.5, whereas it was 88.5 for the $F_{2:4}$ progeny. We then assessed the consistency of these values among environments using an agglomerative hierarchical (Ward's method based on averaged values) cluster analysis. Environments which show similar GxE deviation patterns across the lines will cluster together. This is an adaptation of

Table 3
Estimates of heritability and variance components for grain yield (Mg/ha, adjusted to 15% moisture) evaluated with 112 random $F_{2:4}$ and TC progeny from the maize cross B73xMo17 grown at several central U.S. cornbelt environments in 1987 and 1988.

Progeny	h^2	$\sigma^2_l \pm SE$	$\sigma^2_{le} \pm SE$	σ^2_ϵ
$F_{2\cdot4}$	.74	0.296 ± 0.54	$0.334 \pm .038$	$0.386 \pm .024$
TC	.56	$0.162 \pm .034$	$0.159 \pm .044$	$1.022 \pm .051$

the method proposed by Corsten and Denis (1990) where we used BLUP of GxE deviations rather than the actual yield values.

The cluster of TC progeny (Figure 5) indicates that the estimated GxE patterns among lines were most similar among Ankeny, Winterset and Kellog in 1987 and the patterns from the remaining environments were relatively dissimilar to the first group and to each other. The cluster of environments found on GxE deviations for $F_{2.4}$ progeny (Figure 6) indicated a distinct pattern associated with stressful conditions of 1988, i.e., the $F_{2.4}$ progeny responded to the stressful environments of 1988 with a unique pattern of GxE deviations. Because both types of progeny responded differently to the stressful environments, i.e., those from 1988, we might expect unique QTL from 1988 environments.

These data provided a unique opportunity to investigate a model in which the environments within 1987 can be treated as random representatives of the central U.S. cornbelt model and classify year effects based on heat and drought stress as fixed effects. The latter model has the form:

$$y_{hijkm} = \mu + S_h + e(s)_{j/h} + r(se)_{k/hj} + q_m + l(q)_{i/m} + qs_{mh} + qe(s)_{mj/h} + l(qes)_{i/mjh} + \epsilon_{hijkm}, \qquad \{5\}$$

where S_h takes on the values 0 or 1 and is a fixed effect describing whether the lines are grown in a normal environment or a stressful environment, qs_{mh} represents a fixed effect QTL that responds to heat and drought stress differently than to normal environmental conditions, and $qe(s)_{mj/h}$ is the interactive effect of the mth QTL with the jth environment of the hth class of stress. The remaining terms were defined in $\{1\}$ and $\{4\}$.

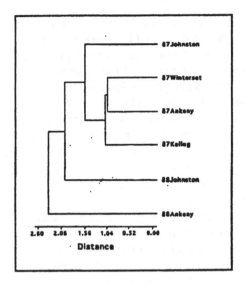

Figure 5. Dendrogram of environments in which 112 topcross (TC) progeny from the cross V73xMo17 were clustered based in an agglomerative hierarchical clustering of BLUP of GxE effects. Environments which have the same pattern of predicted GxE effects exhibited among the lines will cluster together.

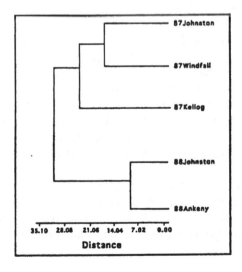

Figure 6. Dendrogram of environments in which 112 $F_{2:4}$ progeny from the cross B73xMo17 were clustered based on an agglomerative hierarchical clustering of BLUP of GxE effects. Environments which have the same pattern of predicted GxE effects exhibited among the lines will cluster together.

C. CRITERIA FOR IDENTIFICATION OF QTL AND QXE

A linkage map based on 96 RFLP markers that were selected for uniform coverage of ~20cM/interval was constructed using MAPMAKER (Lander et al., 1987) and a mapping protocol similar to that outlined by Landry et al. (1991). We then searched the genome for grain yield (Mg/ha) QTL using IM, as implemented in the publicly available software MAPMAKER/QTL v. 0.9 (Lincoln and Lander, 1989) , and MT analyses, as implemented in SAS' Mixed procedure (Wolfinger, 1992) to search the genome for QTL and QxE effects. Because each of the 112 F_{24} and TC progeny were grown in replicated plots at each environment, the averaged grain yield values, i.e., model {3}, were used for IM analyses, whereas the measured grain yield values *per se*, i.e., model {4}, were used for the MT analyses.

Prior to analyses by either technique, we randomly assigned the realized values for grain yield to each of the 112 lines, 100 times and obtained the maximum (LOD or F) value from scanning the genome for each randomization. We decided that an appropriate threshold was a value of the statistic (LOD or F) associated with no more than 10% of the randomizations.

D. RESULTS

1. Comparison of results from IM and MT analyses.

As with other studies in which these techniques have been compared, (Stuber et al., 1993; Darvasi et al., 1993), we found the inferences and estimates of QTL effects to be similar, (Figure 7 and Tables 4 and 5). Inferences about QTL based on thresholds determined from permutation tests were slightly more powerful (discussed below) using the MT analyses of combined data (models {4} or {5}) than using IM analyses of averaged data (model {3}) for both types of progeny. This is not a comparison of IM vs. MT analyses as much as it is a comparison of the models used to analyze the data.

Because IM of combined data has not yet been implemented in readily available software, it is possible to compare IM with MT analyses only when averaged data are used in both techniques. For TC progeny, one would expect a functional relationship between the F and LOD statistics, i.e. $F = 2 \ln 10(LOD)$ (Knapp et al., 1992). As with backcross progeny, only two genotypic classes are included in the analysis of TC progeny. The heterozygous class is not included in analysis of TC progeny because it provides no information about the underlying genetic effects (Beavis et al., 1994). Empirical data may not produce results that fit the relationship unless inferences about missing genotypic data are

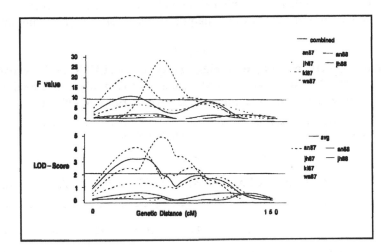

Figure 7. Comparison of MT analyses (F values) with IM analyses (LOD scores) for topcross (TC) yield QTL on chromosome 7 in each of six environments and for the combined (MT) or averaged (IM) data. The horizontal line represents the threshold value for declaring presence of a QTL using the test statistic of the respective analysis technique.

similarly derived.   IM, as implemented in MAPMAKER/QTL, utilizes information from flanking markers to infer missing genotypic information, whereas our implementation of MT analyses using SAS' MIXED procedure does not. The relationship between LOD scores and F statistics for progeny with three genotypic classes, such as $F_{2.4}$ lines, is not known. Darvasi et al. (1993) have shown that for genomes with markers spaced at intervals less than every 20 cM, IM is only slightly more powerful at detecting QTL than MT analyses. Analyses of these data affirm their results for both types of progeny (data not shown).

Because analyses of combined data {4} or {5} were more informative than analyses of averaged data {3}, we realized slightly increased power with MT analyses. As statistical models that allow analysis of combined data are implemented in IM and composite IM software, it may be that we will realize additional slight improvements in the power to detect QTL in these data. Since inferences about QTL from both types of analyses were essentially the same and inferences about QxE were possible only with the MT analyses, further results from the MT analyses are reported herein.

2. Identified QTL and QxE.

Inferential statistics based on grain yield from TC progeny exceeded threshold values in four of the six environments (Figure 8). No QTL were identified at Johnston in 1987 or 1988. A total of eight QTL was identified and none was identified consistently in all environments. The QTL on chromosome 3 was associated with elevated test statistics at most of the environments and it exceeded the threshold in the combined analysis. There was no evidence for QxE effects on chromosome 3. Previous analysis of this genomic region using IM on

Table 4

Estimated genetic effects (combining ability) of yield QTL in TC progeny from B73xMo17. Estimates were obtained using MT-analyses of a model {4} in which environments were treated as random effects.

Location of QTL		Estimated genetic effects						
Chromo-some	Marker	Combined	An87	Jh87	Kl87	Ws87	An88	Jh88
3	UMC60	-0.21*	-0.15	-0.25	-0.32*	-0.22	0.00	-0.19
5	bnl 5.71	0.14	0.17	0.13	0.30*	0.09	-0.13	0.17
7	bnl15.21	-0.22*	-0.25*	-0.13	-0.43*	-0.23	-0.18	-0.07
8	bnl 12.3	-0.18*	-0.18	-0.13	-0.23	-0.32*	-0.12	-0.11
9	CSS1	0.09	0.07	-0.03	-0.08	0.06	0.44*	0.10

*Indicates that the associated 'F' statistic is significant ($P<0.1$).

the averaged yield values did not detect a significant LOD score (Beavis et al., 1994), although the LOD scores were elevated. Thus, for this region of the genome, MT analyses of the combined data were slightly more powerful at detecting a QTL than IM analyses of averaged data. The QTL on chromosome 5 exceeded the threshold only at Ankeny in 1987 and there was little evidence for QxE in this genomic region. Segregation on the short arm of chromosome 7 was strongly associated with yield at Kellog and Ankeny in 1987, which showed some evidence for a QTL at Winterset in 1987, but little evidence for the expression of yield QTL at the remaining three environments. This QTL was significant in the combined analysis and showed significant evidence for QxE effects. The estimated additive effects (Table 4) indicate that the B73 allele combined better with the tester allele than the Mo17 allele at all of the environments, but the estimated magnitudes were not large in the Johnston environments. Thus, there was no evidence for cross-over QxE. The QxE that was detected was due to differences in the magnitude of the estimated QTL effects. A QTL was identified on the long arm of chromosome 8 in the combined analysis and at Winterset in 1987. F values were elevated on chromosome 8 in analyses of all of the 1987 environments, but QxE effects were not significant. A QTL was identified on the

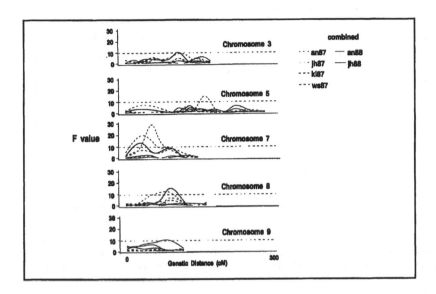

Figure 8. Plots of F values from MT analyses of topcross (TC) yield QTL. Analyses were conducted on data from each of six environments and on data combined across environments. Only chromosomes in which F values exceeded the threshold in at least one of the analyses are shown. The horizontal line represents the threshold value for declaring presence of a QTL in the combined analysis using model {4}.

Table 5

Estimated additive genetic effects of yield QTL in $F_{2:4}$ progeny from B73xMo17.[a]

Location of QTL		Estimated genetic effects					
Chromo-some	Marker	Combined	An87	Jh87	Wn87	An88	Jh88
1	php1122	-0.08	-0.08	-0.26	-0.50*	-0.01	-0.01
1	bn18.10	0.25*	0.47*	0.10	0.18	0.34*	0.26
9	php10005	-0.32	-0.59*	-0.50*	-0.38	-0.33	-0.12

* Indicates that the associated 'F' statistic is significant (P<0.1).
[a] Estimated dominance effects are not reported because they were not significant in most cases. Estimates were obtained using MT-analysis of a model {4} in which environments were treated as random effects.

142

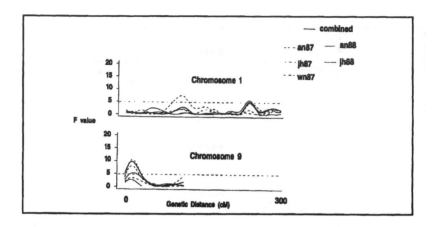

Figure 9. Plots of F values for MT analyses of $F_{2.4}$ yield QTL. Analyses were conducted on data from each of six environments and on data combined across environments. Only chromosomes in which F values exceeded the threshold in at least one of the analyses are shown. The horizontal line represents the threshold value for declaring presence of a QTL in the combined analysis using model {5}.

long arm of chromosome 9 only at Ankeny in 1988 but the QxE effects in this genomic region were not significant.

Inferential statistics on yield QTL from $F_{2.4}$ progeny exceeded threshold values in four of the five environments (Figure 9). No QTL were identified at Ankeny in 1988. Three QTL were identified and none of these had F values that consistently exceeded the threshold values in all environments. Based on the characterization of lines and environments, we anticipated that QTL for $F_{2.4}$ progeny identified in stress environments would be similar to each other and different from QTL identified in non-stress environments. The actual results did not meet the anticipated results. All of the QTL identified in the combined analysis of $F_{2.4}$ progeny were identified in both stress and non-stress environments. Two QTL were identified on chromosome 1: one was closely linked to php1122 and the other was closely linked to bnl8.10. The former was significant only at Windfall, Indiana in 1987. This QTL was not detected in the combined analyses using either model {4} or {5}, nor were QxE or QTLxStress interactions significant in this region of the genome. The QTL linked to bnl8.10 was significant at Ankeny in 1987 and 1988 as well as at Johnston in 1988. This QTL was significant in the combined analysis using both models {4} and {5}, but LOD scores from IM of averaged values did not exceed the threshold (Table 4) (Beavis et al., 1994). Thus, for this region of the genome, the MT analysis, based on model {4}, was slightly more powerful at detecting QTL. Combined analyses based on model {4} showed no evidence for significant QxE effects in the region of bnl8.10. A third QTL was identified on chromosome 9. Analyses at three environments, Johnston 1987 and 1988 as well as Ankeny 1988, exceeded the threshold. Both combined analyses, models {4} and {5}, as well as IM analyses of averaged phenotypic values (Beavis et al., 1994), detected large significant QTL effects on the short arm of chromosome 9. The combined analyses based on model {4} in which environments were treated as a random sample from all possible central U.S. cornbelt environments did detect significant QxE effects (P< 0.1) in this region of the genome. The estimated genetic effects (Table 5) indicate that the B73 allele provided positive effects at all environments, although the estimated effects were fairly small at Johnston in 1988. There was little evidence for dominance at any of the environments. Thus, there was no evidence for cross-over QxE. The QxE that was detected was due to changes in magnitude.

IV. DISCUSSION

A. EVIDENCE FOR QxE ?

There have been several reported experiments in which QTL have been evaluated in multiple environments (Guffy et al., 1989; Paterson et al., 1991; Stuber et al., 1992; Zehr et al., 1992; Bubeck et al., 1993; Hayes et al., 1993; Schon et al., 1994). Most of these evaluated QxE in a qualitative manner as the consistency, or lack thereof, of test statistics across the test environments (Guffy, 1989; Paterson et al., 1991; Stuber et al., 1992; Bubeck et al., 1993; Schon et al., 1994). Only one reported study of maize QTL has shown large differences in plots of test statistics among environments (Bubeck et al., 1993). The study by Paterson et al. (1991) on agronomic traits in tomato also found large differences on LOD plots among environments. There is a tendency to consider differences in plots of QTL test statistics from different environments as evidence of QxE. However, differences in plots of QTL test statistics may not be associated with statistically significant QxE. The challenge is to determine if the differences are of biological interest or simply due to sampling variation.

Studies by Zehr et al. (1992) and Hayes et al. (1993) tested QxE using inferential statistics, although neither of these studies estimated Type I error rates on a genome basis. Thus, it is difficult to assess the levels of significance of the identified QxE from these studies. Zehr et al. (1992) found significant evidence for yield QxE but at only two of eight QTL. The study by Hayes et al. (1993) on barley did not find much evidence for QxE for most agronomic traits but did find eight genomic regions with significant QxE for grain yield.

A common feature of the three studies (Paterson et al., 1991; Bubeck et al., 1993; Hayes et al., 1993) where evidence for QxE was strongest is that wide crosses were used to generate the segregating populations and the traits were evaluated in very diverse environments. As a contrast, the experiment that we reported herein was conducted using adapted germplasm in diverse environments. The environments in 1987 were very different from those of 1988, but B73 and Mo17, although from different heterotic pools, are both adapted to the central U.S. cornbelt and the F_1 hybrid was, at one time, the most widely grown hybrid in the U.S. cornbelt.

Of the five QTL identified using either TC or $F_{2.4}$ progeny in the combined analyses, all were associated with elevated test statistics in several environments; i.e., inferences from the combined analyses were supported by inferences from multiple environments. The QTL on chromosomes 3 and 8 in the TC progeny and the QTL on the long arm of chromosome 1 in the $F_{2.4}$ progeny are supported with elevated F values in at least three environments and the large effect QTL on chromosome 7 in the TC progeny and the large effect QTL on chromosome 9 in the $F_{2.4}$ progeny showed large effects at two of the environments and fairly strong evidence at a third environment.

Yet for all five of the QTL that we identified, there was at least one environment in which there was little or no evidence for a QTL. We identified three additional QTL that had significant test statistics in a single environment but were not associated with elevated test statistics in the remaining environments. Yield QTL for TC progeny on chromosomes 5 and 9 and yield QTL for $F_{2.4}$ progeny near the centromere of chromosome 1 were significant in one environment with little evidence for QTL in the remaining environments. None of these was significant in a combined analysis.

Despite the inconsistencies in QTL plots from one environment to the next, we found significant QxE effects at only two: chromosome 7 in the TC progeny and chromosome 9 in the $F_{2.4}$ progeny. In both of these cases, a significant large effect QTL was detected in at least two of the environments but also showed only slight evidence for expression in at least two of the remaining environments. It may be that there are QxE effects in other regions of the genome but the statistical tests were not sufficiently powerful to detect them. This illustrates a recurring theme in QTL analyses. If hypothesis tests are conducted in a statistically responsible manner on data derived from exploratory QTL experiments, there may not be sufficient power to detect real QxE (or QTL) effects.

It is important to keep in mind that these experimental data represent the initial phase of a QTL research project. The goal of the experiment was exploratory: to identify regions of the genome that co-segregate with the expression of harvestable grain yield and several other quantitative traits in several environments that represent the central U.S. cornbelt. These types of experiments typically evaluate small numbers of progeny (<500 at a limited number of environments) from a population in which linkage disequilibrium has been maximized. As such, the number of DNA preps and field plots are minimized as are the number of marker loci needed to saturate (1 marker per 20 cM) the genome. Consequently, there are not a large number of informative, i.e. recombinant, individuals evaluated in these experiments. Thus, power, precision and accuracy are all compromised especially if there are a large number of QTL segregating in the population (Beavis, 1994). Despite the weakness of exploratory QTL experiments, if the frequency of false positives is controlled, some of the QTL can be mapped and some of the QxE effects can be detected. Depending upon the ultimate goals of the project, e.g., study of gene expression, map-based cloning or marker-aided selection, follow-up experiments can be designed to increase the frequency of informative individuals and informative environments to improve the power, accuracy and precision of the information about the QTL underlying the expression of the trait.

B. SUGGESTIONS

The results of the QTL and QxE analyses conducted in our example, as well as those reported in the literature, have been somewhat unsatisfying. However, the results indicate at least two areas for further development of data analyses: 1) the need to extend IM and composite IM to multiple environments; and 2) the need to integrate recently developed GxE analysis techniques with IM techniques.

Those familiar with QTL analyses will note that we did not report results from analyses based on multiple QTL models. As noted above, the algorithms for implementing multiple QTL models are still being investigated and the most recent developments, such as composite IM (Zeng, 1993), are not yet available in widely distributed software. Caution should used in extending these methods to multiple environments where inferences about QxE will be taken beyond the environmental conditions of the experiment as the models are GLMM. As such, considerable statistical theory will need to be developed and implemented into efficient computing algorithms. We are aware of some effort in this area at North Carolina State University (Zeng, personal communication) and anticipate that they will provide the QTL community with thorough and statistically responsible developments.

To our knowledge, no effort has been devoted to integration of recent GxE analyses which investigate the non-additive nature of GxE effects (Gollob, 1968; Mandel, 1971; Gauch, 1988; Zobel et al., 1988) with QTL analyses. We looked for explanations of the results of cluster analyses of GxE in the results of QTL analyses among environments. Based on the cluster analyses of the BLUP estimates of GxE (Figures 5 and 6), we expected to find differences between yield QTL identified in stress environments from those identified in non-stress environments. In particular, we expected to find either unique sets of yield QTL in the stress environments or reverse estimates of genetic effects between stress and non-stress environments, i.e., cross-over effects (Baker, 1988). For the $F_{2.4}$ progeny, two of the three QTL were detected in both stress and non-stress environments, but there was no evidence for cross-over QTL effects, nor was there significant evidence for QTLxStress interaction effects in the genome. For the TC progeny, we identified only one QTL in the stress environments and found no significant evidence for QTLxStress interaction effects in the genome.

146

We do not have an explanation for why the QTL analyses did not reveal patterns associated with stress, but a conjecture worth pursuing is that the analysis of grain yield values *per se* has too much noise and is not focused on what has been referred to as the important GxE patterns in the data (Gauch, 1988).

Consider the simple hypothetical example given in the introduction. If the GxE component of yield for the lines was controlled by a single locus, then a cluster analysis of the GxE component would reveal two groups. A principal component analysis would reveal a single component. That is, one subset of the segregating lines would have similar BLUPs of GxE effects across the environments and the remaining subset of lines would have the opposite BLUPs of GxE effects across the environments. Each of the lines could be coded as belonging to one of the groups and then the locus responsible for such a clustering of lines could be mapped in a co-segregation analysis, such as that given in Table 1.

In this hypothetical example, there are no other loci involved in the expression of the traits and the trait is measured without error, whereas traits, such as yield, are not measured without error and there are likely to be numerous loci with genes that are affected by environment. But, of course, it is for such traits that QTL analyses are being developed. The question is what is an appropriate metric to capture the pattern revealed in the cluster analysis of the GxE effects? One metric to consider is the Additive Main effects and Multiplicative Interactions (AMMI) (Mandel, 1971; Gauch, 1988). The idea is to use AMMI to model the expression of the quantitative trait for a line in an environment, then apply QTL analyses to the predicted values based upon a model with a reduced number of estimated multiplicative parameters underlying the GxE effects. We are currently investigating this approach. One of the statistical issues in such an approach is that AMMI models were developed under the assumption that the GxE effects are fixed; i.e., they are estimated using the ANOVA method. Can BLUPs of GxE effects in the GLMM context of QTL analyses be used in estimating the multiplicative parameters of the model and what are the limitations on inferences drawn from use of these models?

REFERENCES

Asins M.J., and E.A. Carbonell. 1988. Detection of linkage between restriction fragment length polymorphisms and quantitative traits. *Theoretical and Applied Genetics* 76:623-626.
Baker, R.J. 1988. Tests for crossover genotype-environmental interactions. *Canadian Journal of Plant Science* 68:405-410.
Beavis, W.D. 1994. The power and deceit of QTL experiments: Lessons from comparative QTL studies. in *Proceedings of the Forty-ninth Annual Corn and Sorghum Industry Research Conference*. American Seed Trade Association, Washington, D.C.
Beavis W.D., O.S. Smith, D. Grant, and R. Fincher. 1994. Identification of quantitative trait loci using a small sample of topcrossed and F₄ progeny from Maize. *Crop Science* 34:882-896.
Bubeck, D.M., M.M. Goodman, W.D. Beavis, and D. Grant. 1993. Quantitative trait loci controlling resistance to gray leaf spot in maize. *Crop Science* 33:838-847.
Churchill, G.A., and R.W. Doerge. 1994. Empirical threshold values for quantitative trait

mapping. *Genetics* 138:963-971.

Corsten, L.C.A, and J.B. Denis. 1990. Structuring interaction in two-way Tables by clustering. *Biometrics* 46:207-215.

Cowen, N.M. 1988. The use of replicated progenies in marker-based mapping of QTLs. *Theoretical and Applied Genetics* 75:857-862

Cowen, N.M. 1989. Multiple linear regression analysis of RFLP data sets used in mapping QTLs. p.113-116. in T. Helentjaris and B. Burr (eds) *Development and Application of Molecular Markers to Problems in Plant Genetics.* Cold Spring Harbor Laboratory, Cold Spring Harbor, NY.

Darvasi, A, A. Weinreb, V. Minke, J.I. Weller, and M. Soller. 1993. Detecting marker-QTL linkage and estimating QTL gene effect and map location using a saturated genetic map. *Genetics* 134:943-951.

Edwards, M.D., C.W. Stuber, and J.F. Wendel. 1987. Molecular-marker-facilitated investigation of quantitative-trait loci in maize: I Numbers, genomic distribution and types of gene action. *Genetics* 115:113-125.

Fisher, R.A. 1918. The correlation between relatives on the supposition of Mendelian inheritance. *Transactions Royal Society Edinburgh* 52:399-433.

Gauch, H.G. 1988. Model selection and validation for yield trials with interaction. *Biometrics* 44:705-715.

Gollob, H.F. 1968. A statistical model which combines features of factor analytic and analysis of variance techniques. *Psychometrika* 33:73-115.

Good, P. 1994. *Permutation tests: A practical guide to resampling methods for testing hypotheses.* Springer Verlag, NY.

Guffy, R.D., C.W. Stuber, and M.D. Edwards. 1989. Dissecting and enhancing heterosis in corn using molecular markers. p 99-120. in *Proceedings of the 25th Annual Corn Breeders School,* Champaign, Illinois.

Haley, C.S., and S.A. Knott. 1992. A simple method for mapping quantitative trait loci in line crosses using flanking markers. *Heredity* 69:315-324.

Hayes, P.M., B.H. Liu, S.J. Knapp, F. Chen, B. Jones, T. Blake, J. Franckowiak, D. Rasmusson, M. Sorrels, S.E. Ullrich, D. Wesenberg, and A. Kleinhofs. 1993. Quantitative trait locus effects and environmental interaction in a sample of North American barley germplasm. *Theoretical and Applied Genetics* 87:392-401.

Hocking, R.R. 1985. *The Analysis of Linear Models.* Brooks/Cole Publ Monterey, California, USA.

Hoisington, D.A., M.G. Neuffer, and V. Walbot. 1982. Disease lesion mimics in maize I. Effect of genetic background, temperature, developmental age and wounding on necrotic spot formation with *Les 1. Developmental Biology* 93:381-388.

Jansen, R.C. 1992. A general mixture model for mapping quantitative trait loci using molecular markers. *Theoretical and Applied Genetics* 85:252-260.

Jansen, R.C. 1993. Interval mapping of multiple quantitative trait loci. *Genetics* 135:205-211.

Jansen, R.C. 1994. Controlling the type I and type II errors in mapping quantitative trait loci. *Genetics* 138:871-881.

Kang, M.S. 1990. Understanding and utilization of genotype-by-environment interaction in plant breeding. p 52-68. in M.S. Kang (ed) *Genotype-By-Environment Interaction and Plant Breeding* LSU Ag Center, Baton Rouge, Louisiana, USA.

Knapp S.J., W.C. Bridges, and D. Birkes. 1990. Mapping quantitative trait loci using molecular marker linkage maps. *Theoretical and Applied Genetics* 79:583-592.

Knapp, S.J., W.C. Bridges, and B.H. Liu. 1992. Mapping quantitative trait loci using nonsimultaneous and simultaneous estimators and hypothesis tests. p 209-237. in J.S. Beckman and T.S. Osborn (eds) *Plant Genomes: Methods for Genetic and Physical Mapping.* Kluver, Dardrecht, The Netherlands.

Knapp, S.J. 1994. Mapping quantitative trait loci. (in press) in R.L. Phillips and I.K. Vasil (eds) *DNA markers in plants,* Kluwer, Dordrecht, The Netherlands.

Lander, E.S., P. Green, J. Abrahamson, A. Barlow, M.L. Daly, S.E. Lincoln, and L. Newburg. 1987. MAPMAKER: An interactive computer package for constructing primary gene linkage maps of experimental and natural populations. *Genomics* 1:174-181.

148

Lander, E.S. , and D. Botstein. 1989. Mapping Mendelian factors underlying quantitative traits using RFLP linkage maps. *Genetics* 121:185-199.

Landry, B.S., N. Hubert, T. Etoh, J.J. Harada, and S.E. Lincoln. 1991. A genetic map for *Brassica napus* based on restriction fragment length polymorphisms detected with expressed DNA sequences. *Genome* 34:543-552.

Lincoln, S.E., and E.S. Lander. 1990. Mapping genes controlling quantitative traits using MAPMAKER/QTL. Tech Rpt of Whitehead Inst for Biomed Res. Cambridge, Massachusetts, USA.

Langrdige, J. and B. Griffing. 1959. A study of high temperature lesions in *Arabidopsis thaliana*. *Molecular and General Genetics* 194:15-23.

Lehman, E.C. 1986. *Testing statistical hypotheses*. 2nd ed, John Wiley and Sons, NY.

Mandel, J. 1971. A new analysis of variance model for non-additive data. *Technometrics* 13:1-18.

Martinez, O. and R.N. Curnow. 1992. Estimating the locations and the size of the effects of quantitative trait loci using flanking markers. *Theoretical and Applied Genetics* 85:480-488.

Miller, A.J. 1990. *Subset Selection in Regression*. Chapman and Hall, London, UK.

Moreno-Gonzales, J. 1992. Genetic models to estimate additive and non-additive effects of marker-associated QTL using multiple regression techniques. *Theoretical and Applied Genetics* 85:435-444.

Paterson, A.H., S. Damon, J.D. Hewitt, D. Zamir, H.D. Rabinowitch, S.E. Lincoln, E.S. Lander, and S.D. Tanksley. 1991. Mendelian factors underlying quantitative traits in tomato: Comparison across species, generations, and environments. *Genetics* 127:181-197.

Patterson, H.D., and R. Thompson. 1971. Recovery of inter-block information when block sizes are unequal. *Biometrika* 58: 545-554.

Rasmussen, J.M. 1935. Studies on the inheritance of quantitative characters in *Pisum* I. Preliminary note on the genetics of the time flowering. *Hereditas* 20: 161-180.

Rodolphe, F., and M. Lefort. 1993. A multi-marker model for detecting chromosomal segments displaying QTL activity. *Genetics* 134:1277-1288.

SAS. 1990. *SAS/STAT Users Guide. Volume 2, Version 6.0, Fourth Edition*. SAS Institute Inc. Cary, North Carolina, USA.

Sax, K. 1923. The association of size differences with seed-coat pattern and pigmentation in *Phaseolus vulgaris*. *Genetics* 8:552-560.

Schactman, D.P. and J.I. Schroeder. 1994. Structure and transport mechanism of a high-affinity potassium uptake transporter from higher plants. *Nature* 370:655-658.

Schon, C.C., A.E. Melchinger, J. Boppenmaier, E. Brunklaus-Jung, R.G. Herrmann, and J.F. Seitzer. 1994. RFLP mapping in maize: Quantitative trait loci affecting testcross performance of elite European flint lines. *Crop Science* 34:378-389.

Searle, S.R. 1971. *Linear Models*. John Wiley and Sons, NY.

Searle, S.R. 1987. *Linear Models for Unbalanced Data*. John Wiley and Sons, NY.

Searle, S.R. 1988. Mixed models and unbalanced data: wherefrom, whereat and whereto. *Communications in Statistics: Theory and Methods* 17:935-968.

Soller, M., T. Brody, and A. Genizi. 1976. On the power of experimental design for the detection of linkage between marker loci and quantitative loci in crosses between inbred lines. *Theoretical and Applied Genetics* 47:35-39.

Soller, M., and J.S. Beckman. 1990. Marker-based mapping of quantitative trait loci using replicated progenies. *Theoretical and Applied Genetics* 80:205-208.

Stam, P. 1991. Some aspects of QTL analysis. in *Proceedings of the Eighth Meeting of the Eucarpia Section of Biometrics in Plant Breeding*. Brno, July, 1991.

Stuber, C.W., S.E. Lincoln, D.W. Wolff, T. Helentjaris, and E.S. Lander. 1992. Identification of genetic factors contributing to heterosis in a hybrid from two elite maize inbred lines using molecular markers. *Genetics* 132:823-839.

Thoday, J.M. 1961. Location of polygenes. *Nature* 191:368-370.

Utz, H.F., and A.E. Melchinger. 1994. Comparison of difficult approaches to interval mapping of quantitative trait loci. In *9th Meeting Eucarpia Section on Biometrics in Plant Breeding*. Waginen, The Netherlands.

Weller, J.I. 1986. Maximum likelihood techniques for the mapping and analysis of quantitative trait loci with the aid of genetic markersl. *Biometrics* 42:627-640.

Weller, J.I. 1992. Statistical methodologies for mapping and analysis of quantitative trait loci. p. 81-207. in J.S. Beckman and T.S. Osborn (eds) *Plant Genomes: Methods for Genetic and Physical Mapping.* Kluver, Dardrecht, The Netherlands.

Wolfinger, R.D., R.D. Tobias, and J. Sall. 1991. Mixed Models: A Future Direction. *Proceedings of the Sixteenth Annual SAS Users Group Conf:* 1380-1388.

Wolfinger, R.D. 1992. The Mixed Procedure. In *SAS technical report P-229, SAS/STAT Software: Changes and Enhancements, Release 6.07.*

Zehr, B.E., J.W. Dudley, J. Chojecki, M.A. Saghai-Maroof, and R.P. Mowers. 1992. Use of RFLP markers to search for alleles in a maize population for improvement of an elite hybrid. *Theoretical and Applied Genetics* 83:903-911.

Zeng, Z-B. 1993. Theoretical basis of precision mapping of quantitative trait loci. *Proceedings of the National Academy of Science USA* 90:10972-10976.

Zeng, Z-B. 1994. Precision mapping of quantitative trait loci. *Genetics* 136:1457-1468.

Zobel, R.W., M.J. Wright, and H.G. Gauch. 1988. Statistical analysis of a yield trial. *Agronomy Journal* 80:388-393.

Chapter 6

ANALYSIS OF GENOTYPE-BY-ENVIRONMENT INTERACTION AND PHENOTYPIC STABILITY

H.-P. Piepho[1]

I. INTRODUCTION

In the past, a major concern in agricultural research has been to develop high yielding crop cultivars. Recently, however, stable and sustainable yields under varying environmental conditions have constantly been gaining importance over increased yields. In face of an ever-growing over-production in the industrialized countries, it is to be expected that this trend will continue in the future. Stable yields also play a major role in the developing countries. A main strategy among small-scale subsistence farmers, particularly in marginal areas, is risk-minimization. In these areas, stable yields are the key to sustainable food supplies.

In agricultural research, different concepts and definitions of stability have been developed mainly for application in plant breeding programs and in the evaluation of yield trials (Lin et al., 1986; Westcott, 1986; Becker and León, 1988). The interest of plant breeders in stability stems from the need to develop well-buffered cultivars. The term stability refers to the behavior of a crop in varying environments. The breeders' aim is to develop cultivars that are stable across a range of environments. Environments may be locations or years or combinations of both.

Two different approaches to assessing stability may be distinguished: the static concept and the dynamic concept (Becker and León, 1988). According to the static concept (Type 1 statistics in Lin et al., 1986), maximum stability occurs when the yield of the genotype under consideration is constant across environments, i.e., stability in the sense of homeostasis. According to the dynamic concept (Type 2 statistics in Lin et al., 1986), a genotype is regarded as stable if its performance in different environments is close to what can be expected from the potentials of those environments. Maximum stability occurs if the difference between a genotype's yield and the environmental index, commonly defined as the mean of all tested genotypes, is constant across environments. Whenever this difference is not the same in all environments, the corresponding genotype is said to interact with environments. Thus, if one follows the dynamic concept, the goal of breeding stable genotypes may be translated as the goal of minimizing genotype-by-environment interactions. It is this goal with which we will be concerned in this article, and consequently, we will be dealing with stability

[1]Institut fuer Nutzpflanzenkunde, Universitaet-Gesamthochschule Kassel, D-37213 Witzenhausen, Germany

152

measures that quantify a genotype's contribution to the overall genotype-by-environment interaction.

The dynamic approach regards interactions as random unpredictable fluctuation (noise). In certain cases, however, one may seek to further analyse the interactions and extract predictable information from it (pattern). This leads to the regression approach first suggested by Yates and Cochran (1938) and further elaborated by Finlay and Wilkinson (1963), Eberhart and Russell (1966) and Perkins and Jinks (1968) [see also Piepho (1993a)]. Recent develoments comprise application of a multiplicative interaction model, which was first introduced by Mandel (1961, 1969, 1971) and Gollob (1968), and has been introduced in the agricultural context as AMMI (Additive Main Effects Multiplicative Interaction) model (Gauch, 1988, 1992). These models are appropriate, if one is interested in predicting genotypic yields in specific environments, for which yield trial data are available. A further advantage of these models is that they may be used for modeling and understanding interaction (Gauch, 1992).

If, on the contrary, one is interested in genotypes that perform well in a larger, well defined region, of which only a small sample of environments has been tested, one cannot predict interaction for each environment of that region. Then, from a practical point of view, all interaction becomes unpredictable noise, and it is reasonable to minimize genotype-by-environment interaction, which is in accordance with the dynamic concept. A similar argument is put forward by Eskridge (1990): "In situations where there are sufficient funds and economic justification to breed for a particular environment, stability is irrelevant and yield in that environment is paramount. However, if cultivars are being selected for a large group of environments then stability and mean yield across all environments are of major importance and yield in a specific environment is of marginal importance."

The aims of this contribution are (a) to review measures for assessing stability in the dynamic sense, i.e., in terms of genotype-by-environment interaction, and to discuss their relative merits, (b) to suggest statistical tests, which aid the breeder in deciding whether observed stability differences are significant or merely due to sampling variation, (c) to show how to estimate stability from problem data, i.e., data with missing values, unequal numbers of replications per environment, or heterogeneous error variances, and (d) to propose a procedure for partitioning genotype-by-environment interaction in genotype-by-year-by-location data.

II. MODEL AND STABILITY STATISTICS

The analysis of phenotypic stability is often based on a two-way mixed model of the form (Shukla, 1972a)

$$x_{ijm} = \mu + \alpha_i + \beta_j + (\alpha\beta)_{ij} + e_{ijm}$$

where x_{ijm} is the m-th replicate (m=1,...,L) of the phenotypic value of genotype i (i=1,...,K) in environment j (j=1,...,N), μ is the overall mean, α_i is the fixed effect of genotype i, β_j is the random effect of environment j, $(\alpha\beta)_{ij}$ is the random interaction effect of genotype i and environment j, and e_{ijm} is the experimental error associated with x_{ijm}. It is assumed that all random effects are independently distributed with zero mean.

Stability statistics are usually based on means, for which the model is

$$x_{ij} = \mu + \alpha_i + \beta_j + v_{ij} \qquad [2.1]$$

where $x_{ij} = \Sigma_m x_{ijm}/L$ and $v_{ij} = (\alpha\beta)_{ij} + \Sigma_m e_{ijm}/L$. Note that with the means model it is no longer possible to distinguish between genotype-environmental interaction and error.

Strictly speaking, the model used by Shukla (1972a) is appropriate only for data from yield trials laid out as a completely randomized (CR) design. A much more common design prevailing in most multilocation trials is the randomized complete block (RCB) design, for which the linear model is

$$x_{ijm} = \mu + \alpha_i + \beta_j + (\alpha\beta)_{ij} + \tau_{m(j)} + e_{ijm}$$

where $\tau_{m(j)}$ is the effect of the m(j)-th block within the j-th environment. The corresponding means model is given by

$$x_{ij} = \mu + \alpha_i + \beta'_j + v_{ij}$$

where $\beta'_j = \beta_j + \Sigma_{m(j)}\tau_{m(j)}/L$. The only difference to Eq. [2.1] lies in the addition of the mean of blocks to the environmental effect. β'_j may simply be regarded as a modified environmental effect. The difference between RCB and CR design is irrelevant for most procedures described in this paper (except for those in section F). The reason is that most procedures are concerned with an analysis of estimates of v_{ij} and that they are independent of the magnitudes of β'_j or β_j. If data are from incomplete blocks (e.g. a lattice design), an analysis of means x_{ij} by the methods to be described in this paper must be regarded as approximative.

A genotype is regarded as stable, if it has a small variance of the effects v_{ij}, i.e. a small "stability variance." This term for $\sigma^2_i = \text{Var}(v_{ij})$ was coined by Shukla (1972a). If we assume that all genotypes have a common error variance $\sigma^2_e = \text{Var}(e_{ijm})$, stability differences result merely from differences of a genotype's interaction variance $\Theta^2_i = \text{Var}[(\alpha\beta)_{ij}]$. The stability variance can then be expressed as $\sigma^2_i = \Theta^2_i + \sigma^2_e/L$. Maximum stability occurs when $\Theta^2_i = 0$ and hence $\sigma^2_i = \sigma^2_e/L$, i.e., when the variability of v_{ij} effects is minimal. The statistical design has an influence on the magnitude of σ^2_e and hence of σ^2_i (σ^2_e and σ^2_i will tend to be smaller for an RCB design than for a CR design), but it does not influence contrasts among the stability variances of any two genotypes. Clearly, $\sigma^2_r - \sigma^2_s = \Theta^2_r - \Theta^2_s$ for any r,s=1,...,K, which is independent of σ^2_e.

In what follows, we will consider several well known and some new stability statistics that may be interpreted as measures of the spread of v_{ij}'s. The rationale for introducing new measures is the aim for statistical procedures that are robust to changes in the parent distribution of the v_{ij} effects, which is often assumed normal, but may, at times, depart from normality (Piepho, 1992b). Another intention is to find measures, for which statistical tests are available. The measures are discussed in this and the next section, while the tests will be dealt with in section D.

The statistics to be described momentarily are based on the observed residuals

$$\hat{v}_{ij} = x_{ij} - \bar{x}_{i.} - \bar{x}_{j} + \bar{x}_{..}$$

where

$$\bar{x}_{i.} = \sum_{j=1}^{N} x_{ij}/N, \quad \bar{x}_{j} = \sum_{i=1}^{K} x_{ij}/K, \quad and \quad \bar{x}_{..} = \sum_{i=1}^{K}\sum_{j=1}^{N} x_{ij}/NK.$$

The ecovalence (Wricke, 1965) is given by

$$W_i = \sum_{j=1}^{N} \hat{v}_{ij}^2$$

Shukla's unbiased estimator of σ^2_i is (Shukla, 1972a)

$$\hat{\sigma}_i^2 = \frac{1}{(K-1)(K-2)(N-1)}\left[K(K-1)W_i - \sum_{r=1}^{K} W_r\right] \qquad [2.2]$$

This estimator is a MINQUE (Minimum Norm Quadratic Unbiased Estimator) of σ^2_i (Rao, 1970). It should be noted that the MINQUE is equivalent to W_i for ranking purposes, as are the measures suggested by Plaisted and Petersen (1959) and Plaisted (1960) (Lin et al., 1986). In the sequel we will, therefore, confine our attention to the MINQUE.

The MINQUE of σ^2_i may be shown to be equivalent to Grubbs estimator of the variance of measurement errors (Piepho, 1992b), which is based on the mixed linear model [1] and has application in comparisons of the precision of measurement instruments (Grubbs, 1948). For the latter field of application, Jaech (1985) proposed a Maximum-Likelihood-Estimator (MLE) of σ^2_i, given by the following equations (for K>3):

$$\tilde{\sigma}_i^2 = \frac{(N-1)(b_0 b_{1i} - b_{2i})}{N b_{0i}^2} - \frac{1}{b_{0i}}, \qquad [2.3]$$

$$b_{0i} = \sum_{s \neq i}^{K} \frac{1}{\sigma_s^2} \qquad\qquad [2.4]$$

$$b_{1i} = \sum_{s \neq i}^{K} \frac{V_{i \cdot s}^2}{\sigma_s^2} \qquad\qquad [2.5]$$

$$b_{2i} = \sum_{s \neq i}^{K} \sum_{\substack{r > s \\ s \neq i}}^{K} \frac{V_{s \cdot r}^2}{\sigma_s^2 \sigma_r^2}, \qquad\qquad [2.6]$$

and

$$V_{s \cdot r}^2 = \left[\sum_{j=1}^{N} y_{srj}^2 - \left(\sum_{j=1}^{N} y_{srj} \right)^2 / N \right] / (N-1)$$

with $y_{srj} = x_{sj} - x_{rj}$. An iterative solution is obtained by assigning starting values to $\sigma_1^2, \ldots, \sigma_{K-1}^2$ (e.g. the MINQUE's) and solving the above system of equations for σ_K^2. Then with the current estimates of σ_s^2 (s unequal to i) one computes estimates of the other σ_i^2 (i<K). This procedure is repeated until all estimates converge to a predetermined level of accuracy. This estimate has recently been suggested as an estimate of the stability variance in the analysis of stability (Piepho, 1992b and 1993b). In the sequel it will be referred to as MLE.

To compute Huehn's nonparametric stability measures, the residuals are ranked within environments. Let the resulting ranks (ranging from 1 to K) be denoted by r_{ij}. Then, Gini's mean difference of ranks $S_i^{(1)}$ and the rank variance $S_i^{(2)}$ are computed as (Nassar and Huehn, 1987; Huehn, 1990)

$$S_i^{(1)} = \frac{2 \sum_{j<j'}^{N} |r_{ij} - r_{ij'}|}{N(N-1)} \quad \text{and} \quad S_i^{(2)} = \frac{\sum_{j=1}^{N} (r_{ij} - \bar{r}_{i.})^2}{(N-1)}, \quad \text{where} \quad \bar{r}_{i.} = \sum_{j=1}^{N} r_{ij}/N.$$

Instead of ranking within environments, one may rank the residuals in the whole data set, which implies assigning ranks from 1 to NK. Gini's mean difference and the variance of these ranks are denoted by $P_i^{(1)}$ and $P_i^{(2)}$, respectively (Piepho, 1992b). In analogy to Wricke's ecovalence one may compute the sum of absolute values of the residuals (Piepho and Lotito, 1992):

$$L_i = \sum_{j=1}^{N} |\hat{v}_{ij}|$$

Furthermore, the absolute values of the residuals may be transformed to ranks. The rank sum is then a measure of the variability of the residuals. If ranking is done within environments, the rank sum of the i-th genotype is denoted by R_i (Piepho and Lotito, 1992). For ranking over all residuals, the rank sum is designated by $P_i^{(3)}$ (Piepho, 1992b). For clarity it is useful to distinguish among the following three groups: Group 1: Measures based on v_{ij}-values (MINQUE and MLE of σ_i^2; L_i). Group 2: Measures based on ranking of v_{ij}-values within environments ($S_i^{(1)}$; $S_i^{(2)}$; R_i). Group 3: Measures based on ranking v_{ij}-values across the whole data set ($P_i^{(1)}$; $P_i^{(2)}$; $P_i^{(3)}$). The relation between measures of group 2 and group 3 is summarized in Table 1.

Table 1

Relation between measures based on rankings within environments (group 2) and measures based on rankings across whole data set (group 3)

Ranking of	Measure of rank dispersion	Stability measure	
		Group 2	Group 3
v_{ij}	Gini's mean absolute difference	$S_i^{(1)}$	$P_i^{(1)}$
	Variance	$S_i^{(2)}$	$P_i^{(2)}$
$|v_{ij}|$	Rank sum	R_i	$P_i^{(3)}$

III. RANKING ABILITY - SIMULATIONS

Since all statistics described in Section II measure the variability of the v_{ij}'s, they lend themselves to inferences on the parameter values of the stability variances σ_i^2. In practical situations, namely in breeding programmes, selection of the best genotypes is the prime objective. Selection involves ranking of genotypes. One may view the stability rank order of a set of genotypes as given by the rank order of the stability variances σ_i^2. To assess this stability rank order, the genotypes are grown in different environments. In statistical terms, we thus obtain realisations of the random effects v_{ij}, which are estimated to compute the stability measures given in section B. For selection purposes we seek the stability measure, which best reflects the "true" stability rank order given by the rank order of the σ^2_i, i.e. which has the best ranking ability. The concordance between true rank order and the rank order displayed by the stability measure under consideration may be quantified by Spearman's rank correlation. In practice, σ^2_i is unknown. In a Monte Carlo simulation, however, we may assign defined values of σ^2_i to each genotype such that the true stability rank order is known a priori.

A common assumption in mixed linear models is that random effects are normally distributed. In the particular case at hand, this leads to the assumption that the distribution of the v_{ij}-effects is normal. In practice, however, it is difficult to verify this assumption, and there is some evidence that v_{ij}'s may at times be nonnormally distributed (Piepho, 1992b). In fact, statistical procedures for the stability variance are very sensitive to violations of the normality assumption (Brindley and Bradley, 1985).

It is to be expected that the distribution of v_{ij}-effects has an influence on the ranking ability of the different measures of stability. This influence was studied by means of a Monte Carlo experiment, in which v_{ij} effects were generated from different distributions (Piepho, 1992b and 1994d). We quantified the concordance between true stability rank order and the estimated rank order, as given by the stability measure estimates, by Spearman's rank correlation (r_s), averaged over 1000 runs of the Monte Carlo experiment. In the following the term "ranking ability" refers to this average of r_s obtained in the Monte Carlo experiment.

It is noteworthy that the Monte Carlo experiment did not identify one single stability measure that was best with all distributions. The results suggest that, given a normal distribution of v_{ij} effects, it is best to estimate stability by the MINQUE of σ^2_i (or equivalently by Wricke's ecovalence) if the number of environments is small. With a larger number of environments ($N \geq 10$) the MLE of σ^2_i is preferable. The situation does not change dramatically under mild departures from normality, and differences in ranking ability tend to be small. The Monte Carlo simulation has clearly demonstrated, however, that in some cases, namely for longer-tailed distributions of v_{ij}, it may be worthwhile to use one of the more robust measures L_i, $S_i^{(1)}$, $S_i^{(2)}$, $P_i^{(1)}$, and $P_i^{(2)}$, while R_i and $P_i^{(3)}$ tended to have the lowest ranking ability within groups 2 and 3, respectively.

Inspection of expansive data sets has revealed that the departure from the normality assumption is usually not dramatic, so use of the MINQUE or, when the number of environments is not too small, the MLE is probably adequate in most cases. As a diagnostic tool, it is useful to investigate the shape of the distribution of the v_{ij}-effects. Procedures for this problem are discussed by Piepho (1992b).

Regarding ranking ability, the choice of the best performing stability parameter seems considerably less critical than the choice of the number of environments and genotypes. This result in part explains the discouraging empirical finding that repeatability of stability estimates tends to be rather low (Weber and Wricke, 1987, Becker and Léon, 1988). The simulation results suggest that low repeatability may be partly due to statistical errors in the stability estimates, which are best reduced by including in the analysis as many genotypes and environments as possible. This conjecture is corroborated by results of Becker (1987), who demonstrated for parameters of the regression approach by Eberhart and Russell (1966) that heritability may be improved by increasing the number of years and the number of locations.

Stability statistics may also be used for statistical testing. The parametric procedures that are available for MINQUE and the MLE have been shown to be very sensitive to departures from normality, while tests based on measures L_i, $S_i^{(1)}$, and $S_i^{(2)}$ are rather robust (Piepho, 1992b; also see next section). Thus, the latter measures may be preferable for testing purposes in case of departures from normality, and the simulation results have shown that little, if any, information is lost with regard to ranking if one uses these measures in place of MLE and MINQUE.

IV. TESTS

Stability statistics are subject to sampling variation. This variation may be considerable particularly if the number of environments is small, rendering stability estimates rather unreliable. A high sampling variation in the stability estimates is associated with relatively low ranking ability.

In order to decide whether observed stability differences between genotypes are statistically significant or solely due to sampling variation, it is advisable to perform statistical tests. First, one may want to test the global null hypothesis that all genoypes are equally stable, i.e.,

$$H_0: \sigma_i^2 = \sigma^2 ; \quad i = 1, 2 ..., K$$

In case the global H_0 is rejected, one is usually interested in pairwise comparisons among specific genotypes, e.g., between new cultivars or lines and a check variety:

$$H_0: \sigma_i^2 = \sigma_{i'}^2 ; \quad i \neq i ; \quad i = 1, 2 ..., K$$

A. GLOBAL TESTS

1. Parametric global tests

Shukla (1972a) suggested application of his proposed sphericity test (Shukla, 1972b). Unfortunately this test is inoperative when K>N, i.e., when the number of genotypes exceeds that of environments (Piepho, 1992b), and this is the most common situation in plant breeding programmes.

Shukla (1982) suggested a chi-squared appoximation to the distribution of

$$T_{sh} = K \ln(K^{-1} \sum_{k=1}^{K} W_k) - \sum_{k=1}^{K} \ln(W_k)$$

which he showed by Monte Carlo simulation to be more accurate than that by Johnson (1962). The test statistic q_{sh} may be computed as (see Gill, 1984, Piepho; 1992a, 1993c):

$$q_{sh} = (K-1)(N-1)T_{sh}/c,$$

where

$$c = K-1 + \frac{K^2-1}{3K(N-1)} - \frac{2(K^4-1)}{15K^3(N-1)^3} - \frac{(N-1)\binom{K}{2}}{(K-1)^2(1+K(N-1)/2)}$$

$$- \frac{4(N-1)\binom{K}{3}}{(K-1)^3}\left[\frac{1}{1+K(N-1)/2} - \frac{1}{2+K(N-1)/2}\right]$$

Under the global H_0, q_{sh} is approximately chi-squared distributed with K-1 degrees of freedom (df).

According to Anscombe (1981: 345-348) the statistic

$$T_A = \frac{A\sum_{i=1}^{K}\left(\hat{\sigma}_i^2/\hat{\sigma}^2 - 1\right)^2}{K \, Var\left(\hat{\sigma}_i^2/\hat{\sigma}^2\right)}$$

with

$$Var\left(\hat{\sigma}_i^2/\hat{\sigma}^2\right) = \frac{2(K-1)^2}{(v+2)(K-2)} \quad , \quad \hat{\sigma}^2 = \sum_{i=1}^{K} \hat{\sigma}_i^2/K$$

$$A = \frac{(v+4)(v+6)}{(N-1)[v-2/(K-2)]} \quad , \quad v = (K-1)(N-1)$$

may be regarded as a chi-squared variable with A df. Anscombe (1981) qualifies his test as a "rough test".

Brindley and Bradley (1985) suggested the test statistic

$$U_B = \frac{1}{2}(N-1)(K-1)(K-2)\ln(T_B)$$

with

$$T_B = \frac{2K \sum_{i<l}^{K} \hat{\sigma}_i^2 \hat{\sigma}_l^2}{(K-1)(\sum_{i=1}^{K} \hat{\sigma}_i^2)^2},$$

which under the global H_0 is approximately chi-squared with K-1 df.

The tests by Anscombe (1981), Shukla (1982), and Brindley and Bradley (1985) have been shown by Monte Carlo simulation to have accurate empirical significance levels if v_{ij}'s are normally distributed, but they are very sensitive even to mild departures from the normal distribution (Piepho, 1992b).

Jaech (1985) suggested a large sample likelihood-ratio test for the homogeneity of σ_i^2. Unfortunately, this test leads to very liberal empirical significance levels for N<20 (Piepho, 1992b). Recently, Mudholkar and Sarkar (1992) proposed a modification of the test by Shukla (1972b), which is applicable for arbitrary N and K. This test involves computation of linear orthogonal contrasts. Unfortunately, there is some ambiguity associated with this test, since the choice of contrasts is not unique (Piepho, 1993c).

In summary, we would suggest the tests by Anscombe (1981), Shukla (1982), and Brindley and Bradley (1985) in case v_{ij}'s are normally distributed. These tests should not be used, however, if one cannot rule out departures from normality.

2. Global tests based on ranks

In order to obtain procedures that are not sensitive to nonnormality, it is worthwhile to use rank transforms of the observed residual. Nassar and Huehn (1987) and Huehn and Nassar (1989, 1991) (also see Chapter by M. Huehn in this volume) developed rank tests based on $S_i^{(1)}$ and $S_i^{(2)}$. Under the global H_0, the statistic

$$S^{(q)} = \sum_{i=1}^{K} Z_i^{(q)} \quad with \quad Z_i^{(q)} = \frac{[S_i^{(q)} - E(S_i^{(q)})]^2}{Var(S_i^{(q)})} \qquad (q = 1, 2)$$

is approximately chi-squared with K df. $E(S_i^{(q)})$ and $Var(S_i^{(q)})$ are, respectively, the expected value and variance of $S_i^{(q)}$. Huehn and Nassar (1989) give the following explicit formulae:

$$E(S_i^{(1)}) = \frac{K^2-1}{3K} \qquad\qquad [4.1]$$

$$E(S_i^{(2)}) = \frac{K^2-1}{12} \qquad\qquad [4.2]$$

$$Var(S_i^{(1)}) = \frac{(K^2-1)[(K^2-4)(N+3)+30]}{45K^2N(N-1)} \qquad [4.3]$$

$$Var(S_i^{(2)}) = \frac{(K^2-1)[2(K^2-4)(N-1)+5(K^2-1)]}{360N(N-1)} \qquad [4.4]$$

These rank-based procedures are approximate, since the estimated residuals are not stochastically independent. Extensive simulations have shown that the empirical significance level of these tests is not independent of the parent distribution of the residuals, and for highly non-normal distributions the empirical α may be rather inaccuate (Piepho, 1992b). Thus, it would be an over-statement to qualify any of these procedures as non-parametric or distribution free. What can be said is that these procedures are much more robust to departures from normality than the parametric tests given in section D.1.a. Simulations have also shown that even for a parent normal distribution the $S_i^{(2)}$-test is somewhat too liberal for small N and N<<K, which is unfortunate, since this is a very common constellation.

In order to obtain independent rankings, one may do the following (Piepho 1995b): (1) Randomly select pairs of environments (columns j and j', say) and compute differences $y_{id}=x_{ij}-x_{ij'}$. This leads to D=N/2 new columns. When N is odd, (N-3)/2 pairs are obtained. From the remaining three environments (columns j, j' and j", say) compute the contrasts $y_{id}=2x_{ij}-x_{ij'}-x_{ij''}$, leading to a total of D=(N-1)/2 new columns. (2) Compute ranks (denoted as r_{id}) within the D new columns. Note that these rankings are stochastically independent. Also, the variability of ranks depends only on the residuals v_{ij}. (3) Based on ranks r_{id}, the statistics $S_i^{(q)}$ (q=1,2) may be computed and tested by the Huehn-Nassar procedure (N must be replaced by D in eqs. [4.1] to [4.4]).

Even for independent rankings, the asymptotic distribution of $S_i^{(q)}$ (q=1,2) is not chi-squared with K df, because of the correlation among $S_i^{(q)}$'s. One could improve the approximation by incorporating the correlations in the derivation of the test statistic, but exact correlations are cumbersome to derive. Matters simplify for the following statistic:

$$U_i = \frac{\sum_{d=1}^{D}[r_{id}-\tfrac{1}{2}(K+1)]^2}{D}$$

Both $S_i^{(2)}$ and U_i are variances of ranks. With $S_i^{(2)}$ the rank mean is estimated from the data, while U_i exploits the fact that under H_0 the expected rank mean is known to be $\tfrac{1}{2}(K+1)$ for every i. Under H_0 the statistic

$$U = \frac{180D}{K(K+1)(K^2-4)} \sum_{i=1}^{K} U_i^2 - \frac{5D\,(K^2-1)(K-1)}{4(K^2-4)}$$

is asymptotically chi-squared with K-1 degrees of freedom. This approximation is satisfactory for moderate K and N. Better approximations for small K and N are considered in Piepho (1995b). Extensive tables with exact critical values are being prepared.

Recently, Piepho (1994f) has suggested the following distribution-free procedure. Assume that $K \geq 3$ and $N \geq 4$ and that under H_0, v_{ij} are identically (not necessarily normally) independently distributed. Then the global H_0 may be tested as follows: (1) Arrange the x_{ij}-data in a two-way table of K rows and N columns. (2) Randomly split the data into two subsets of N/2 columns if N is even or of (N-1)/2 and (N+1)/2 rows, if N is odd, keeping the original columns intact. (3) For each of the two subsets, estimate the row variances, using the MINQUE given in [2.2]. Within each subset rank these row variances. (4) Under H_0, each rank order in one of the subsets is equally likely, given the rank order in the other subset. Thus, we may test H_0 by testing Spearman's rank correlation computed from these two rank orders. If the rank correlation is statistically significant, the null hypothesis of equal stability variances is rejected.

It is conjectured that this test and the U-test will often be less powerful than the parametric tests and Huehn-Nassar tests. But they have the advantage of being distribution-free. A closer investigation of the power of the various rank tests is in preparation.

3. Robust global tests

Levene (1960) suggested to test the global H_0 by subjecting estimates of either $|v_{ij}|$ (L_1-test) or $(v_{ij})^2$ (L_2-test) to a one-way ANOVA. The rationale of this procedure is the well-documented high robustness of the ANOVA F-test to nonnormality. This robustness may be exploited here by transforming what is basically a scale (variance) problem into a location (means) problem. Levene's tests have good power compared to the rank tests, but unfortunately they are too liberal for small N and N<K as are most of the rank-tests. One reason is that the one-way ANOVA requires independent errors, while estimates of $|v_{ij}|$ and $(v_{ij})^2$ are not stochastically independent, even though v_{ij}'s are independent (Piepho, 1992b).

Recently, Piepho (1995a) has suggested a modification of the L_2-test, which takes into account correlations among

$$t_{ij} = \hat{v}_{ij}^{2}$$

Under H_0,

$$T_2 = \frac{\sum\limits_{i=1}^{K}(W_i/N)^2 - \left(\sum\limits_{i=1}^{K}W_i/N\right)^2/K}{a(K-1)}$$

with

$$a = a_1\hat{\sigma}^4 + a_2\hat{\mu}_4$$

$$a_1 = 2(N-1)(K-2)n^{-1}N^{-1} - 3a_2$$

$$a_2 = (K-2)^2(N-1)^2n^{-2}N^{-1}$$

$$v = (K-1)(N-1), \quad n = KN$$

$$\hat{\mu}_4 = (c_1b_2 - c_2b_1)^{-1}\left[b_2\sum\limits_{i=1}^{K}\sum\limits_{j=1}^{N}t_{ij}^2 - c_2\left(\sum\limits_{i=1}^{K}W_i\right)^2\right]$$

$$\hat{\sigma}^4 = (c_1b_2 - c_2b_1)^{-1}\left[-b_1\sum\limits_{i=1}^{K}\sum\limits_{j=1}^{N}t_{ij}^2 + c_1\left(\sum\limits_{i=1}^{K}W_i\right)^2\right]$$

$$b_1 = vNn^{-2}(N-1)(K^2-3K+3)$$

$$b_2 = vK^{-1}(N+1)(K-1) - 3b_1$$

$$c_1 = (K^2-3K+3)(N^2-3N+3)v/n^2$$

$$c_2 = 3(v^2/n - c_1)$$

approximately follows an F-distribution with (K-1) and K(N-1) degrees of freedom.

Simulations by Piepho (1995a) for normal distribution of v_{ij} indicate that the L_1-test and the L_2-test are quite liberal when K>N. The T_2-test suffers from the same defect, but deviation from the nominal α is much less severe than for the L_1-test and the L_2-test. For a nominal α=.05, K=4 and N=4, e.g., the empirical α for the L_1-test, the L_2-test, and the T_2-test were, respectively, .0574, .0806, and .0598, while for K=30 and N=4, the empirical α's were .3340, .3352, and .1386, respectively.

In extensive simulations the L_1-test was shown to have better power than the $S_i^{(1)}$- and $S_i^{(2)}$-tests across a range of different distributions, while being more sensitive to nonnormality (Piepho, 1992b). The L_2-test and the T_2-test were not included in these simulations, but it is expected that their power is comparable to that of the L_2-test. In summary it is suggested to use the T_2-test in place of the L_1- and L_2-tests.

B. PAIRWISE COMPARISONS

1. Parametric tests for pairwise comparisons

For these tests it is assumed that v_{ij}'s are normally distributed. Shukla's (1972b) test (henceforth referred to as S-test) for comparing two genotypes is based on linear contrasts

$$z_{1j} = (x_{ij} - x_{i'j}) / \sqrt{2}$$

and

$$z_{2j} = (x_{ij} + x_{i'j} - 2\bar{x}_{(K-2)j}) / \sqrt{6}$$

where

$$\bar{x}_{(K-2)j} = \sum_{i \neq i, i'} x_{ij} / (K-2)$$

It can be shown that

$$Var(z_{1j}) = (\sigma_i^2 + \sigma_{i'}^2)/2$$

$$Var(z_{2j}) = \left[\sigma_i^2 + \sigma_{i'}^2 + 4\bar{\sigma}_{K-2}^2/(K-2)\right]/6$$

$$Cov(z_{1j}, z_{2j}) = (\sigma_i^2 - \sigma_{i'}^2)/\sqrt{12}$$

where

$$\overline{\sigma}^2_{K-2} = \sum_{l \neq i, i'} \sigma_l^2 / (K-2)$$

Under the null hypothesis of equal variances σ^2_i and $\sigma^2_{i'}$, the correlation of z_{1j} and z_{2j} is zero. The null hypothesis may thus be tested with the sample correlation coefficient r_z by noting that $r_z(1-r_z^2)^{-\frac{1}{2}}(N-2)^{\frac{1}{2}}$ is distributed as t with N-2 degrees of freedom.

The test by Maloney and Rastogi (1970) (henceforth referred to as MR-test) is constructed in a similar fashion. Under the null hypothesis the correlation of sums ($s_j = x_{ij} + x_{i'j}$) and differences ($d_j = x_{ij} - x_{i'j}$) of the observations of genotype i and i' is zero. We have

$$Var(d_j) = \sigma_i^2 + \sigma_{i'}^2$$

$$Var(s_j) = \sigma_i^2 + \sigma_{i'}^2 + 4\sigma_\beta^2 \quad where \quad \sigma_\beta^2 = Var(\beta_j)$$

$$Cov(s_j, d_j) = \sigma_i^2 - \sigma_{i'}^2$$

The sample correlation r_{sd} of s_j and d_j is tested as above: under H_0, $t_{sd} = r_{sd}[1 - r^2_{sd}]^{-\frac{1}{2}}(N-2)^{\frac{1}{2}}$ is distributed as t with N-2 degrees of freedom.

Johnson's (1962) test (henceforth referred to as the J-test) uses the ratio $J = W_i / W_{i'}$. Under H_0 this ratio approximately has a central F-distribution with f, f' degrees of freedom, where $f = (f-2p^2)/(1-p^2)$, $f = N-1$, $p = -(2R-1)[(K-2)R+1]^{-1}$, $R = \sigma_i^2/\sigma^2 = \sigma_{i'}^2/\sigma^2$, and $\sigma^2 = \Sigma_i \sigma_i^2 / K$. Johnson also gives the exact distribution of J. If R is close to unity and K is large, p is numerically small, and there will probably be little error in using the naive approximation taking $f=f=N-1$. The problem in practice, however, is that R is usually unknown, and p cannot be properly determined. On the global null hypothesis we may take $p = -(K-1)^{-1}$ (Johnson, 1962). But this is of little use, since we are interested in individual comparisons only if the global null hypothesis was rejected. If we take $f=f$, the error is largest when $R \ll 1$, i.e., when the two stability variances to be compared are much smaller than the mean of all variances, and in stability analysis this comparison is usually the most interesting one.

Because of the ambiguity in determining the degrees of freedom for the J-test, we will consider a lower bound on the degrees of freedom f'. f' is a strictly monotonically increasing function of p^2 on [0,1) (Note that $p^2 \in [0,1]$). The minimum possible value of p^2 is zero, so f' will assume a minimum for $p^2=0$, in which case $f'=f=N-1$. The F-test based on (N-1, N-1) degrees of freedom will tend to be conservative. In the sequel, this test is denoted as the J_c-test, where c stands for "conservative".

It may be shown (Piepho, 1994b) that in terms of power the S-test is preferable to the J_c-test. A comparison of the S-test to the MR-test reveals that the power of both tests depends on the population value of the correlations between z_{1j} and z_{2j} and between $_j s$ and $_j d$. The test with the larger absolute value of this correlation will have greater power in a given situation. So the S-test will be more powerful if

$$\sigma_\beta^2 > \bar{\sigma}_{K-2}^2/(K-2) = (K\sigma^2 - \sigma_l^2 - \sigma_{l'}^2)/(K-2)^2 \qquad [4.5]$$

where

$$\sigma^2 = \sum_{i=1}^{K} \sigma_i^2/K$$

For given σ^2 the expression on the right side of the above inequality will approach a maximum as σ_l^2 and $\sigma_{l'}^2$ tend to zero. So the S-test will always be more powerful than the MR-test if

$$\sigma_\beta^2 > K\sigma^2/(K-2)^2 \qquad [4.6]$$

It is, therefore, suggested that σ_β^2 and $\hat\sigma$ be estimated for a dataset under investigation to evaluate Eq.[4.6]. The S-test is preferable if Eq.[4.6] holds. Otherwise one may estimate σ_l^2 and $\sigma_{l'}^2$ and scrutinize Eq.[4.5]. If Eq.[4.5] holds, the S-test should be used, otherwise the MR-test is expected to give better power. The choice based on evaluation of Eqs.[4.5] and [4.6] may not always lead to the most powerful test, since the variance component estimates are subject to sampling errors. Nevertheless, this seems better than to randomly select one of the tests, in which case the probability of choosing the more powerful one is 50%. With the procedure suggested above, this probability will certainly be larger than 50%.

We have investigated 24 extensive datasets from German registration trials and found that Eq.[4.6] was valid in all cases. Eqs.[4.5] and [4.6] also show that the power of the S-test tends to increase with the number of genotypes included in the analysis. This is not the case for the MR-test.

The three parametric tests for pairwise comparisons are rather sensitive to departures from the normal distribution, as are the parametric global tests (Piepho, 1992b).

2. Rank tests/robust tests for pairwise comparisons

Nassar and Huehn (1987) proposed to use standard procedures for pairwise comparisons based on the fact that $S_i^{(q)}$ (q=1,2) is asymptotically normally distributed. Since $Var(S_i^{(q)})$ is known, the df are infinite. So following this suggestion, one may perform the standard normal test with

$$u = \frac{S_i^{(q)} - S_{i'}^{(q)}}{\sqrt{2\,Var(S_i^{(q)})}}$$

where $Var(S_i^{(q)})$ (q=1,2) are as given in [4.3] and [4.4].

It should be stressed that the formulae for $Var(S_i^{(q)})$ (q=1,2), and hence the test, are (approximately) valid only under the global null hypothesis, whereas one is usually interested in pairwise comparisons when the global null hypothesis has been rejected. Simulations have shown that the above test does not generally control the nominal α when the global null hypothesis is incorrect (Piepho, 1992b).

V. ESTIMATING THE STABILITY VARIANCE WHEN SOME G x E COMBINATIONS ARE MISSING

The procedures for estimating stability outlined so far assume that a complete two-way table of the environmental means of K genotypes tested in N environments is available. In practice, however, one is often faced with two-way tables in which one or more cells are empty, e.g., when some genotypes have not been tested in all environments. Piepho (1994a) proposed two procedures for estimating σ^2_i, when some cells are empty. These will be depicted briefly.

Shukla's estimate of σ^2_i from a complete two-way table can be shown to be identical to Grubbs' estimate of the variance in errors of measurement (Piepho, 1992). For the first genotype Grubbs' estimate is given by

$$\check{\sigma}_1^2 = (K-1)^{1} \left[\sum_{r=2}^{K} V_{1r}^2 - (K-2)^{-1} \sum_{1 < s \leq r}^{K} V_{sr}^2 \right]$$

where

$$V_{s \cdot r}^2 = \left[\sum_{j=1}^{N} y_{srj}^2 - \left(\sum_{j=1}^{N} y_{srj} \right)^2 / N \right] / (N-1)$$

with $y_{srj} = x_{sj} - x_{rj}$ (Grubbs, 1948). Estimates for the other genotypes are obtained by an obvious rotation of subscripts. Grubbs estimate is based on the method of moments, in which sample moments are equated to population moments (Jaech, 1985). We have

$$E[V_{s \cdot r}^2] = \sigma_s^2 + \sigma_r^2 \qquad (s = 1,...,K-1; \ r \neq s)$$

In order to estimate σ^2_i, $E[V^2_{s \cdot r}]$ is replaced by $V^2_{s \cdot r}$. The system of equations to be solved for σ^2_i is then given by

$$V^2_{s,r} = \sigma^2_s + \sigma^2_r \qquad (s = 1,...,K-1; \; r \neq s)$$

There are K(K-1)/2 different equations in K unknowns, so that for K>3 there are more equations than there are unknowns. Grubbs' estimates are the least squares solutions of these equations (Jaech, 1985). Formally the system of equations can be represented as

$$Q\sigma = V \qquad\qquad [5.1]$$

where **Q** is a K(K-1)/2 x K matrix with elements 0 and 1 that picks the appropriate σ^2_i's, σ is a K dimensional vector of σ^2_i's and **V** is a K(K-1)/2 dimensional vector of $V^2_{s,r}$'s. **Q'Q** has full rank and can thus be inverted. The solution of [5.1] is

$$\check{\sigma} = (Q'Q)^{-1}Q'V \qquad\qquad [5.2]$$

The method of moments may also be employed when some data are missing. For two genotypes s and r we can compute $V^2_{s,r}$ as long as they are grown together in at least two environments. If this is the case we will say that the two genotypes s and r are connected. In order to obtain a unique solution of Eq. [5.1], we require that there be at least K connected pairs of genotypes, since we need at least as many equations as there are unknowns. Also, each genotype must be connected to at least one other genotype. Thus, the method may break down in some instances when many data are missing.

In the unbalanced case, another estimate of σ^2_i can be obtained by the MINQUE principle of estimation (Rao, 1970). It has already been pointed out that for balanced data Shukla's estimator is a MINQUE of σ^2_i. Rao (1970) provides a computational procedure for MINQUE in the general case, which can be used in data sets with empty cells. Computational details are described in Piepho (1994a).

Stability estimates obtained from sparse two-way tables should be used with some caution due to the large sampling variation associated with such estimates. Piepho (1994a) gives general formulae for the sample variance of the MINQUE and Grubbs' estimates, which may be a useful guide for interpretation.

VI. UNEQUAL NUMBER OF REPLICATIONS PER ENVIRONMENT AND HETEROGENEOUS ERROR VARIANCES

So far we have assumed that the number of replicates is the same in each environment and that the error variance is constant for each genotype. If any of these two assumptions is violated, estimating stability by the above procedures is not adequate. It may then be more appropriate to estimate $\Theta^2_i = Var[(\alpha\beta)_{ij}]$, rather than $\sigma^2_i = Var[v_{ij}]$, for each genotype. In case the yield trials in each environment were laid out as a randomized complete block (RCB) design, MINQUE estimates of Θ^2_i, as suggested by Deutler (1991), may be used (for details see Piepho, 1994e). Since the RCB design is commonly used in many yield trials, we will confine our discussion to this design.

To allow for heterogeneous error variances, we will denote the error variance of the i-th genotype as σ^2_{ei}. It is assumed that in the j-th environment the yield trial was laid out as a RCB design with L_j replications. The observation in the j-th environment (j=1,..,N) of the m-th replicate (m=1,...,L_j) of the i-th genotype (i=1,...,K) will be represented by x_{ijm}. Unbiased estimates of σ^2_{ei} and Θ^2_i are then given by

$$\hat{\sigma}^2_{ei} = \frac{K(K\text{-}1)U_i - \sum_{r=1}^{K} U_r}{\left(\sum_{j=1}^{N} L_j - N\right)(K\text{-}1)(K\text{-}2)}$$

where

$$U_i = \sum_{j=1}^{N}\sum_{m=1}^{L_i}(x_{ijm} - \bar{x}_{ij.} - \bar{x}_{.jm} + \bar{x}_{.j.})^2$$

and

$$\hat{\theta}^2_i = S^2_{vi} - \frac{N}{\sum_{j=1}^{N} L_j}\hat{\sigma}^2_{ei}$$

where

$$S^2_{vi} = \frac{NK(K\text{-}1)Z_i - N\sum_{r=1}^{K} Z_r}{\sum_{j=1}^{N} L_j(N\text{-}1)(K\text{-}1)(K\text{-}2)}$$

$$Z_i = \sum_{j=1}^{N} L_j(\bar{x}_{ij.} - \bar{x}_{i..} - \bar{x}_{.j.} + \bar{x}_{...})^2$$

These estimates of Θ^2_i and σ^2_{ei} are MINQUE (Minimum Norm Quadratic Unbiased Estimators) as is Shukla's estimator of the stability variance (Shukla, 1972a). In case of heterogeneous error variances and/or unequal numbers of replications per environment Θ^2_i should be preferred to Shukla's stability variance.

VII. PARTITIONING DIFFERENT SOURCES OF G x E INTERACTION

Assessment of the stability variance is usually based on multilocation trials, which yield a two-way table of genotypes and locations. This approach has the limitation of ignoring genotype-by-year interactions, which may have a large influence on stability. Usually, however, the plant breeder is interested in stability across locations as well as across years, while farmers are mainly interested in stability across years. Therefore, three types of interaction are of interest: (1) Genotype-by-location interaction, (2) Genotype-by-year interaction, and (3) Genotype-by-location-by-year interaction.

If Shukla's estimate of the stability variance is computed from genotype-by-location data of one year, it does not include genotype-by-year interaction. This may be rather misleading as the following hypothetical example shows: Suppose that based on genotype-by-location data, a certain genotype was shown in different years to have the smallest stability variance of all genotypes. It is possible that despite a low stability variance, this genotype has the highest genotype-by-year interaction. This type of interaction is not detected by the stability variance as estimated from two-way tables of genotype-by-locations. This means that the genotype may be highly unstable across years, even though its stability variance is small in different years.

In order to include all types of interaction, it has been suggested to consider each location-year combination as a "macroenvironment" (Eberhart and Russell, 1966; Becker and Léon, 1988). The stability variance may then be computed from a two-way table of genotypes and macroenvironments. This estimate will contain all three interaction components. An undesirable property of this estimate is that the three components are weighted differently depending on the number of years and the number of genotypes involved. Suppose that genotypes were tested for C years in each of N locations and with L replications. It may be shown that in this case the stability variance has expectation

$$\frac{C(N\text{-}1)}{CN\text{-}1} Var_i(\alpha\beta) + \frac{N(C\text{-}1)}{CN\text{-}1} Var_i(\alpha\tau) + Var_i(\alpha\beta\tau) + \sigma_e^2/L$$

where $Var_i(\alpha\beta)$ = Genotype-by-location interaction variance of i-th genotype, $Var_i(\alpha\tau)$ = Genotype-by-year interaction variance of i-th genotype, $Var_i(\alpha\beta\tau)$ = Genotype-by-location-by-year interaction variance of i-th genotype, and σ_e^2 is the common error variance. This does not appear to be a desirable estimate, and in analogy to Shukla's stability variance it seems more appropriate to assess

$$\Gamma_i = Var_i(\alpha\beta) + Var_i(\alpha\tau) + Var_i(\alpha\beta\tau) + \sigma_e^2/L$$

for each component of gentoype-by-environment interaction separately. An estimation procedure for this purpose is proposed by Piepho (1994c):
(1) For each genotype-by-location combination compute the mean across years.

From the resulting K x N two-way table of means compute Shukla's estimate. This estimate will henceforth be denoted as GL_i. (2) Perform the same analysis with the genotype-by-year table for means across locations. By analogy, we obtain estimates GY_i. (3) Compute

$$GLY_i = \sum_{j=1}^{N} \sum_{k=1}^{C} (x_{ijk} - \bar{x}_{ij.} - \bar{x}_{i.k} - \bar{x}_{.jk} + \bar{x}_{i..} + \bar{x}_{.j.} + \bar{x}_{..k} - \bar{x}_{...})^2$$

where x_{ijk} is the mean of the i-th genotype in the j-th location and k-th year. (4) Compute

$$Q_i = \frac{K(K-1)GLY_i - \sum_{r=1}^{K} GLY_r}{(K-1)(K-2)(N-1)(C-1)}$$

Q_i is an unbiased estimator of $Var(\alpha\beta\tau) + \sigma_e^2/L$. Note the resemblance with Shukla's estimate of the stability variance. The unbiased estimators for $Var_i(\alpha\beta)$ and $Var_i(\alpha\tau)$ are, respectively

$$S_i^2(\alpha\beta) = GL_i - C^{-1}Q_i$$

and

$$S_i^2(\alpha\tau) = GY_i - N^{-1}Q_i$$

A stability measure comprising all genotype-environmental interaction may be defined as

$$G_i = S_i^2(\alpha\beta) + S_i^2(\alpha\tau) + Q_i$$

since

$$E(G_i) = \Gamma_i = Var_i(\alpha\beta) + Var_i(\alpha\tau) + Var_i(\alpha\beta\tau) + \sigma_e^2/L$$

as desired. Γ_i will be denoted as the "generalized stability variance", and G_i is an unbiased estimate of Γ_i.

VIII. A CONCLUDING REMARK

Throughout this article we have assumed uncorrelated interaction effects. Whether this assumption is tenable or not will much depend on the population of environments from which the testing environments were sampled and on the set of genotypes investigated.

If the regression approach or an AMMI model with one or more principal component axes (PCA) explains a major portion of the genotype-by-environment interaction sum of squares, this may be seen as an indication that not all pairs of interactions are independent. Shukla (1972a) states that his concept of the stability variance is appropriate when only a small fraction of the G x E interaction sum of squares can be attributed to heterogeneity among regressions. Wricke and Weber (1980) point out that in breeding applications the fraction explained by the regression is usually small. A test of significance for the regression model (Freeman and Perkins, 1971) or AMMI-PCAs (Gauch, 1992; Cornelius, 1993) may be regarded as a preliminary test to the procedures suggested here. Alternatively, significance testing may be done by cross validation procedure to identify significant PCAs in the AMMI model (Gauch and Zobel, 1988; Piepho, 1994g). Nonsignificance of tests for PCAs has been reported by Crossa et al. (1990). Gauch (1992: 111) gives several cases in which one or two significant PCAs were found in an AMMI analysis. If any of these tests is significant, but the proportion of the total interaction sum of squares explained by the heterogeneity of regressions or by the significant AMMI-PCAs is small, then it may still be worthwhile to use the procedures suggested in this article despite minor distortions possibly caused by the correlation among v_{ij}'s. An investigation of the robustness of these procedures to correlations among interaction effects would be quite rewarding, but is beyond the scope of this article.

REFERENCES

Anscombe, F.J. 1981. Computing in Statistical Sciences through AP.Springer Series in Statistics. Springer-Verlag, New York, Heidelberg, Berlin.

Becker, H.C. 1987. Zur Heritabilitaet statistischer Maßzahlen fuer die Ertragssicherheit. *Vortraege fuer Pflanzenzuechtung* 12:134-144.

Becker, H.C., and J. Léon. 1988. Stability analysis in plant breeding. *Plant Breeding* 101:1-23.

Brindley, R.D., and R.A. Bradley. 1985. Some new results on Grubbs' estimates. *Journal of the American Statistical Association* 80: 711-714.

Cornelius, P.L. 1993. Statistical tests and retention of terms in the additive main effects and multiplicative interaction model for cultivar trials. *Crop Science* 33:1186-1193.

Crossa J., H.G. Gauch, and R.W. Zobel. 1990. Additive main effects and muliplicative interaction analysis of two international maize cultivar trials. *Crop Science* 30:493-500.

Deutler, T. 1991. Grubbs-type estimators for reproducibility variances in an interlaboratory test study. *Journal of Quality Technology* 23:324-333.

Eberhart, S.A., and W.A. Russell. 1966. Stability parameters for comparing varieties. *Crop Science* 6:36-40.

Eskridge, K.M. 1990. Selection of stable cultivars using a safety first rule. *Crop Science* 30:369-374.

Finlay, K.W., and G.N. Wilkinson. 1963. The analysis of adaption in a plant breeding programme. *Australian Journal of Agricultural Research* 14:742-754.

Freeman, G.H., and J.M. Perkins. 1971. Environmental and genotype-environmental components of variability. VIII. Relations between genotypes grown in different environments and measures of these environments. *Heredity* 26:15-23.

Gauch, H.G. 1988. Model selection and validation for yield trials with interaction. *Biometrics* 44:705-715.

Gauch, H.G. 1992. *Statistical analysis of regional yield trials.* Elsevier, Amsterdam.

Gauch, H.G., and R.W. Zobel. 1988. Predictive and postdictive success of statistical anylyses of yield trials. *Theoretical and Applied Genetics* 76:1-10.

Gill, J.L. 1984. Heterogeneity of variance in randomized block experiments. *Journal of Animal*

Science 59:1339-1344.

Gollob, H.F. 1968. A statistical model which combines features of factor analytic and analysis of variance techniques. *Psychometrika* 33: 73-115.

Grubbs, F.E. 1948. On estimation of precision of measuring instruments and product variability. *Journal of the American Statistical Association* 43:243-264.

Hochberg, Y., and A.C. Tamhane. 1987. *Multiple comparison procedures.* Wiley, New York.

Huehn, M. 1979. Beitraege zur Erfassung der phaenotypischen Stabilitaet. I. Vorschlag einiger auf Ranginformationen beruhender Stabilitaetsparameter. *EDV in Medizin und Biologie* 10:112-117.

Huehn, M. 1990. Nonparametric measures of phenotypic stability. Part 1: Theory. *Euphytica* 47:189-194.

Huehn, M., and R. Nassar. 1989. On tests of significance for nonparametric measures of phenotypic stability. *Biometrics* 45:997-1000.

Huehn, M., and R. Nassar. 1991. Phenotypic stability of genotypes over environments: On tests of significance for two nonparametric measures. *Biometrics* 47:1096-1097.

Jaech, J.L. 1985. *Statistical analysis of measurement errors.* Wiley, New York.

Johnson, N.L. 1962. Some notes on the investigation of heterogeneity in interactions. *Trabajos de estadistica* XIII:183-199.

Levene, H. 1960. Robust tests for equality of variances. In: I. Olkin, S.G. Ghurye, W. Hoeffding, W.G. Madow, H.B. Mann (eds.). *Contributions to probability and statistics. Essays in honor of Harold Hotelling.* Stanford University Press, Stanford, CA, 278-292.

Lin, C.S., M.R. Binns, and L.P. Levkovitch. 1986. Stability analysis: where do we stand? *Crop Science* 26:894-900.

Maloney, C.J., and S.C. Rastogi. 1970. Significance tests for Grubbs' estimators. *Biometrics* 26:671-676.

Mandel, J. 1961. Non-additivity in two-way analysis of variance. *Journal of the American Statistical Association* 56:878-888.

Mandel, J. 1969. The partitioning of interaction in analysis of variance. *Journal of Research of the National Bureau of Standards B* 73B:309-328.

Mandel, J. 1971. A new analysis of variance model for non-additive data. *Technometrics* 13:1-18.

Miller, R.G. 1981. *Simultaneous statistical inference.* (2nd ed.). Springer Series in Statistics. Springer, New York.

Mudholkar, G.S., and I.C. Sarkar. 1992. Testing homoscedascity in a two-way table. *Biometrics* 48:883-888.

Nassar, R., and M. Huehn. 1987. Studies on estimation of phenotypic stability: Tests of significance for nonparametric measures of phenotypic stability. *Biometrics* 43:45-53.

Perkins, J.M., and J.L. Jinks. 1968. Environmental and genotype-environmental components of variability. III. Multiple lines and crosses. *Heredity* 23:339-356.

Piepho, H.P. 1992a. Comment on Shukla's test for homogeneity of variances in a two-way classification. Letter. *Animal Science* 70:1644-1645.

Piepho, H.P. 1992b. Vergleichende Untersuchungen der statistischen Eigenschaften verschiedener Stabilitaetsmaße mit Anwendungen auf Hafer, Winterraps, Ackerbohnen sowie Futter- und Zuckerrueben. Diss. Kiel, 163 p.

Piepho, H.P. 1993a. Note on bias in estimates of the regression coefficient in the analysis of genotype-environmental interaction. *Heredity* 70:98-100.

Piepho, H.P. 1993b. Use of the maximum likelihood method in the analysis of phenotypic stability. *Biometrical Journal* 35:815-822.

Piepho, H.P. 1993c. Tests for homoscedasticity in a two-way layout. Letter. *Biometrics* 49:1279-1280.

Piepho, H.P. 1994a. Missing observations in the analysis of stability. *Heredity* 72:141-145 (Correction 73 (1994): 58)

Piepho, H.P. 1994b. Tests for pairwise comparisons of variances in a two-way layout for the analysis of phenotypic stability. *Informatik, Biometrie und Epidemiologie in Medizin und Biologie* 25:172-180.

Piepho, H.P. 1994c. Partitioning genotype-environmental interaction in regional yield trials via a generalized stability variance. *Crop Science* 34:1682-1685.

Piepho, H.P. 1994d. Ranking ability of various measures of phenotypic stability. *Informatik, Biometrie und Epidemiologie in Medizin und Biologie* 25:181-189.

174

Piepho, H.P. 1994e. Application of a generalized Grubbs model in the analysis of genotype-environment interaction. *Heredity* 73:113-116.

Piepho, H.P. 1994f. C410. A distribution-free test for homoscedasticity in a two-way layout. *Journal of Statistical Computation and Simulation* 49:223-225.

Piepho, H.P. 1994g. Best linear unbiased prediction for regional yield trials: A comparison to additive main effects multiplicative interaction analysis. *Theoretical and Applied Genetics* 89:647-654.

Piepho, H.P. 1995a. A robust test for homoscedasticity in a two-way layout. *Biometrical Journal* 37:151-160.

Piepho, H.P. 1995b. Distribution-free tests for one-way homoscedasticity in a two-way layout. submitted

Piepho, H.P., and S. Lotito. 1992. Rank correlation among parametric and nonparametric measures of phenotypic stability. *Euphytica* 64:221-225.

Plaisted, R.L. 1960. A shorter method for evaluating the ability of selections to yield consistently over locations. *American Potato Journal* 37:166-172.

Plaisted, R.L., and L.C. Peterson. 1959. A technique for evaluating the ability of selections to yield consistently in different locations or seasons. *American Potato Journal* 36:381-385.

Rao, C.R. 1970. Estimation of heteroscedastic variances in linear models. *Journal of the American Statistical Association* 65:161-172.

Shukla, G.K. 1972a. Some statistical aspects of partitioning genotype-environmental components of variability. *Heredity* 29:237-245.

Shukla, G.K. 1972b. An invariant test for homogeneity of variances in a two-way classification. *Biometrics* 28:1063-1072.

Shukla, G.K. 1982. Testing the heterogeneity of variances in a two-way classification. *Biometrika* 69:411-416.

Weber, W.E., and G. Wricke. 1987. Wie zuverlaessig sind Schaetzungen von Parametern der phaenotypischen Stabilitaet? *Vortraege zur Pflanzenzuechtung* 12:120-133.

Westcott, B. 1986. Some methods of analysing genotype-environment interactions. *Heredity* 56:243-253.

Wricke, G. 1962. Über eine Methode zur Erfassung der oekologischen Streubreite in Feldversuchen. *Zeitschrift fuer Pflanzenzuechtung* 47:92-96.

Wricke, G., and W.E. Weber. 1980. Erweiterte Analyse von Wechselwirkungen in Versuchsserien. In: Koepke, W. and Überla, K. (eds.). Biometrie - heute und morgen. Springer, Berlin.

Yates, F., and W.G. Cochran. 1938. The analysis of groups of experiments. *Journal of Agricultural Science* 28:556-580.

Chapter 7

USING THE SHIFTED MULTIPLICATIVE MODEL CLUSTER METHODS FOR CROSSOVER GENOTYPE-BY-ENVIRONMENT INTERACTION[1]

J. Crossa,[2] P. L. Cornelius,[3] and M. S. Seyedsadr[4]

I. INTRODUCTION

Genotype x environment interaction is the differential response of genotypes to differing environments and comprises the following types of interaction: 1) crossover interaction (COI) or genotypic rank changes across environments, the most crucial interaction in crop improvement and production (Baker, 1988, 1990), 2) non-COI or scale changes among environments and 3) a combination of both.

Statistical tests originally proposed for identifying and quantifying COI in medical trials (Azzalini and Cox, 1984; Gail and Simon, 1985) have been used to analyze genotype x environment interaction (Baker, 1988; 1990). Gregorius and Namkoong (1986) defined "separability" of genotypic effects and environmental effects from one another. One property of such separability is the absence of genotypic rank changes (i.e., absence of COIs). They proposed a joint multiplicative operator $[\gamma(g)\varepsilon(e)]$ for describing patterns of genotypic response $[\gamma(g)]$ to environmental effects $[\varepsilon(e)]$ such that all variability would be attributed to $\gamma(g)\varepsilon(e)$ plus a constant. However, no specific statistical methods were proposed.

The shifted multiplicative model (SHMM) developed by Seydesadr and Cornelius (1992), when restricted to only one multiplicative term (SHMM$_1$), in the context of genotype x environment interaction, postulates proportionality of genotypic and environmental effects and is equivalent to Gregorius and Namkoong's (1986) multiplicative operator. The SHMM model, supplemented by other statistical methods, has been a powerful analytical tool for determining separability of genotypic effects from environmental effects within subsets of environments or cultivars, i.e., clustering sites without genotypic rank change (Cornelius et al., 1992; Crossa et al., 1993) or clustering genotypes without genotypic rank change (Cornelius et al., 1993). These

[1] The investigation reported in this paper (No. 94-3-63) is in connection with a project of the Kentucky Agricultural Experiment Station and is published with the approval of the Director.
[2] Senior Biometrician, Biometrics and Statistics Unit, International Maize and Wheat Improvement Center (CIMMYT), Lisboa 27, Apdo. Postal 6-641, 06600, México D.F., México.
[3] Professor of Agronomy and Statistics, Department of Agronomy, University of Kentucky, Lexington, KY 40546-0091, USA.
[4] Senior Research Biostatistician, Amgen, Inc., Biostatistics 17-1-A-393, 1840 DeHavilland Drive, Thousand Oaks, CA 91320-1789, USA.

clustering methods use a SHMM-based distance measure between environments or genotypes.

In this monograph we review and apply SHMM-based methods for clustering genotypes into groups in which COI are negligible. We also describe and compare alternative constrained non-COI solutions for subsets of sites and genotypes.

II. SHMM AND ITS RELATIONSHIP TO COI

The general t-term SHMM is $\bar{y}_{ij.} = \beta + \sum_{k=1}^{t} \lambda_k \alpha_{ik} \gamma_{jk} + \epsilon_{ij}$ (Seyedsadr and Cornelius, 1992), where, for a two-way table (or subtable) of g genotypes and e environments, $\bar{y}_{ij.}$ is the mean of the ith genotype in the jth environment, β is the shift parameter, λ_k is the scale parameter for the kth multiplicative term which allows the genotype and site parameters, α_{ik} and γ_{jk}, respectively, to be subject to normalization constraints $\sum_i \alpha_{ik}^2 = \sum_j \gamma_{jk}^2 = 1$, and ϵ_{ij} is the error associated with $\bar{y}_{ij.}$. It is assumed that the ϵ_{ij} are NID $(0, \sigma^2/n)$, where n is the number of replicates in a site and σ^2 is the within site error variance (assumed homogeneous). Multiplicative terms are orthogonal to one another, i.e., $\sum_i \alpha_{ik} \alpha_{ik'} = \sum_j \gamma_{jk} \gamma_{jk'} = 0$ for $k \neq k'$. The first multiplicative term is called "primary effects", second term "secondary effects", etc. Least squares solutions for $\hat{\lambda}_k$, $\hat{\alpha}_{ik}$ and $\hat{\gamma}_{jk}$ are obtained by singular value decomposition (SVD) of the matrix $Z = [z_{ij}] = [\bar{y}_{ij.} - \hat{\beta}]$. However, $\hat{\beta} = \bar{y}_{...} - \sum_{k=1}^{t} \hat{\lambda}_k \bar{\hat{\alpha}}_k \bar{\hat{\gamma}}_k$, where $\bar{\hat{\alpha}}_k = g^{-1} \sum_i \hat{\alpha}_{ik}$, $\bar{\hat{\gamma}}_k = e^{-1} \sum_j \hat{\gamma}_{jk}$, which depends on the SVD. Consequently, SHMM analysis requires an iterative algorithm [except that if the number of multiplicative terms (t) is one less than the smaller dimension of the two-way table of data, the solution can be obtained by transformation of a model which does have a closed form solution]. Computational algorithms are described by Seyedsadr and Cornelius (1992).

The SHMM model with one multiplicative term (SHMM₁) is $\bar{y}_{ij.} = \beta + \lambda \alpha_{i1} \gamma_{j1} + \epsilon_{ij}$, which postulates that only primary effects exist, i.e., after subtracting the shift parameter there is proportionality of genotypic and environmental effects. Mandel (1961) obtained a reparameterization of his "concurrent regression" model equivalent to SHMM₁ although he did not call it a "shifted multiplicative model".

Sufficient conditions for a non-COI SHMM₁ (Cornelius et al., 1992) are: 1) SHMM₁ is an adequate model for fitting the data and 2) estimated primary effects of sites ($\hat{\gamma}_{j1}$) have the same sign. If this holds, then the fitted SHMM₁ predicts proportionality of genotypic differences from one site to another.

When predicted values of SHMM₁ are plotted against primary effects of sites, $\hat{\gamma}_{j1}$, the graph consists of a set of concurrent regression lines, one for each genotype. A set of regression lines are concurrent if they all intersect at one point (Mandel, 1961). For SHMM$_1$, with $\hat{\gamma}_{j1}$ as the absissa, intersection is at the point (0, $\hat{\beta}$). If this point is outside (to the left or right) of the region that contains the data, and, furthermore, SHMM$_1$ does not show lack of fit, then the genotypic effects are "separable from environmental effects" in the sense of Gregorius and Namkoong (1986). Data in which genotype and site effects are nearly additive will have the SHMM$_1$ intersection point very far removed (left or right) from the region containing the data, producing regression lines which appear almost parallel. Conversely, if the SHMM$_1$ intersection point is within the region containing the data, the graph displays complete rank reversal of genotypes for sites whose positions (as determined by their $\hat{\gamma}_{j1}$ values) on the horizontal axis fall on opposite sides of the origin, thus giving a SHMM$_1$ which displays COI. In general, the existence of large COIs in the data will result in the fitted SHMM$_1$ displaying COI and/or significant lack of fit of SHMM$_1$. Graphs of four hypothetical SHMM$_1$ response patterns are shown by Cornelius et al. (1992) (their Figures 1A-D). The four cases correspond to: 1) complete separability, 2) genotypic effects separable from environmental effects, 3) environmental effects separable from genotypic effects and 4) genotypic and environmental effects inseparable.

III. SHMM CLUSTERING METHODS

SHMM analysis for identifying subsets of sites without COI was first employed by Cornelius et al. (1992). The authors presented an exploratory method in which sites were grouped by inspecting the magnitude and sign of primary, secondary, tertiary, etc., effects of sites from a SHMM analysis of the entire data set. Separate SHMM analyses were performed on subsets until clusters were found to which SHMM$_1$ adequately fit the data. This approach was characterized as a "step-down" method because it begins with all sites as a group and successively finds smaller subsets. It showed that SHMM could be a useful basis for clustering, but the scheme is not easily automated on a computer. Later, other SHMM-based methods for clustering sites or genotypes with negligible COI were developed (Crossa et al., 1993; Cornelius et al., 1993; Crossa and Cornelius, 1993; Crossa et al., 1995).

A. SHMM CLUSTERING METHOD FOR GROUPING SITES OR GENOTYPES

The main principle in SHMM cluster methodology is that the variation owing to secondary and higher effects (tertiary, quaternary, etc.) that exist in the entire data set will be recovered as primary effects in smaller clusters or as differences among the clusters. The SHMM cluster method of sites searches for subsets of sites with negligible genotypic rank change by defining the distance between two sites as the residual sum of squares, RSS(SHMM$_1$), after least squares fitting of SHMM$_1$ (Crossa et al., 1993). If SHMM$_1$ shows COI, i.e., has $\hat{\gamma}_{j1}$ which are not all of like sign, the distance should be RSS(SHMM$_1$) for a SHMM$_1$ constrained to be a non-COI solution. Methods for obtaining constrained solutions are reviewed later in this chapter. For a pair of sites, calculation of distances is facilitated by the fact that SREG$_1$ [the sites, i.e., columns, regression model of Bradu and Gabriel (1978) with only one multiplicative term] can be reparameterized to SHMM$_1$ (Crossa et al. 1993; Seyedsadr and Cornelius, 1992). Consequently, for an unconstrained solution for a pair of sites, RSS(SHMM$_1$)=RSS(SREG$_1$).

Once the distance matrix has been obtained, a complete linkage (farthest neighbor) cluster analysis is computed and a dendrogram generated. In order to identify groups of sites as large as possible to which SHMM$_1$ gives an adequate fit, a dichotomous splitting procedure is used. First, SHMM analysis is performed on the entire data set. If SHMM$_1$ is not adequate, SHMM analysis is done on each of the last two clusters joined in the dendrogram. If SHMM$_1$ does not give an adequate fit, the next step is to move down to the next branch of the dendrogram, then to next branch, etc., until subsets of sites are found to which SHMM$_1$ does provide an adequate fit.

This clustering approach has been characterized as a "step-up" method because, beginning with individual sites, successively larger sets are amalgamated. However, since dichotomous splitting will move in the opposite direction (from all sites as a group to smaller subsets), it would be more appropriate to refer to it as a "step-up-step-down" method. It could, however, be accomplished as a strictly "step-up" method by analyzing each cluster as soon as it is formed and invoking suitable stopping rules.

Cornelius et al. (1993) adapted the SHMM cluster method for sites to the problem of grouping genotypes without significant genotypic rank change. Unconstrained SHMM$_1$ solutions for pairs of genotypes are obtained simply by exchanging the roles of genotypes and sites. Thus, for an unconstrained solution for a pair of genotypes, the distance measure RSS(SHMM$_1$)=RSS(GREG$_1$) (Cornelius et al., 1993) [GREG$_1$ is the genotype, i.e., rows, regression model of Bradu and Gabriel (1978) with only one multiplicative term]. However, constrained solutions do not derive simply by exchanging roles because it is still the genotypes with respect to which the solution must be non-COI. Constrained non-COI SHMM$_1$ solutions do not exist in closed form for pairs of genotypes. Once the distances among all pairs of genotypes are computed, a dendrogram is obtained and subsets of genotypes, to which SHMM$_1$ is adequate to describe the data, are identified in the same manner as when sites are clustered.

B. LACK OF FIT OF SHMM$_1$

A cluster of sites or genotypes is judged acceptable if SHMM analysis does not show significant evidence for multiplicative terms beyond primary effects in a non-COI solution. Test criteria used have usually been the F_1 and F_{GH1} tests of the sum of squares due to the sequentially fitted multiplicative terms (Cornelius et al., 1992; Cornelius, 1993). F_1 and F_{GH1} employ method of moments approximations to obtain approximate F statistics. The F_{GH1} test of secondary effects is conservative if primary effects are small, but otherwise is the preferred procedure. Simulation results show that F_1 tests of secondary effects are very liberal if primary effects are non-null (which is usually the case). Thus, nonsignificance of secondary effects by the F_1 test implies adequacy of SHMM$_1$, and significance of secondary effects by the F_{GH1} test implies inadequacy. When primary effects are large, significance of secondary effects by the F_1 test should not be taken very seriously unless F_1 shows still more significant terms (e.g., tertiary and quaternary). In practice, we draw our conclusions primarily from the F_{GH1} test.

Another test of lack of fit of SHMM$_1$ is the F_R test of the residuals (Cornelius et al., 1992), which assigns *ge-g-e* df to RSS(SHMM$_1$), where g and e are the numbers of genotypes and sites, respectively, in the cluster being analyzed. Thus for testing SHMM$_1$ residuals, $F_R = n[\text{RSS}(\text{SHMM}_1)]/(ge\text{-}g\text{-}e)s^2$ which, under the hypothesis of no lack of fit, and if primary effects are large, is distributed approximately as central F with *ge-g-e* and f df, where f is the df of the pooled error mean square s^2. If F_R is significant, then SHMM$_1$ is not adequate to fit the data. Although F_R is a simple test procedure for choosing a truncated model, it has less power for detecting the need of another multiplicative term than the F_{GH1} test. A more detailed review of statistical tests for multiplicative models is given in Cornelius et al. (1995). F_2 and F_{GH2} can be used in lieu of F_1 and F_{GH1}, respectively.

IV. UNCONSTRAINED AND CONSTRAINED SOLUTIONS

A. UNCONSTRAINED NON-COI SHMM$_1$ SOLUTIONS

When SHMM$_1$ is an adequate model to fit the data for a cluster and all primary effects of sites have the same sign, the solution is an unconstrained non-COI SHMM$_1$ solution. The unconstrained SHMM$_1$ solution exists in closed form if the cluster contains only two sites or only two genotypes, but not otherwise (Seyedsadr and Cornelius, 1992). As already mentioned, the unconstrained non-COI RSS(SHMM$_1$) is equal to RSS(SREG$_1$) for a cluster containing only two sites and is equal to RSS(GREG$_1$) for a cluster containing only two genotypes, provided that the like-sign condition for primary effects of sites holds in each of these cases.

B. CONSTRAINED NON-COI SHMM$_1$ SOLUTIONS

Often SHMM$_1$ is an adequate model to fit the data, but primary effects of sites show different signs. In such cases, the point of intersection, $(0, \hat{\beta})$, of the concurrent regression lines lies in the midst of the plotted data points, and, in order to force all variation due to COIs into the residuals, it is necessary to compute a constrained solution by choosing a shift parameter value that forces the smallest or the largest $\hat{\gamma}_{j1}$ to be equal to zero [the proper choice is the one that gives the smallest RSS(SHMM$_1$)]. Two types of constrained solutions can be computed: 1) least squares (LS) non-COI SHMM$_1$ solutions and 2) singular value decomposition (SVD) non-COI SHMM$_1$ solutions.

1. Constrained LS Non-COI Solution for Two Sites

Suppose that the unconstrained SHMM$_1$ solution for the pair of sites h and m shows COI and that $\Sigma_i(\bar{y}_{ih.}-\bar{y}_{.h.})^2 < \Sigma_i(\bar{y}_{im.}-\bar{y}_{.m.})^2$. Let $\hat{\beta} = \bar{y}_{.h.}$,

$$\hat{\lambda}_1 = [\Sigma_i(\bar{y}_{im.}-\bar{y}_{.h.})^2]^{1/2}, \; \hat{\gamma}'_1=(\hat{\gamma}_{h1},\hat{\gamma}_{m1}) = (0,1), \text{ and } \hat{\lambda}_1\hat{\alpha}_1 = \begin{bmatrix} \bar{y}_{1m.}-\bar{y}_{.h.} \\ \bar{y}_{2m.}-\bar{y}_{.h.} \\ \cdot \\ \cdot \\ \cdot \\ \bar{y}_{gm.}-\bar{y}_{.h.} \end{bmatrix}$$

where $\hat{\alpha}_1$ is a vector with i^{th} element $\hat{\alpha}_{i1}$. Then

$$\hat{Y} = \hat{\beta}J + \hat{\lambda}_1\hat{\alpha}_1\hat{\gamma}'_1 = \bar{y}_{.h.}J + \begin{bmatrix} \bar{y}_{1m.}-\bar{y}_{.h.} \\ \bar{y}_{2m.}-\bar{y}_{.h.} \\ \cdot \\ \cdot \\ \cdot \\ \bar{y}_{gm.}-\bar{y}_{.h.} \end{bmatrix} [\,0,1] = \begin{bmatrix} \bar{y}_{.h.} & \bar{y}_{1m.} \\ \bar{y}_{.h.} & \bar{y}_{2m.} \\ \cdot & \cdot \\ \cdot & \cdot \\ \cdot & \cdot \\ \bar{y}_{.h.} & \bar{y}_{gm.} \end{bmatrix}$$

where J is a $g \times 2$ matrix of ones. This is the closed form constrained LS solution given by Crossa et al. (1993), for which

$$RSS(SHMM_1) = \Sigma_i(\bar{y}_{ih.}-\bar{y}_{.h.})^2. \tag{1}$$

Note that a non-COI SHMM$_1$ can also be obtained by interchanging the roles of sites h and m, giving $\hat{\beta} = \bar{y}_{.m.}$, $\hat{\lambda}_1 = [\Sigma_i(\bar{y}_{ih.}-\bar{y}_{.m.})^2]^{1/2}$, $\hat{\gamma}'_1=(\hat{\gamma}_{h1},\hat{\gamma}_{m1}) = (1,0)$, $\hat{\lambda}_1\hat{\alpha}_{i1} = \bar{y}_{ih.}-\bar{y}_{.m.}$, and $RSS(SHMM_1) = \Sigma_i(\bar{y}_{im.}-\bar{y}_{.m.})^2$. But this cannot be the LS solution if its RSS is greater than (1). Thus the constrained LS RSS(SHMM$_1$) for a pair of sites h and m is $\min[\Sigma_i(\bar{y}_{ih.}-\bar{y}_{.h.})^2, \Sigma_i(\bar{y}_{im.}-\bar{y}_{.m.})^2]$.

2. Constrained SVD Non-COI Solution for Two Sites

Recall our definition of matrix Z as an array of values $[\bar{y}_{ij.}-\hat{\beta}]$. For unconstrained least squares SHMM$_1$ solutions, the t multiplicative terms are given by the first t components of the SVD of Z. The residual sum of squares is

$$RSS(SHMM_t) = trace(\mathbf{Z'Z}) - \Sigma_{k=1}^t L_k \qquad (2)$$

where L_k is the kth largest eigenvalue of $\mathbf{Z'Z}$. For $SHMM_1$, (2) reduces to

$$RSS(SHMM_1) = trace(\mathbf{Z'Z}) - L_1. \qquad (3)$$

But (3) does not give the same result as (1) [the constrained LS $RSS(SHMM_1)$ for a pair of sites]. However, there usually exists a constrained $SHMM_1$ for which (3) does hold. The matrix $\mathbf{Z}$ for two sites, h and m, and g genotypes is

$$\mathbf{Z} = \begin{bmatrix} \bar{y}_{1h.} - \hat{\beta} & \bar{y}_{1m.} - \hat{\beta} \\ \bar{y}_{2h.} - \hat{\beta} & \bar{y}_{2m.} - \hat{\beta} \\ \cdot & \cdot \\ \cdot & \cdot \\ \cdot & \cdot \\ \bar{y}_{gh.} - \hat{\beta} & \bar{y}_{gm.} - \hat{\beta} \end{bmatrix}.$$

The symmetric matrix $\mathbf{Z'Z}$ is

$$\mathbf{Z'Z} = \begin{bmatrix} \Sigma_i(\bar{y}_{ih.} - \hat{\beta})^2 & \Sigma_i(\bar{y}_{ih.} - \hat{\beta})(\bar{y}_{im.} - \hat{\beta}) \\ \Sigma_i(\bar{y}_{ih.} - \hat{\beta})(\bar{y}_{im.} - \hat{\beta}) & \Sigma_i(\bar{y}_{im.} - \hat{\beta})^2 \end{bmatrix}.$$

In order for this to give $\hat{\mathbf{\gamma}}_1 = (0,1)$ or $(1,0)$, $\mathbf{Z'Z}$ must be a diagonal matrix, i.e., $\Sigma_i(\bar{y}_{ih.} - \hat{\beta})(\bar{y}_{im.} - \hat{\beta}) = 0$. Thus, the constrained LS solution will satisfy (3) if and only if $\Sigma_i(\bar{y}_{ih.} - \bar{y}_{.h.})(\bar{y}_{im.} - \bar{y}_{.m.}) = 0$, i.e., if and only if there is no correlation of genotypic differences in site h with genotypic differences in site m (an event with vanishingly small probability). We refer to a constrained solution which does satisfy (3), i.e., for which $\Sigma_i(\bar{y}_{ih.} - \hat{\beta})(\bar{y}_{im.} - \hat{\beta}) = 0$, as a constrained SVD solution. There are two possible solutions:

$$1)\ \hat{\mathbf{\gamma}}_1' = (\hat{\gamma}_{h1}, \hat{\gamma}_{m1}) = (0,1),\ \hat{\lambda}_1 = [\Sigma_i(\bar{y}_{im.} - \hat{\beta})^2]^{1/2},\ \hat{\lambda}_1\hat{\mathbf{\alpha}}_1 = \begin{bmatrix} \bar{y}_{1m.} - \hat{\beta} \\ \bar{y}_{2m.} - \hat{\beta} \\ \cdot \\ \cdot \\ \cdot \\ \bar{y}_{gm.} - \hat{\beta} \end{bmatrix},$$

$$\hat{Y} = \begin{bmatrix} \hat{\beta} & \bar{y}_{1m.} \\ \hat{\beta} & \bar{y}_{2m.} \\ \cdot & \cdot \\ \cdot & \cdot \\ \cdot & \cdot \\ \hat{\beta} & \bar{y}_{gm.} \end{bmatrix} \text{ and } RSS(SHMM_1) = \Sigma_i(\bar{y}_{ih.} - \hat{\beta})^2;$$

2) $\hat{\gamma}'_1 = (\hat{\gamma}_{h1}, \hat{\gamma}_{m1}) = (1,0)$, $\hat{\lambda}_1 = [\Sigma_i(\bar{y}_{ih.} - \hat{\beta})^2]^{1/2}$, $\hat{\lambda}_1 \hat{\alpha}_1 = \begin{bmatrix} \bar{y}_{1h.} - \hat{\beta} \\ \bar{y}_{2h.} - \hat{\beta} \\ \cdot \\ \cdot \\ \cdot \\ \bar{y}_{gh.} - \hat{\beta} \end{bmatrix},$

$$\hat{Y} = \begin{bmatrix} \bar{y}_{1h.} & \hat{\beta} \\ \bar{y}_{2h.} & \hat{\beta} \\ \cdot & \cdot \\ \cdot & \cdot \\ \cdot & \cdot \\ \bar{y}_{gh.} & \hat{\beta} \end{bmatrix} \text{ and } RSS(SHMM_1) = \Sigma_i(\bar{y}_{im.} - \hat{\beta})^2.$$

For either solution, $\hat{\beta}$ must be such that $\Sigma_i(\bar{y}_{ih.} - \hat{\beta})(\bar{y}_{im.} - \hat{\beta}) = 0$, i.e.,
$\Sigma_i[(\bar{y}_{ih.} \bar{y}_{im.}) - (\bar{y}_{ih.} + \bar{y}_{im.})\hat{\beta} + \hat{\beta}^2] = 0$ for which
$\hat{\beta} = 0.5\{(\bar{y}_{.h.} + \bar{y}_{.m.}) \pm [(\bar{y}_{.h.} + \bar{y}_{.m.})^2 - 4\Sigma_i \bar{y}_{ih.} \bar{y}_{im.}/g]^{1/2}\}$. Provided that the quantity within the square brackets is nonnegative, the latter equation gives two real solutions, $\hat{\beta}_1$ and $\hat{\beta}_2$, say. For either solution, RSS=min$[\Sigma_i(\bar{y}_{ih.} - \hat{\beta})^2, \Sigma_i(\bar{y}_{im.} - \hat{\beta})^2]$. Thus, the constrained SVD RSS(SHMM$_1$)=min$[\Sigma_i(\bar{y}_{ih.} - \hat{\beta}_1)^2, \Sigma_i(\bar{y}_{im.} - \hat{\beta}_1)^2, \Sigma_i(\bar{y}_{ih.} - \hat{\beta}_2)^2, \Sigma_i(\bar{y}_{im.} - \hat{\beta}_2)^2]$. Note that for whichever site is constrained (h or m, i.e., $\hat{\gamma}_{h1} = 0$ or $\hat{\gamma}_{m1} = 0$), the distance is the sum of squared differences between the genoype means at that site and the $\hat{\beta}$ value.

3. Constrained SVD Non-COI Solution for More than Two Sites

Constrained solutions for any set with more than two sites do not exist in closed form (Cornelius et al., 1993). If site h is chosen to have its primary effect equal to zero, the constrained SVD solution requires that $\hat{\lambda}_1 \hat{\gamma}_{h1} = \Sigma_i \hat{\alpha}_{i1} (\bar{y}_{ih.} - \hat{\beta}) = 0$ for which $\hat{\beta} = \Sigma_i \hat{\alpha}_{i1} \bar{y}_{ih.} / \Sigma_i \hat{\alpha}_{i1}$ (Cornelius et al., 1993). Thus, β is iteratively estimated using this equation and α_{i1} estimated by SVD of the matrix Z (or by computing the eigenvectors of ZZ'). Note that Z changes in each iteration. The constraint $\Sigma_i \hat{\alpha}_{i1} (\bar{y}_{ih.} - \hat{\beta}) = 0$ forces the vector of deviations of the genotype means from $\hat{\beta}$ in site h to be orthogonal to the vector of primary effects of genotypes $\hat{\alpha}_{i1}$. This orthogonality constraint is required by the SVD solution, but not by the LS solution.

4. Constrained LS Non-COI Solution for More than Two Sites

For any SHMM$_1$ solution such that $\hat{\gamma}_{h1} = 0$,

$$RSS(SHMM_1) = \Sigma_i (\bar{y}_{ih.} - \hat{\beta})^2 + \Sigma_{j \neq h} \Sigma_i (\bar{y}_{ij.} - \hat{\beta} - \hat{\lambda}_1 \hat{\alpha}_{i1} \hat{\gamma}_{j1})^2. \qquad (4)$$

The constrained SVD solution of Cornelius et al. (1993) does not, in fact, minimize (4), and, consequently, is not a constrained LS solution. With the constrained LS solution, $\hat{\gamma}_{j1} = 0$, but other parameters in the multiplicative term are estimated by SVD of $Z_{(-h)}$, where $Z_{(-h)}$ is obtained from Z by deleting the hth column. The solution for $\hat{\beta}$ is $\hat{\beta} = \bar{y}_{...} - [(e-1)/e] \hat{\lambda}_1 \bar{\alpha}_1 \bar{\gamma}_1$ where $\bar{y}_{...}$ is the grand mean for the genotypes and sites in the cluster, e is the number of sites in the cluster and $\bar{\gamma}_1$ is the mean of the $e-1$ unconstrained (i.e., nonzero) $\hat{\gamma}_{j1}$ values. A Newton-Raphson algorithm for computing this solution will be given later. With $Z_{(-h)}$ in (3) substituted for Z (and L$_1$ now defined as the largest eigenvalue of $Z'_{(-h)} Z_{(-h)}$ or of $Z_{(-h)} Z'_{(-h)}$) gives the second term in (4), but $\Sigma_i (\bar{y}_{ih.} - \hat{\beta})^2$ must be added to obtain the constrained RSS(SHMM$_1$).

5. When Attempted Non-COI Solution Still Shows COI

Suppose, on the basis of the unconstrained solution, that we have determined that the constraint $\hat{\gamma}_{h1} = 0$ should be imposed, but subsequently obtain a constrained solution with $\hat{\gamma}_{m1} < 0$, $\hat{\gamma}_{h1} = 0$, $\hat{\gamma}_{j1} > 0$ (or, if the SHMM$_1$ point of intersection is to be moved to the right boundary of the data, rather than the left, $\hat{\gamma}_{j1} < 0$, $\hat{\gamma}_{h1} = 0$, $\hat{\gamma}_{m1} > 0$ might result) for some $m \neq h$ and all $j \neq m$ or h. The solution, while it satisfies the constraint $\hat{\gamma}_{h1} = 0$, is still not a non-COI SHMM$_1$, since the nonzero $\hat{\gamma}$ values are not all of like sign. Our strategy for resolving this problem depends on which kind of constrained solution is sought. If a constrained SVD solution is sought, usually a non-COI SHMM$_1$ can be obtained by a second attempt in which the constraint will be $\hat{\gamma}_{m1}=0$ instead of $\hat{\gamma}_{h1}=0$. We have not attempted such a strategy for constrained LS solutions because it seems likely to result in $\hat{\gamma}_{h1}<0$, $\hat{\gamma}_{m1}=0$, $\hat{\gamma}_{j1}>0$ for $j \neq m$ or h, still not a non-COI solution. Rather, as a second attempt to obtain a LS non-COI solution, we impose the constraints $\hat{\gamma}_{h1}=\hat{\gamma}_{m1}=0$, i.e., choose $\hat{\beta}$ to minimize

$$\Sigma_i[(\bar{y}_{ih.}-\hat{\beta})^2 + (\bar{y}_{im.}-\hat{\beta})^2] + \Sigma_{j\neq h,m}\Sigma_i(\bar{y}_{ij.}-\hat{\beta}-\hat{\lambda}_1\hat{\alpha}_{i1}\hat{\gamma}_{j1})^2.$$

We acknowledge that imposing the double constraint without actually verifying that neither single-constraint solution will give a non-COI SHMM$_1$ is not strictly correct procedure. But we believe that in practice it will make little difference and, consequently, have not undertaken the task of programming to explore all such "boundary solutions".

6. Algorithm for Constrained LS Non-COI Solutions

Constrained LS SHMM$_1$ solutions may be computed by a modification of the Newton-Raphson algorithm given by Seyedsadr and Cornelius (1992) for computation of unconstrained SHMM solutions. Suppose, in an analysis of a set of g genotypes and e environments, primary effects must be forced to zero for a set $\tilde{S}$ of $\tilde{e}$ sites, but unconstrained in the complementary set S of $e-\tilde{e}$ sites. The residual sum of squares to be minimized is $\Sigma_{j\in\tilde{S}} \Sigma_i(\bar{y}_{ij.}-\hat{\beta})^2 + \Sigma_{j\in S} \Sigma_i(\bar{y}_{ij.}-\hat{\beta}-\hat{\lambda}_1\hat{\alpha}_{i1}\hat{\gamma}_{j1})^2$ for which the solution for $\hat{\beta}$ is

$$\hat{\beta} = \bar{y}_{...}-[(e-\tilde{e})/e]\hat{\lambda}_1\bar{\hat{\alpha}}_1\bar{\hat{\gamma}}_1,$$

where $\bar{\hat{\gamma}}_1 = (e-\tilde{e})^{-1}\Sigma_{j\in S} \hat{\gamma}_{j1}$. Put $f(\hat{\beta}) = \hat{\beta} - \bar{y}_{...} + [(e-\tilde{e})/e]\hat{\lambda}_1\bar{\hat{\alpha}}_1\bar{\hat{\gamma}}_1 = 0$. To iteratively solve for $\hat{\beta}$, let $\hat{\beta}^{(w)}$ be the wth iteration estimate and put $\hat{\beta}^{(w+1)} = \hat{\beta}^{(w)} + u^{(w)}$ where

$$u^{(w)} = \frac{f(\hat{\beta}^{(w)})}{[\partial f(\hat{\beta})/\partial\hat{\beta}]_{\hat{\beta}=\hat{\beta}^{(w)}}},$$

$$\frac{\partial f(\hat{\beta})}{\partial(\hat{\beta})} = 1-\frac{(e-\tilde{e})^2}{e}[\bar{\hat{\gamma}}_1^2-g\Sigma_{k=2}^p\frac{(\hat{\lambda}_1\bar{\hat{\alpha}}_1\bar{\hat{\gamma}}_k + \hat{\lambda}_k\bar{\hat{\alpha}}_k\bar{\hat{\gamma}}_1)^2}{\hat{\lambda}_k^2-\hat{\lambda}_1^2}] \qquad (5)$$

where $p=\min(g, e-\tilde{e})$, $\hat{\lambda}_k$ is the kth singular value, and $\bar{\hat{\alpha}}_k$ and $\bar{\hat{\gamma}}_k$ are the means of elements $\hat{\alpha}_{ik}$ and $\hat{\gamma}_{jk}$ of the kth left and right singular vectors, respectively, of $Z_{(-\tilde{S})}$, where $Z_{(-\tilde{S})}$ is obtained from Z by deleting the columns for sites contained in the constrained set $\tilde{S}$. The first component of the SVD of $Z_{(-\tilde{S})}$ will give the parameter estimates $\hat{\lambda}_1$, $\hat{\alpha}_{i1}$ and $\hat{\gamma}_{j1}$ for the unconstrained sites. If $\tilde{e} = 0$ (i.e., if the set $\tilde{S}$ is empty), the Newton-Raphson algorithm described above is precisely as given by Seyedsadr and Cornelius (1992) for $t=1$.

An alternative form for the derivative given by (5) is

$$\frac{\partial f(\beta)}{\partial \hat{\beta}} = 1 - \frac{g(e-\tilde{e})}{e}[\bar{\alpha}_1^2 - (e-\tilde{e})\Sigma_{k=2}^p \frac{(\hat{\lambda}_1\bar{\alpha}_k\bar{\gamma}_1 + \hat{\lambda}_k\bar{\alpha}_1\bar{\gamma}_k)^2}{\hat{\lambda}_k^2 - \hat{\lambda}_1^2}] \quad (6)$$

Using (5), one can avoid having to compute the $\hat{\alpha}_{ik}$ in every iteration by exploiting the relationship, $\hat{\lambda}_k\bar{\alpha}_k = \Sigma_{j\in S}\hat{\gamma}_{jk}(\bar{y}_{.j} - \hat{\beta})$. Using (6), one can avoid having to compute the $\hat{\gamma}_{jk}$ in every iteration by exploiting the relationship, $\hat{\lambda}_k\bar{\gamma}_k = \Sigma_i\hat{\alpha}_{ik}(\bar{y}_{i.} - \hat{\beta})$, where $\bar{y}_{i.}$ is the mean for the ith genotype over the sites with nonzero (i.e., unconstrained) $\hat{\gamma}_{j1}$. If $e-\tilde{e} < g$, (5) is expedient, but (6) is expedient if $e-\tilde{e} > g$.

Strategies suggested by Seyedsadr and Cornelius (1992) to improve convergence properties of the algorithm for unconstrained SHMM are appropriate here also, in particular, halving or quartering $u^{(w)}$ if the full increment fails to reduce RSS, replacing the denominator of $u^{(w)}$ with its absolute value if it is negative and jump-shifting to a large value of opposite sign if $\hat{\beta}$ seems to be going toward $\pm\infty$. The interested reader should see Seyedsadr and Cornelius (1992) for details.

7. Which Constrained Solution to Use

The constrained SVD solution always gives a larger RSS(SHMM$_1$) than the constrained LS solution; thus it imposes the greater penalty on a set which shows COI in its unconstrained solution. However, whether this will lead to more effective clustering of sites or genotypes into non-COI groups has not been critically investigated.

When clustering sites or genotypes, we suggest that constrained non-COI SHMM$_1$ solutions be used in a consistent manner, i.e., if SVD solutions were used to obtain distance measures for pairs of sites (or genotypes) requiring constrained solutions, then also use SVD solutions in any analyses of larger groups requiring constrained solutions. In the sequel, we will denote constrained RSS(SHMM$_1$) obtained by the two methods as RSS(SHMM$_1$)$_{SVD}$ and RSS(SHMM$_1$)$_{LS}$. The unconstrained RSS(SHMM$_1$) will be denoted as RSS(SHMM$_1$)$_{UN}$.

V. EXAMPLE

Data from a CIMMYT maize international yield trial were used to illustrate the use of SHMM clustering and to compare results of the two methods of obtaining constrained solutions. The trial (EVT12, 1988) consisted of 10 genotypes tested in a randomized complete block design with four replicates at each of 47 international sites.

A. RESULTS

SHMM analysis of variance on the entire data set indicated that the first three multiplicative terms were significant by the F_{GH1} test (the first five multiplicative terms were significant by the F_1 test). This indicates that there is much disproportionality of effects in the complete data set, and suggests a need to split the total set of genotypes (or sites) into non-COI subsets. Therefore, we used SHMM clustering to group genotypes into subsets that have non-significant COI.

Distance values after multiplication by 10^{-3} are presented in Table 1. A total of 38 out of 45 pairs of genotypes required constrained solutions (Table 1); therefore, their constrained $RSS(SHMM_1)_{SVD}$ and $RSS(SHMM_1)_{LS}$ were computed. Genotypes 2 and 4 had the smallest distance $RSS(SHMM_1)_{UN} = 5155$ and did not display COI (Figure 1). The pair with next smallest distance was $\{6,7\}$ with $RSS(SHMM_1)_{UN} = 6441$ (Table 2), but did show COI. However, its $RSS(SHMM_1)_{SVD} = 6442$ and $RSS(SHMM_1)_{LS} = 6441$ (Table 1) which imposed only a small penalty for the pair's COI $SHMM_1$ pattern. The pair $\{1,6\}$, with larger $RSS(SHMM_1)_{UN}$ of 9960, received a much larger penalty for its COI $SHMM_1$ pattern (Figure 2) with $RSS(SHMM_1)_{SVD} = 11039$ and $RSS(SHMM_1)_{LS} = 10641$. For this pair, $RSS(SHMM_1)_{SVD}$ was 10.8% larger than $RSS(SHMM_1)_{UN}$, $RSS(SHMM_1)_{LS}$ was 6.8% larger than $RSS(SHMM_1)_{UN}$ and $RSS(SHMM_1)_{SVD}$ was 3.7% larger than $RSS(SHMM_1)_{LS}$ (Table 2). Pair $\{1,6\}$, in fact, showed the greatest increase in distance imposed by a constrained SVD solution, and also the greatest difference between $RSS(SHMM_1)_{SVD}$ and $RSS(SHMM_1)_{LS}$. Its constrained SVD solution is shown as Figure 3. As is always true, the constrained SVD solution imposed a greater penalty than the constrained LS solution.

For most pairs of genotypes requiring constrained solutions, the estimated shift parameter obtained by the constrained SVD solution was a more extreme value (to the lower left or the upper right, depending on the boundary of the scatter of data to which the concurrence point needed to be moved) than the estimated shift parameter obtained by the constrained LS solution. For example, for $\{1,6\}$, $\hat{\beta}$ values for the constrained SVD and LS solutions were 1877 and 1915, respectively.

Table 1
RSS(SHMM$_1$)$_{SVD}$ (above diagonal) and RSS(SHMM$_1$)$_{LS}$ (below diagonal) distances for 45 pairs of genotypes.[a]

Genotype

	1	2	3	4	5	6	7	8	9	10

-------------------------------------- × 10^{-3} --

	1	2	3	4	5	6	7	8	9	10
1		10962	9486	8074	11536	11039	7567	7391[b]	13304	9232
2	10962		9974	5155[b]	9038	12144	10598	7241[b]	7421	11175
3	9486	9974		11057	11531	10571	11659	11919[b]	12064	10346[b]
4	8074	5155[b]	11057		10592	14739	11368	7911[b]	9127	9286
5	11536	9036	11525	10592		16302	14566	9749	11610	15161
6	10641	2142	10454	14738	16277		6442	10662	12670	13768
7	7542	10586	11654	11349	14522	6441		7582	10860	12468
8	7391[b]	7241[b]	11919[b]	7911[b]	9748	10661	7576		7841	15128[b]
9	13294	7416	12063	9122	11563	2670	10860	7839		14474
10	9232	11175	10346[b]	9286	15157	13767	12462	15128[b]	14473	

[a]Significant (P<0.05) lack of fit of a non-COI SHMM$_1$ to a pair of genotypes is indicated if the distance is greater than 10826.
[b] RSS(SHMM$_1$)$_{UN}$

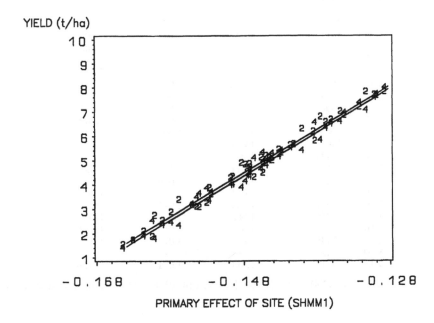

Figure 1. Unconstrained fit of SHMM$_1$ to genotypes 2 and 4.

Table 2

$RSS(SHMM_1)_{UN}$ and ratios of unconstrained and SVD and LS constrained $SHMM_1$ solutions for 45 pairs of genotypes.

Pair	$RSS(SHMM_1)_{UN}$	$\dfrac{RSS(SHMM_1)_{SVD}}{RSS(SHMM_1)_{UN}}$	$\dfrac{RSS(SHMM_1)_{LS}}{RSS(SHMM_1)_{UN}}$	$\dfrac{RSS(SHMM_1)_{SVD}}{RSS(SHMM_1)_{LS}}$
	------ 10^{-3} ------			
1 2	10960	1.000	1.000	1.000
1 3	9293	1.021	1.021	1.000
1 4	8074	1.000	1.000	1.000
1 5	11502	1.003	1.003	1.000
1 6	9960	1.108	1.068	1.037
1 7	6891	1.098	1.095	1.003
1 8[a]	7391	---	---	---
1 9	12604	1.056	1.055	1.001
1 10	9042	1.021	1.021	1.000
2 3	9961	1.001	1.001	1.000
2 4[a]	5155	---	---	---
2 5	8736	1.034	1.034	1.000
2 6	11912	1.020	1.019	1.000
2 7	10362	1.023	1.022	1.001
2 8[a]	7241	---	---	---
2 9	7207	1.030	1.029	1.001
2 10	11155	1.002	1.002	1.000
3 4	10983	1.007	1.007	1.000
3 5	11101	1.039	1.038	1.001
3 6	10321	1.024	1.013	1.011
3 7	11518	1.012	1.012	1.000
3 8[a]	11919	---	---	---
3 9	11910	1.013	1.013	1.000
3 10[a]	10346	---	---	---
4 5	10541	1.005	1.005	1.000
4 6	14116	1.044	1.044	1.000
4 7	10660	1.066	1.065	1.002
4 8[a]	7911	---	---	---
4 9	8496	1.074	1.074	1.000
4 10	9264	1.002	1.002	1.000
5 6	15253	1.069	1.067	1.002
5 7	13580	1.073	1.069	1.003
5 8	9308	1.047	1.047	1.000
5 9	10608	1.095	1.090	1.004
5 10	14627	1.037	1.036	1.000
6 7	6441	1.000	1.000	1.000
6 8	10543	1.011	1.011	1.000
6 9	12670	1.000	1.000	1.000
6 10	13620	1.011	1.011	1.000
7 8	7474	1.014	1.014	1.001
7 9	10856	1.000	1.000	1.000
7 10	12275	1.016	1.015	1.000
8 9	7745	1.012	1.012	1.000
8 10[a]	15128	---	---	---
9 10	14318	1.011	1.011	1.000

[a] Pairs {1,8}, {2,4}, {2,8}, {3,8}, {3,10}, {4,8} and {8,10} do not require constrained solutions.

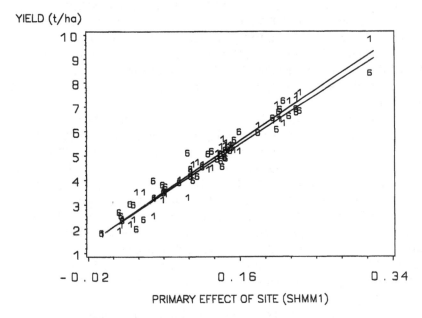

Figure 2. Constrained SVD fit of SHMM$_1$ to genotypes 1 and 6.

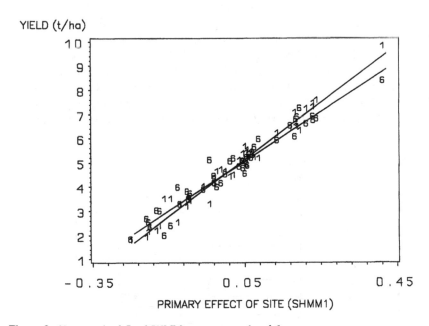

Figure 3. Unconstrained fit of SHMM$_1$ to genotypes 1 and 6.

The mean difference between $RSS(SHMM_1)_{UN}$ and $RSS(SHMM_1)_{SVD}$ was 311, with a minimum of 0.025 for {1,4} and a maximum of 1080 for {1,6}. The mean difference between $RSS(SHMM_1)_{UN}$ and $RSS(SHMM_1)_{LS}$ was 291, with a minimum of 0.013 for {1,4} and a maximum of 1024 for {5,6}. The mean difference between $RSS(SHMM_1)_{LS}$ and $RSS(SHMM_1)_{SVD}$ was 197, with a minimum of 0.001 for {6,9} and a maximum of 399 for {1,6}. As expected, the inequality

$$RSS(SHMM_1)_{SVD} \geq RSS(SHMM_1)_{LS} \geq RSS(SHMM_1)_{UN}$$

held for all pairs of genotypes requiring constrained solutions. Pairs {1,6}, {1,7} and {5,9} showed the three largest $RSS(SHMM_1)_{SVD}/RSS(SHMM_1)_{UN}$ ratios (1.108, 1.098 and 1.095, respectively) (Table 2). The pairs with the three largest $RSS(SHMM_1)_{LS}/RSS(SHMM_1)_{UN}$ ratios were {1,7}, {5,9}, and {4,9}, closely followed by {1,6} (1.095, 1.090, 1.074 and 1.068, respectively). The two largest $RSS(SHMM_1)_{SVD}/RSS(SHMM_1)_{LS}$ ratios were for pairs {1,6} and {3,6} (1.037 and 1.011, respectively) and these were the only pairs for which $RSS(SHMM_1)_{SVD}$ was more than 1% greater than $RSS(SHMM_1)_{LS}$. Genotype pair {1, 6} had the largest ratio as well as the largest difference between $RSS(SHMM_1)_{SVD}$ and $RSS(SHMM_1)_{UN}$ and between $RSS(SHMM_1)_{SVD}$ and $RSS(SHMM_1)_{LS}$.

Cluster analysis using both SVD and LS distance matrices produced dendrograms which were identical with respect to the sequence in which genotypes were joined (Figure 4). The dichotomous splitting and SHMM analysis left two final groups having four genotypes each, G1={2,4,5,9}, G3={1,6,7,8}, and one group with two genotypes, G2={3,10}. The lack of fit of $SHMM_1$ was assessed based on F_{GH1} tests (Table 3). As a consequence of the relationship $RSS(SHMM_1)_{SVD} \geq RSS(SHMM_1)_{LS}$, any cluster judged acceptable using the constrained SVD will also be acceptable under the constrained LS solution. The effect of constrained solutions (SVD and LS) on the SHMM analysis of variance is that part of the variability explained by primary effects in the unconstrained solution is re-allocated to secondary or

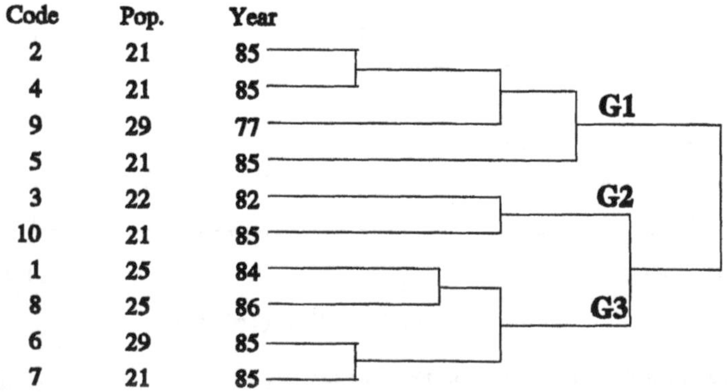

Figure 4. Dendrogram from cluster analysis of 10 genotypes with distance defined as residual sum of squares from fitting $SHMM_1$ subject to a no-COI constraint.

higher effects when a non-COI constraint is imposed on the $SHMM_1$ solution (Table 3). However, for the example, the composition of the final groups was the same for either method of obtaining constrained $SHMM_1$ solutions in analyzing the clusters.

Table 3
SHMM analysis of variance mean squares (10^{-3}) for groups of genotypes using unconstrained and constrained SVD and LS $SHMM_1$ solutions[a].

Source of Variation	Unconstrained	Constrained SVD	LS
		-------- G1 --------	
Primary effects	7552[b,c]	7540[b,c]	7540[b,c]
Secondary effects	198	210[b]	210[b]
Tertiary effects	169	169	169
Quaternary effects	113	113	113
Error	191	191	191
		-------- G3 --------	
Primary effects	6741[b,c]	6735[b,c]	6736[b,c]
Secondary effects	193[b]	199[b]	199[b]
Tertiary effects	144	144	144
Quaternary effects	133	133	133
Error	163	163	163
		-------- G2 --------	
Primary effects	4227[b,c]	--	--
Secondary effects	230[b]	--	--
Error	180	--	--

[a] Mean squares shown for primary, secondary, tertiary and quaternary effects are "pseudo mean squares" (Cornelius, 1993) for F_{GH1} tests.
[b] Significant (P<0.05) by the F_1 test.
[c] Significant (P<0.05) by the F_{GH1} test.

G1 had 10 sites with negative $SHMM_1$ primary effects (Figure 5) and 37 positive in its unconstrained solution; thus a constrained solution (Figure 6) was required. Similarly, G3 had 7 sites with negative primary effects (Figure 7) and 40 positive, also requiring a constrained solution (Figure 8). For G2, regression lines almost coincided and primary effects of sites were all of like sign (Figure 9). The number of sites with $\hat{\gamma}_{j1} = 0$ in constrained LS $SHMM_1$ solutions (not shown) for G1 and G3 was 2 and 1, respectively.

When the unconstrained solution has its crossover point located near the middle of the region that contains the data, the constrained solution will force a lot of variability into the residual that was previously captured in the primary

192

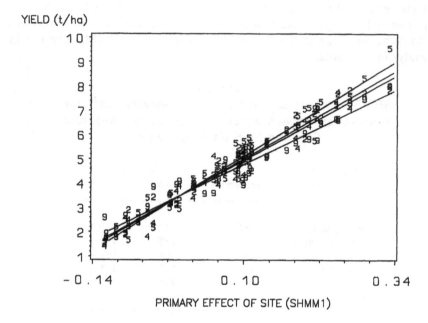

Figure 5. Unconstrained fit of SHMM$_1$ to Gl={2, 4, 5, 9} genotypes.

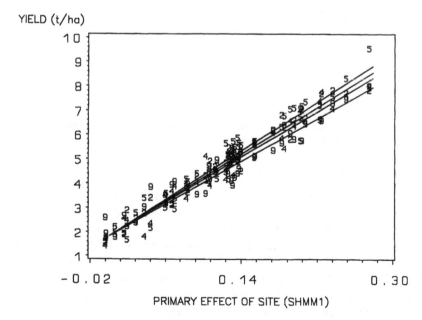

Figure 6. Constrained SVD fit of SHMM$_1$ to Gl={2, 4, 5, 9} genotypes.

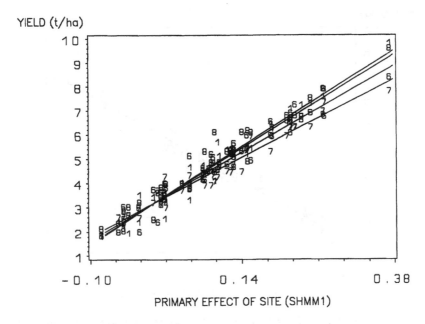

Figure 7. Unconstrained fit of SHMM₁ to G3={1, 6, 7, 8} genotypes.

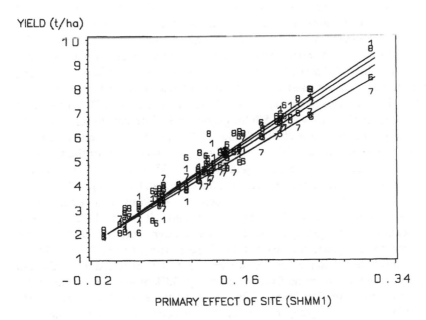

Figure 8. Constrained SVD fit of SHMM₁ to G3={1, 6, 7, 8} genotypes.

YIELD (t/ha)

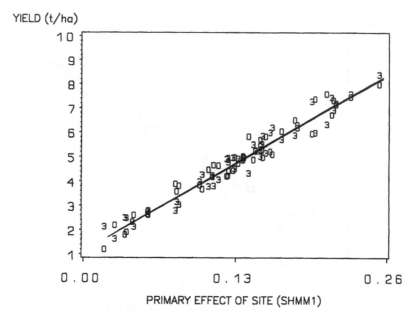

PRIMARY EFFECT OF SITE (SHMM1)

Figure 9. Unconstrained fit of SHMM$_1$ to G2={3, 10} genotypes. Genotype 10 is represented by zeros.

effects by the unconstrained solution. For clusters G1 and G3, the point of concurrence of the unconstrained solution was located to the lower left side of the data (Figures 5 and 7); therefore, only a small amount of variability existing in the unconstrained primary effects is re-allocated to the residual.

The number of observed genotype rank changes (COI) within and among final clusters was assessed by the joint t-test described in Cornelius et al. (1992). The percentage of total 2 × 2 interactions having COI pattern (not necessarily significant) was 71.5% for genotypes in different clusters and 2.2% for genotypes in the same cluster (Table 4). Clusters G1, G2, and G3 showed 34, 33, and 10 significant COIs, respectively. The magnitude of the COIs for genotypes in different subsets ranged from -12165 to 12147 kg ha^{-1} with a mean absolute value of 3468 kg ha^{-1}. On the other hand, the range of the COIs for genotypes in the same subset was much less, -7185 to 10268 kg ha^{-1}, with a mean absolute value of 1990 kg ha^{-1}. Among all of the 14053 possible interactions among genotypes in the same cluster, only 77 (0.55%) were significant COIs; since the test has 5% Type I error rate per 2 × 2 subtable tested, this is less than the number of significant COIs expected under the null hypothesis that there are no COIs. Conversely, 38% of the among-cluster 2 × 2 interactions were significant COIs. These results showed that the SHMM clustering effectively grouped genotypes with negligible COI into the same cluster.

Table 4

All possible 2 x 2 interactions (I), number of crossover interactions (COI) and significant COI obtained by the joint t-test (P<0.05) for genotypes in different clusters and in the same cluster.

	All Interactions	All COI	Significant COI[a]
Among clusters	34592	24726	13026
Within clusters	14053	310	77
G1	6486	141	34
G2	1081	144	33
G3	6486	25	10

[a] The joint t-test is the test for COI which controls the Type I error rate per 2 × 2 subtable tested (Cornelius et al., 1992).

The results further suggest that the choice of SVD or LS methods for obtaining constrained $SHMM_1$ solutions is not of great practical importance. $RSS(SHMM_1)_{SVD}$ was usually only slightly larger than $RSS(SHMM_1)_{LS}$ and the dendrogram obtained and final clusters judged satisfactory were identical. However, examples could occur where a cluster judged unacceptable for the non-COI SVD solution might be acceptable if the non-COI LS solution were used. The possibility of the two methods giving different dendrograms in specific examples is also not ruled out. Therefore, we suggest that constrained distance matrices and dendrograms based on both SVD and LS solutions be obtained and results examined. If dendrograms are the same (as in the previous example), either solution may be used. On the other hand, if dendrograms are different, then both should be subjected to dichotomous splitting until final groups are obtained. To assess whether SVD or LS constrained solutions produced the best result, the joint t-test (Cornelius et al., 1992; Crossa et al., 1993) may be used to evaluate how many COIs are found in the final groups produced by each method.

VI. OTHER SHMM-BASED METHODS

Crossa and Cornelius (1993) and Crossa et al. (1995) proposed SHMM-based "building block" and "fusion" methods for clustering genotypes or sites. These schemes recompute distance measures at each clustering step and, in so doing, make use of "trial residual mean squares" (TRMS) and "optimal residual mean squares" (ORMS). For any cluster containing g genotypes and e sites, $ORMS = RSS(SHMM_1)/(ge-g-e)$ and $RSS(SHMM_1)$ is from the optimal non-COI $SHMM_1$ solution (either type of constrained solution, if such is needed, may be used). A TRMS is by the same formula as ORMS except that the RSS results from a non-optimal fit. In particular, a TRMS results when an unclustered item (genotype or site depending on which is being clustered) is put into a previously formed cluster, but $\hat{\beta}$ is held at the optimal non-COI solution for the previous cluster. The rationale for a TRMS is that TRMS $\geq$ ORMS is easier to compute than ORMS and should give a close approximation to the ORMS for any item which is a good candidate for addition to the cluster.

The building block method begins by finding the best pair to initiate a cluster. Then it uses TRMS to find the best (or at least a "good") choice of an item to add to the cluster. SHMM analysis of the enlarged cluster is then done to find its optimal $\hat{\beta}$ for a non-COI $SHMM_1$ and to evaluate lack of fit. The process is repeated using a new set of TRMS values to identify another item to add to the cluster, then analysis of that new cluster, etc. The cluster is built up in this fashion until it cannot be adequately modeled by a non-COI $SHMM_1$. Once that happens, a new cluster is initiated using a different initial pair. Building of the second cluster can be restricted to items not contained in the first cluster, or items can be allowed to enter more than one cluster. The building block method holds promise as a sort of multiple comparison procedure for "separability", but its usage for that purpose awaits further research.

The fusion method uses the building block strategy, but allows new clusters to be initiated before previously formed clusters are completed and also provides for fusing clusters. It does not allow an item to belong to more than one cluster. TRMS for adding items to previously formed clusters will be used, but it is also necessary to define a TRMS for fusing (i.e., merging) two clusters. Suppose Clusters 1 and 2 are candidates for fusion with one another. Let $\hat{\beta}_1$ and $\hat{\beta}_2$ be the optimal β values for Clusters 1 and 2, respectively. Cluster 1 is said to be forced into Cluster 2 if $\hat{\beta}$ for the fused pair of clusters is put equal to $\hat{\beta}_2$. Denote the resulting TRMS as TRMS1.2. Analogously define TRMS2.1 for forcing Cluster 2 into Cluster 1. Also compute TRMS1#2, defined as the TRMS when $\hat{\beta}$ for the fused clusters is the weighted average $(w_1\hat{\beta}_1 + w_2\hat{\beta}_2)/(w_1 + w_2)$ where w_1=1/TRMS2.1 and w_2=1/TRMS1.2. Then define the TRMS for fusing the two clusters as min(TRMS1.2, TRMS2.1, TRMS1#2).

SHMM fusion begins by forming the best cluster of two items, but thereafter it uses TRMS to identify the best action among (1) possible additions of an item to a cluster already formed and (2) possible fusions of pairs of clusters. Then the ORMS is computed for the action so identified. If this ORMS is less than the ORMS for forming a new cluster from two as yet unclustered items, the action identified by the smallest TRMS is taken. Otherwise, the action taken is to initiate a new cluster with the best available pair of unclustered items. Note that the ORMS value for the latter action is always available, because these will have been computed and stored for all possible pairs in order to find the best pair to initiate the first cluster. The process is terminated whenever the best action found fails to produce a cluster which can be adequately modeled by a non-COI $SHMM_1$. Thus, the end result of each SHMM fusion "cycle" is one of four possible actions: (1) initiation of a new cluster, (2) addition of an unclustered item to an existing cluster, (3) fusion of two clusters or (4) termination of the process.

Further details concerning building block and fusion methods, including flow charts, are given by Crossa et al. (1995). In their example, SHMM fusion of environments (combinations of years and irrigation levels) resulted in the same clusters as obtained by SHMM clustering by complete linkage. But such may not always happen and further research is needed to examine which method will tend to cluster more sites or genotypes and/or to leave a lower proportion of significant COI in the final groups.

VII. SOFTWARE

Software to accomplish SHMM clustering of either genotypes or environments is not available all in one program. A SAS® program for obtaining the dendrogram for clustering environments using constrained LS $SHMM_1$ solutions is given by Crossa et al. (1993). A modification of that program allowing the user to choose either constrained LS or constrained SVD solutions and a SAS® program for obtaining the dendrogram for clustering genotypes using only constrained SVD solutions can be obtained by sending an e-mail address or a diskette along with their request to the second author.

Computation of the cluster analysis by the SHMM fusion method uses several programs. SAS® programs for (1) finding the best unclustered individual (smallest TRMS) to force into a previously formed cluster using that cluster's estimated shift value and (2) computing the TRMS for fusing two clusters can also be obtained from the second author if an e-mail address or a diskette is provided.

SHMM analysis of the clusters is done by the FORTRAN program EIGAOV. Readers interested in becoming authorized users of EIGAOV should address inquiries to the second author.

VIII. ACKNOWLEDGMENT

The authors wish to express their appreciation to Jane Johnson for her assistance in formatting this chapter.

REFERENCES

Azzalini, A., and D.R. Cox. 1984. Two new tests associated with the analysis of variance. *Journal of the Royal Statistical Society Series B.* 46:335-343.

Baker, R.J. 1988. Tests for crossover genotype-environmental interactions. *Canadian Journal of Plant Science.* 68:405-410.

Baker, R.J. 1990. Crossover genotype-environmental interaction in spring wheat. p.42-51. In: M.S. Kang (Ed.) *Genotype-by-environment Interaction and Plant Breeding.* Dept. Agronomy, Louisiana Agric. Expt. Stn., LSU Agricultural Center, Baton Rouge, Louisiana.

Bradu, D., and K.R. Gabriel. 1978. The biplot as a diagnostic tool for models of two-way tables. *Technometrics* 20:47-68.

Cornelius, P.L., M. Seyedsadr, and J. Crossa. 1992. Using the shifted multiplicative model to search for "separability" in crop cultivar trials. *Theoretical and Applied Genetics* 84:161-172.

Cornelius, P.L. 1993. Statistical tests and retention of terms in the additive main effects and multiplicative interaction model for cultivar trials. *Crop Science* 33:1186-1193.

Cornelius, P.L., D.A. Van Sanford and M.S. Seyedsadr. 1993. Clustering cultivars into groups without rank-change interaction. *Crop Science* 33:1193-1200.

Cornelius, P.L., J. Crossa and M.S. Seyedsadr. 1995. Statistical tests and estimators of multiplicative models for genotype-by-environment interaction. Chapter 8, this volume.

Crossa, J., P.L. Cornelius, M. Seyedsadr, and P. Byrne. 1993. A shifted multiplicative model cluster analysis for grouping environments without genotypic rank change. *Theoretical and Applied Genetics* 85:577-586.

Crossa, J., and P.L. Cornelius. 1993. Recent developments in multiplicative models for cultivar trials. In: D.R. Buxton et al. (Eds.) *International Crop Science I.* Crop Science Society of America, Madison, Wisconsin. (p. 571-577).

Crossa, J., P.L. Cornelius, K. Sayre and J.I. Ortiz-Monasterio R.A. 1995. Shifted multiplicative model fusion method for grouping environments without cultivar rank change. *Crop Science.* 35:54-62.

Gail, M., and R. Simon. 1985. Testing for qualitative interactions between treatment effects and patient subsets. *Biometrics* 41:361-372.

Gregorius, H.R., and G. Namkoong. 1986. Joint analysis of genotypic and environmental effects. *Theoretical and Applied Genetics* 72:413-422.

Mandel, J. 1961. Nonadditivity in two-way analysis of variance. *Journal of the American Statistical Association* 56:878-88.

Seyedsadr, M. and P.L. Cornelius. 1992. Shifted multiplicative models for nonadditive two-way tables. *Communications in Statistics Simulation and Compututation* 21:807-832.

Chapter 8

STATISTICAL TESTS AND ESTIMATORS OF MULTIPLICATIVE MODELS FOR GENOTYPE-BY-ENVIRONMENT INTERACTION[1]

P. L. Cornelius,[2] J. Crossa,[3] and M. S. Seyedsadr[4]

I. INTRODUCTION

Applications of multiplicative models to the analysis of agricultural experiments involving different cultivars of a crop species go back at least as far as Fisher and MacKenzie (1923) who used a multiplicative model to analyze a factorial arrangement of 12 potato (*Solanum tuberosum* L.) varieties and six manurial treatments. Finlay and Wilkinson (1963) and Eberhart and Russell (1966) introduced "stability analysis" that uses a model in which the data on each cultivar is regressed on an environmental productivity index, estimated as the main effect of the environment, thus introducing a term into the model which is multiplicative, i.e., the product of a cultivar regression coefficient times an environment parameter, both of which are estimated from the data. With the exception of estimation of variance components due to deviations of the cultivar yields from their regressions, analysis using this model had previously been developed by Mandel (1961) to provide a more general test for interaction in unreplicated two-way tables than Tukey's (1949) test of nonadditivity.

Gollob (1968) and Mandel (1969, 1971) introduced a factor analytic (FANOVA) model for studying interaction in complete two-way layouts. Mandel (1971) perceived that there were also potential applications to models that would omit the main effects of one or both factors, allowing such effects to be absorbed into multiplicative terms. Gabriel (1978) showed the connection between least squares fitting of a multiplicative model and singular value decomposition (SVD) of a matrix. The stability analyses of Eberhart and Russell (1966) and Finlay and Wilkinson (1963) do not use the least squares solution of Gabriel and, consequently, do not give a least squares solution for simultaneously estimating the environmental indices and the coefficients for the cultivar regressions on those indices.

[1] The investigation reported in this paper (No. 93-3-51) is in connection with a project of the Kentucky Agricultural Experiment Station and is published with approval of the Director.
[2] Professor of Agronomy and Statistics, Department of Agronomy, University of Kentucky, Lexington, KY 40546-0091, U.S.A.
[3] Senior Biometrician, Biometrics Unit, International Maize and Wheat Improvement Center (CIMMYT), Apdo. Postal 6-641, 06600 Mexico. D.F., Mexico
[4] Senior Research Biostatistician, Amgen, Inc., Biostatistics 17-1-A-393, 1840 DeHavilland Drive, Thousand Oaks, CA 91320-1789, U.S.A.

Zobel et al. (1988), Gauch(1988) and Gauch and Zobel (1988) renamed Gollob's FANOVA model as the Additive Main Effects and Multiplicative Interaction (AMMI) model and developed application to crop cultivar trials. In so doing, they found that Gollob's test procedures applied to the entire data set, if used as criteria for model choice, would generally retain more multiplicative interaction terms in the model than could be shown by cross validation to improve predictive accuracy. Consequently, they recommended that cross validation be used to determine the number of terms to retain. Cross validation was by a data-splitting method that used N_m (say) replicates of data for modeling and the remaining $N_v = n - N_m$ replicates for validation. Cornelius (1993) concluded that, if such cross validation is done as the criterion for model choice, then N_m should always be $n-1$.

Cross validation, however, does not necessarily lead to the best model obtainable from the complete data set. Cornelius (1993) showed that the apparent propensity of the Gollob test to overfit the model could be at least partially explained by liberality of the Gollob test used and suggested that choosing a model on the basis of a test procedure which does control Type I error rates would result in a model that, if not actually better, at least would not be much worse than could be expected of a model chosen by cross validation.

Note that if all multiplicative terms are retained, this produces predicted values identical to the raw cell means. An implication of more reliable prediction by parsimonious multiplicative models, strongly argued by Gauch (1988) and Gauch and Zobel (1988) and with which we agree, is that, even in the presence of interaction, from the data one can extract estimates of the realized performance levels of the cultivars in the environments, where tested, that are better estimates than the raw cell means. However, it would be quite understandable if plant breeders and other practitioners well schooled in conventional data analysis techniques (e.g., analysis of variance and mean separation), having been taught that the cell means are the "best unbiased estimates", were to view this claim with astonishment and skepticism.

In this chapter, we review, update and unify recent developments in multiplicative model methodology. We first summarize results on hypothesis testing. Then we discuss our work on "shrinkage" estimates of multiplicative models, which we believe can eliminate the need for either cross validation or hypothesis testing as criteria for choosing the number of multiplicative terms. We will use the term "site" as a synonym of "environment." We suppose that there are g cultivars evaluated at e sites. Let $\bar{y}_{ij.}$ and $\bar{\varepsilon}_{ij.}$ denote the means of the raw data and unobservable residual errors, respectively, on the ith cultivar at the jth site. We suppose that the $\bar{\varepsilon}_{ij.}$ are $NID(0, \sigma^2/n)$. The following are five potentially useful multiplicative model forms:

1. Completely Multiplicative Model (COMM):

$$\bar{y}_{ij.} = \sum_k^t \lambda_k \alpha_{ik} \gamma_{jk} + \bar{\varepsilon}_{ij.},$$

2. Shifted Multiplicative Model (SHMM):

$$\bar{y}_{ij.} = \beta + \sum_k^t \lambda_k \alpha_{ik} \gamma_{jk} + \bar{\varepsilon}_{ij.},$$

3. Genotypes (cultivars) Regression Model (GREG):

$$\bar{y}_{ij.} = \mu_i + \sum_k^t \lambda_k \alpha_{ik} \gamma_{jk} + \bar{\varepsilon}_{ij.},$$

4. Sites (environments) Regression Model (SREG):

$$\bar{y}_{ij.} = \mu_j + \sum_k^t \lambda_k \alpha_{ik} \gamma_{jk} + \bar{\varepsilon}_{ij.},$$

5. Additive Main Effects and Multiplicative Interaction Model (AMMI):

$$\bar{y}_{ij.} = \mu + \tau_i + \delta_j + \sum_k^t \lambda_k \alpha_{ik} \gamma_{jk} + \bar{\varepsilon}_{ij.}.$$

In all of these models, the λ_k scale parameters are ordered, i.e., $\lambda_1 \geq \lambda_2 \geq ... \geq \lambda_t \geq 0$, and the α_{ik} and γ_{jk} satisfy normalization constraints, $\sum_i \alpha_{ik}^2 = \sum_j \gamma_{jk}^2 = 1$, and orthogonality constraints, $\sum_i \alpha_{ik}\alpha_{im} = \sum_j \gamma_{jk}\gamma_{jm} = 0$ for $k \neq m$. For the moment, the number t of multiplicative terms will be arbitrary.

COMM is the model used by Fisher and MacKenzie (1923). Cornelius et al. (1992, 1993), Crossa et al. (1993), Crossa and Cornelius (1993) and Crossa et al. (1995a, 1995b) have described use of SHMM for identifying clusters of cultivars or sites in which cultivar rank changes are statistically insignificant. SHMM also has potential use as a model for stability analysis, but such application has not been extensively studied. The "Genotypes Regression Model" is so named because the model expresses the performance of each genotype as its multiple regression on t unknown functions (γ_{jk}) of unidentified characteristics of the sites. Thus, GREG is a reparameterization of the stability analysis model of Finlay and Wilkinson (1963) and Eberhart and Russell (1966), with the first multiplicative term perceived as the cultivar regressions, with coefficients α_{i1} on environmental indices γ_{j1} (the scale parameter λ_1 can be absorbed into either the regression coefficients α_{i1} or the regressors γ_{j1}, or partly into each), and the deviations further modeled as multiplicative components (provided $t > 1$). We use the name "Genotypes Regression" in preference to the more generic "Rows Regression" of Bradu and Gabriel (1978). Similarly, the name "Sites Regression" is in preference to "Columns Regression."

II. LEAST SQUARES ESTIMATES

There is extensive statistical literature on analysis using multiplicative models for nonadditive two-way data, much of which is cited by Seyedsadr and Cornelius (1992). Least squares estimation of the multiplicative terms are through SVD of the two-way array (matrix Z) of data obtained by subtracting the least squares estimates of the model's additive effects from the cell means. Computing software to accomplish SVDs is readily available. SVD expresses a matrix as a sum of orthogonal components, e.g., $z_{ij} = \sum_k^r \hat{\lambda}_k \hat{\alpha}_{ik} \hat{\gamma}_{jk}$, where r is the rank of matrix Z, and the $\hat{\alpha}_{ik}$ and $\hat{\gamma}_{jk}$ satisfy the same normalization and orthogonality constraints as we described for the α_{ik} and γ_{jk} parameters in the multiplicative models. Gabriel (1978) showed that, if the components of the SVD are arranged in decreasing order with respect to the "singular values" $\hat{\lambda}_k$, retaining only the first component gives a "rank one" least squares approximation to Z, retaining two components gives a "rank two" least squares approximation, etc. Thus, for any of the multiplicative model forms, the least squares estimates of the parameters in the multiplicative terms will be given by the first t components of the SVD of the Z matrix appropriate for that model form. For the five model forms we consider, the following are the least squares estimates of the additive effects, and formulas for elements of the appropriate Z matrix.

COMM: $z_{ij} = \bar{y}_{ij.}$,

SHMM: $\hat{\beta} = \bar{y}_{...} - \sum_k^t \hat{\lambda}_k \bar{\hat{\alpha}}_k \bar{\hat{\gamma}}_k$, $z_{ij} = \bar{y}_{ij.} - \hat{\beta}$,

where $\bar{\hat{\alpha}}_k = g^{-1} \sum_i \hat{\alpha}_{ik}$ and $\bar{\hat{\gamma}}_k = e^{-1} \sum_j \hat{\gamma}_{jk}$,

GREG: $\hat{\mu}_i = \bar{y}_{i..}$, $z_{ij} = \bar{y}_{ij.} - \bar{y}_{i..}$,

SREG: $\hat{\mu}_j = \bar{y}_{.j.}$, $z_{ij} = \bar{y}_{ij.} - \bar{y}_{.j.}$,

AMMI: $\hat{\mu} = \bar{y}_{...}$, $\hat{\tau}_i = \bar{y}_{i..} - \bar{y}_{...}$, $\hat{\delta}_j = \bar{y}_{.j.} - \bar{y}_{...}$, $z_{ij} = \bar{y}_{ij.} - \bar{y}_{i..} - \bar{y}_{.j.} + \bar{y}_{...}$.

Given a program for computing SVDs, computation of least squares estimates in COMM, GREG, SREG or AMMI is very straightforward. SHMM is more complicated; the matrix to be subjected to SVD depends on the SVD to be obtained, and an iterative algorithm is required. Moreover, changing the number of multiplicative terms in SHMM will change the least squares estimate of the shift ($\hat{\beta}$), resulting in a different Z matrix to be decomposed with consequent changes in all parameter estimates. Computational algorithms for least squares fitting of SHMM have been described by Seyedsadr and Cornelius (1992).

III. TESTING SIGNIFICANCE OF THE MULTIPLICATIVE TERMS

The sequential sum of squares SS_k due to the kth successively fitted multiplicative term in COMM, GREG, SREG and AMMI is $SS_k = n\hat{\lambda}_k^2$. For SHMM, the sequential sum of squares due to the kth multiplicative term is

$SS_k = n\left[\sum_{m=k}^{p} \hat{\lambda}^2_{m(k-1)} - \sum_{m=k+1}^{p} \hat{\lambda}^2_{m(k)}\right]$, where $p = \min(g,e)$ and $\hat{\lambda}_{m(k)}$ is the mth largest singular value of Z when $t = k$. The squared singular value, $\hat{\lambda}^2_k$, can also be obtained as the kth largest eigenvalue of $Z'Z$ (or of ZZ'). For COMM, GREG, SREG and AMMI, the eigenvalues (and, thus, also the sequential sums of squares) are distributed as the eigenvalues of a p-variate Wishart matrix with q df (Johnson and Graybill, 1972), where p is the smaller, and q the larger, of Mandel's (1971) "row degrees of freedom" and "column degrees of freedom" in matrix Z. This does not hold for sequential sums of squares in SHMM analysis, but, for convenience in the sequel, we will define p and q for SHMM to be the same as for COMM. However, the total df due to multiplicative terms in a saturated SHMM is not pq, but rather $pq-1$. Values of p and q are:

COMM: $p = \min(g,e)$, $q = \max(g,e)$;
SHMM: $p = \min(g,e)$, $q = \max(g,e)$;
GREG: $p = \min(g,e-1)$, $q = \max(g,e-1)$;
SREG: $p = \min(g-1,e)$, $q = \max(g-1,e)$;
AMMI: $p = \min(g-1,e-1)$, $q = \max(g-1,e-1)$.

When an independent error mean square ("pooled error"), s^2, is available, an appropriate statistic for testing the hypothesis H_0: $\lambda_k = 0$ is the ratio SS_k/s^2 or some function of this ratio. For testing the first term in COMM, GREG, SREG or AMMI with $2 \leq p \leq 5$ and $2 \leq q \leq 6$, the ratio SS_1/s^2 can be compared with Boik's (1986) tables of percentiles of the "studentized maximum root", but, to our knowledge, tables of critical values are not available for $p > 5$ and/or $q > 6$ which commonly occur in crop cultivar trials. Moreover, Boik's tables are not appropriate for SHMM analysis and exact P-values are not easily computed in any case. Consequently, it is generally necessary to use approximate procedures. We will review several such procedures, all but one of which is constructed by multiplying (or dividing) SS_k/s^2 by a constant intended to make the resulting test statistic approximately distributed as an F-statistic.

A. GOLLOB'S F-TEST

Gollob's (1968) test assigns degrees of freedom to SS_k by counting the number of independent parameters in the kth multiplicative term. Then an F-test is computed as in analysis of variance procedures for linear models. An analysis of variance F-test assumes that, under the null hypothesis, the numerator and denominator of the F-statistic are each distributed (independently) as a multiple of a chi-square random variable. The Gollob F-test is invalid because the eigenvalues do not have chi-square distributions (Johnson and Graybill, 1972). Cornelius (1993) showed by computer simulation that the Gollob F-tests are extremely liberal and can easily result in two or more too many terms judged significant. Type I error rate was 60%

for the first term in a completely null case with $p = 8$, $q = 19$. Williams and Wood (1993) reached a similar conclusion. The liberality does decrease with successive terms and the test may become conservative for one or more of the late terms in a saturated model. [Cornelius (1993) found power of the 0.05-level Gollob test was only 2% for the eighth term when $p = 8$, $q = 19$ and all $\lambda_k = 5\sigma / \sqrt{n}$; the Type I error rate, if the eighth term was actually null, would surely be still smaller.] Because of the problems in controlling the Type I error rate, we do not recommend the Gollob test.

B. OTHER F APPROXIMATIONS

Cornelius et al. (1992) defined test statistics denoted as F_1, F_2, F_{GH1}, and F_{GH2}. We will characterize the approximations used to obtain F_1 and F_{GH1} as "F_1-type approximations" and those that obtain F_2 and F_{GH2} as "F_2-type approximations." Cornelius (1993) developed Simulation tests and Iterated Simulation tests that also use F_2-type approximations.

1. F_2-Type Approximations

We first describe the F_2-type approximation, which is computationally simpler than the F_1-type. Let f denote the df of the pooled error mean square s^2. Further let $E(\cdot)$ and $V(\cdot)$ denote the expectation and variance of whatever variable, or function of variables, is given as the argument. Now define the following quantities:

$$u_{1k} = E(SS_k/\sigma^2), \text{ given } \lambda_k = 0,$$

$$u_{2k}^2 = V(SS_k/\sigma^2), \text{ given } \lambda_k = 0,$$

$$h_{2k} = 2u_{1k}^2/u_{2k}^2,$$

$$F_{2k} = SS_k/u_{1k}s^2, \text{ df} = (h_{2k}, f).$$

F_{2k} is the F-statistic for an F_2-type test of the kth multiplicative term. The logic for this approximation, first developed by Johnson (1976) for $k = 1$, is that, under the null hypothesis, to a two-moment approximation, SS_k/u_{1k} is approximately distributed as σ^2 times a chi-square random variable with h_{2k} df divided by the df h_{2k}. In effect, what we have are two different values used for the df of SS_k, viz., u_{1k} for the purpose of computing a mean square, and h_{2k} for purposes of referral to tables of critical values or for computing P-values. Different F_2-type statistics differ only with respect to the values used for u_{1k} and u_{2k}. The Gollob F-test is, in fact, an F_2-type approximation wherein u_{1k} is obtained by counting number of independent parameters added to the model and u_{2k}^2 is put equal to $2u_{1k}$ (thus giving $h_{2k} = u_{1k}$). Unfortunately, correct values for u_{1k} and u_{2k} are not so easily obtained.

a. The F_2 Test

The F_2 test (Cornelius et al., 1992, 1993) is a test under the complete null hypothesis (all $\lambda_k = 0$). This test effectively controls Type I error rates for the first term, but simulation results for SHMM analysis (unpublished) have shown that it is liberal for subsequent terms if the first term is nonnull. This undoubtedly holds for the other model forms as well. Like the Gollob tests, the F_2 test can easily lead to two or more too many terms judged statistically significant, and may, in fact, be worse than the Gollob test in this regard. We continue to routinely compute it (or its F_1-type analog) because any terms that it finds nonsignificant can assuredly be considered negligible. Such a statement cannot be made for the Gollob test because the Gollob test can actually become conservative for one or more of the late multiplicative terms. Significant results by the F_2 test for terms beyond the first term are not to be trusted (if the Gollob test is used, even significance of the first term cannot be trusted). The failure to control Type I error rates for terms beyond the first is not the fault of the F_2-type approximation per se, but rather due to use of inaccurate values of u_{1k} and u_{2k} for $k > 1$. Better control of Type I error rates is achieved by F_{GH2} or Simulation tests described in the next two sections (or their F_1-type analogs described later).

b. The F_{GH2} Test

The F_{GH2} test (Cornelius et al., 1992; Cornelius, 1993) uses values of u_{1k} and u_{2k} appropriate under a partial null hypothesis, which assumes that $\lambda_k = 0$, but the true λ values for preceding terms are all very large (relative to error). The test is a synthesis of results of Mandel (1971) concerning properties of the sequential sums of squares (eigenvalues) under the complete null hypothesis, and theoretical results of Goodman and Haberman (1990) concerning asymptotic properties of the residuals as the true λ values for the terms fitted become large relative to the variance of a cell mean (see also Marasinghe, 1985; Schott, 1986). The kth term is tested by an F_2-type test, but with u_{1k} and u_{2k} taken as values appropriate for the first term in a problem with p and q reduced to $p-k+1$ and $q-k+1$. This test effectively controls the Type I error rate at or below the intended significance level, but has low power for detecting the latter terms in a series of terms with small, but nonnull, λ. This may, in fact, be an argument in its favor, because terms with small, albeit nonnull, λ values apparently tend to be so fettered with "noise" that their retention in a working model actually reduces predictive accuracy (Gauch, 1988; Gauch and Zobel, 1988). Cornelius (1993) presented simulation results suggesting that F_{GH2} tests at 0.05-level would often behave quite well as criteria for obtaining a parsimonious model for prediction or estimation purposes.

c. Values of u_{1k} and u_{2k} for F_2 and F_{GH2} Tests

If $p \leq 19$ and $q \leq 99$, approximate values of u_{1k} and u_{2k} can be obtained for F_2 tests of the first three multiplicative terms, or for F_{GH2} tests of all terms, in COMM, GREG, SREG or AMMI analysis, from tables given by Mandel (1971) or by functions obtained by Cornelius (1980) by regression analysis of Mandel's tables. For F_{GH2} tests, always enter the table or use the function for the first multiplicative term, but with the "row degrees of freedom" and "column degrees of freedom" (p and q) sequentially reduced by one for each successive term tested. For SHMM analysis, tables of u_{1k} and u_{2k} are given by Seyedsadr (1987) and functions of p and q which approximate those values for the first five terms under the complete null hypothesis are given by Seyedsadr and Cornelius (1991). Approximating functions for u_{1k} and u_{2k} under the partial null hypothesis, for use in computing F_{GH1} or F_{GH2} tests, adapted from Cornelius (1980) and, for SHMM, from Seyedsadr and Cornelius (1991), are given in the Appendix to this chapter. These functions are not valid for $p-k+1>19$ or for $q-k+1>99$.

d. Simulation And Iterated Simulation Tests

Cornelius (1993) developed a simulation scheme for approximating u_{1k} and u_{2k} under the partial null hypothesis, but relaxing the assumption that previous terms are "large." Generally, for COMM, GREG, SREG or AMMI, $\hat{\lambda}_k^2$ is an upward biased estimate of λ_k^2. A downward adjustment is made to remove the error (approximately) and square root taken to give an adjusted estimate of λ_k; specifically the adjusted value is $\tilde{\lambda}_k = (1 - 1/\ddot{F}_k)^{1/2}\hat{\lambda}_k$, where $\ddot{F}_k = n\hat{\lambda}_k^2/\ddot{u}_{1k}s^2$ and $\ddot{u}_{1k}$ is such that $E(n\hat{\lambda}_k^2) = \ddot{u}_{1k}\sigma^2 + n\lambda_k^2$. Note that, for terms with true $\lambda_k > 0$, $\ddot{F}_k$ is an F_2-type statistic computed with u_{1k} determined under the alternative hypothesis, not under the null hypothesis. If $\ddot{F}_k < 1$, put $\tilde{\lambda}_k = 0$. By simulation, determine u_{1k} and u_{2k} under the assumption $\lambda_m/\sigma = \tilde{\lambda}_m/s$ for $m < k$, and $\lambda_m/\sigma = 0$ for $m \geq k$. This gives u_{1k} and u_{2k} values which can then be used to compute an F_2-type statistic for testing the hypothesis H_0: $\lambda_k = 0$. At the conclusion of the process, a set of simulated $\ddot{u}_{1k}$ values may be obtained under the assumption that $\lambda_k/\sigma = \tilde{\lambda}_k/s$ for all k. Iterated Simulation tests are done by repeating the whole process using these simulated $\ddot{u}_{1k}$ values to obtain new $\tilde{\lambda}_k$ values. To initiate the scheme, one has to have a set of $\ddot{u}_{1k}$ with which to start. For this, we use Gollob's df. [Theorems of Goodman and Haberman (1990) imply that Gollob's df should be a good approximation to $\ddot{u}_{1k}$ for terms that are large relative to error.] If the adjusted $\tilde{\lambda}_k$ are not sequentially decreasing in magnitude, pool as many adjacent $\hat{\lambda}_k^2$ values as necessary to obtain a nonincreasing sequence, and compute a pooled F statistic and a common $\tilde{\lambda}$ value to use for all terms in the pooled set.

Cornelius (1993) describes how to save computing time by exploiting the properties that the expected values of u_{1k}, u_{2k} and $\ddot{u}_{1k}$ do not depend on the true values of the α_{ik} and γ_{jk} parameters (except that they must satisfy the orthonormality constraints) and that, for each of the model forms AMMI, GREG and SREG, sampling properties of the singular values of the $g \times e$ matrix Z are the same as for the singular values of a $p \times q$ matrix of data which has the same true singular values, but from which no additive effects have been removed (thus having no constraints on its rows or columns). This allows the simulation to be done as follows. First, generate a $p \times q$ matrix R of random normal noise. Obtain its SVD to generate a simulated result under the complete null hypothesis. Then add $\tilde{\lambda}_1$ to R_{11} and do the SVD again to obtain a result under the partial null hypothesis H_0: $\lambda_2=0$, given $\lambda_1 = \tilde{\lambda}_1$. Then further addition of $\tilde{\lambda}_2$ to R_{22} followed by SVD gives a result under H_0: $\lambda_3=0$, given $(\lambda_1,\lambda_2)=(\tilde{\lambda}_1,\tilde{\lambda}_2)$. Continue this process, at each step sequentially adding the next $\tilde{\lambda}_k$ to R_{kk}, until this has been done for all nonnull $\tilde{\lambda}_k$.

The Simulation and Iterated Simulation tests have greater power than F_{GH2} while still giving quite satisfactory control of Type I error rates. Simulation tests have not yet been developed for SHMM analysis.

2. F_1-Type Approximations

F_1-type approximations also depend only on the values of SS_k, s^2, f, n, u_{1k} and u_{2k}. Thus, there exists an F_1-type analog of any F_2-type test. Compute

$$v_{1k} = u_{2k}^2 + u_{1k}^2 + (f-4)u_{1k},$$

$$v_{2k} = (f-2)u_{2k}^2 + 2u_{1k}^2,$$

$$g_k = 2+2(f-2)v_{1k}/v_{2k},$$

$$h_{1k} = 2v_{1k}u_{1k}/v_{2k},$$

$$F_{1k} = g_k SS_k / h_{1k} fs^2, \text{ df} = (h_{1k}, g_k).$$

The theoretical basis for this test is that, to a two-moment approximation, $1+SS_k/fs^2$ is distributed as the reciprocal of a Beta random variable with Beta distribution parameters $g_k/2$ and $h_{1k}/2$. The formula for F_{1k} and its df result from inverting the transformation by which a central F-statistic is converted into a Beta random variable (Hogg and Craig, 1978). P-values computed for F_2-type approximations and for their F_1-type analogs are often astonishingly similar, and we see little need to compute both in practice.

Indeed, we would not mention the F_1-type statistics here were it not for the fact that the first author's multiplicative models FORTRAN program EIGAOV does not compute F_2 and F_{GH2} tests, but, instead, computes their F_1-type analogs, F_1 and F_{GH1}, respectively.

3. The F_R Test

The F_R test of Cornelius et al. (1992) provides a simpler test than F_{GH1}, F_{GH2} or the Simulation tests for addressing whether the kth term should be retained in a working model. F_R is constructed as a straightforward F-test of the residual among-cell variation after fitting $k-1$ terms. By results of Goodman and Haberman (1990), the residual sum of squares (on an observation basis) after fitting $k-1$ terms, RSS_{k-1}, is approximately distributed as σ^2 times chi-square with $(p-k+1)(q-k+1)$ df if the terms fitted are sufficiently large. (This result has not been formally proved for SHMM analysis, but the df apparently should be reduced by one in such a case.) Put $F_R = RMS_{k-1}/s^2$, where RMS_{k-1} is the mean square $RSS_{k-1}/[(p-k+1)(q-k+1)]$. If the first $k-1$ terms are large relative to σ^2/n, and the kth term is null, F_R is approximately distributed as central F with $(p-k+1)(q-k+1)$ and f df.

The value $(p-k+1)(q-k+1)$ for the residual df is the same as obtained by allocating Gollob's df to the $k-1$ terms fitted. Note, however, that one is not using the Gollob tests as test criteria, but is assessing the need for each additional term on the basis of significance of residual variation after the previous terms are fitted. Thus, the F_R model building strategy is to conclude that the model requires at least one multiplicative term if the residuals after fitting the additive effects are significant (note that for AMMI this is the F-test for the interaction). Conclude that at least two terms are needed if the residuals after fitting the first term are significant, at least three terms are needed if the residuals after fitting two terms are significant, etc., finally truncating the string of terms retained in the model when residuals are no longer found to be significant.

We have observed that P-values for the test of residuals after fitting $k-1$ terms are usually, but not always, slightly larger (less significant) than for the F_{GH1} and F_{GH2} tests of the kth term. Simulation results of Piepho (1995) for AMMI analyses with $(p,q,n) = (8,19,4)$ indicate that, under the normality and homogeneous variance assumptions, Type I error rates for F_R are very similar to Type I error rates for F_{GH1} and F_{GH2}, with F_R generally having less power for detecting the last nonnull term, but often providing a more powerful lack of fit test for the $(k-1)$-term model when the true number of nonnull terms exceeds k. Piepho (1995) also found F_R to generally be more robust under non-normality or heteroscedasticity, particularly so in cases with large site to site differences among the within-site error variances.

Table 1
**Raw cell means (kg ha^{-1}) for the CIMMYT EVT16B maize yield trial[a]
with cultivars and sites arranged in rank order with respect to their
overall means.**

				Cultivar						
Site	8	2	1	7	3	9	4	5	6	Mean
19	2634	2282	1632	3211	3059	2735	2233	3073	3011	2652
10	3353	2972	3100	2889	2785	2774	2843	2688	3024	2936
7	3088	2515	2832	3537	3529	3061	2998	3556	3949	3229
1	4529	3426	3622	3354	3446	3136	3720	3165	4116	3613
8	4484	5278	6011	4206	4731	3309	2516	2732	2983	4028
2	3393	3919	3728	4767	4082	4500	4539	4079	4878	4209
18	2940	4415	3344	5276	4295	5244	5618	4498	5333	4552
20	4806	4396	4587	4056	5018	4822	4988	5776	5088	4838
17	5243	4211	4415	3862	4749	4989	5161	5454	5807	4877
5	4476	5201	4380	4277	4178	5407	5672	5414	5591	4955
9	4001	4714	4647	4318	5448	5553	4864	5588	5603	4971
15	4321	4126	4601	4346	4537	4889	6331	6328	5961	5049
3	5248	4937	5554	5389	5117	3780	4543	6173	5205	5105
4	4442	4963	4566	4342	5136	5781	6030	5831	5980	5230
11	3713	4349	4433	5052	4526	6430	7117	5995	6150	5307
16	4551	5196	5010	5013	5455	5278	6351	6070	5730	5406
14	5522	5932	5849	5820	5886	6282	6439	6359	6380	6052
6	4986	6036	6437	6745	6459	5610	6678	6882	6916	6305
13	4117	5627	6721	5544	6294	6920	7332	7174	7262	6332
12	5816	7571	6873	7444	7727	8091	8385	8106	7637	7516
Mean	4283	4603	4617	4672	4823	4930	5218	5247	5330	4858

[a] Adapted from Cornelius et al. (1992) and used with permission of Springer-Verlag, Berlin.

IV. HYPOTHESIS TESTING EXAMPLE

The data shown in Table 1 are the raw cell means from the CIMMYT EVT16B maize (*Zea mays* L.) yield trial, which evaluated nine cultivars at 20 sites in randomized complete block designs with $n=4$ replications at each site. The value of s^2 is 602847. We have previously used these data in several published analyses (Cornelius et al., 1992; Crossa and Cornelius, 1993; Cornelius, 1993). Results of Gollob's F, F_2, F_{GH2}, Simulation (F_2-type, not iterated) and F_R tests for AMMI and GREG analyses are shown in Table 2. If the values of u_{1k} and u_{2k} were known precisely, the tests of the first term by F_2, F_{GH2} and Simulation tests would be identical, but here F_{GH2} used u_{1k} and u_{2k} obtained by the approximating functions of Cornelius (1980), but u_{1k} and u_{2k} were obtained directly by simulation for the F_2 tests. The tests of the first

Table 2
Statistical tests of multiplicative terms in AMMI and GREG analyses of the EVT16B data.[a]

$\hat{\lambda}_k$	df	Gollob F		F_2 test				F_{GH2} test				Simulation test				F_R test		
		F	P	u_{1k}	h_{2k}	F	P	u_{1k}	h_{2k}	F	P	u_{1k}	h_{2k}	F	P	df	F	P
							AMMI analysis											
5923	26	8.95	0.000	42.8	76.5	5.43	0.000	42.9	78.2	5.43	0.000	42.8	76.5	5.43	0.000	152	2.73	0.000
3070	24	2.61	0.001	31.9	82.2	1.96	0.000	38.8	68.3	1.61	0.002	38.6	68.5	1.62	0.002	126	1.44	0.004
2552	22	1.96	0.006	24.4	82.0	1.77	0.000	34.8	58.7	1.24	0.116	33.4	61.5	1.29	0.075	102	1.17	0.149
2320	20	1.79	0.020	18.6	73.7	1.92	0.000	30.6	49.3	1.17	0.213	27.3	56.0	1.31	0.074	80	0.95	0.612
1758	18	1.14	0.309	13.8	63.7	1.49	0.012	26.4	40.1	0.78	0.837	21.5	46.7	0.96	0.557	60	0.67	0.974
1313	16	0.72	0.779	9.8	47.3	1.16	0.219	22.0	31.0	0.52	0.986	15.0	41.8	0.76	0.863	42	0.46	0.999
793	14	0.30	0.994	6.6	31.0	0.63	0.941	17.4	21.9	0.24	1.000	9.7	28.1	0.43	0.996	26	0.31	1.000
757	12	0.32	0.986	3.8	17.4	1.00	0.454	12.0	12.0	0.32	0.986	5.3	16.3	0.71	0.783	12	0.32	0.986
							GREG analysis											
16115	27	63.82	0.000	45.2	87.3	38.15	0.000	45.3	84.7	38.06	0.000	45.2	87.3	38.15	0.000	171	12.02	0.000
5218	25	7.23	0.000	34.0	97.9	5.31	0.000	41.3	74.8	4.37	0.000	41.3	78.8	4.38	0.000	144	2.31	0.000
2973	23	2.55	0.000	26.5	91.2	2.21	0.000	37.3	65.2	1.57	0.005	36.9	67.6	1.59	0.003	119	1.28	0.039
2324	21	1.71	0.027	20.6	90.3	1.74	0.027	33.3	55.8	1.08	0.335	31.7	61.5	1.13	0.241	96	0.97	0.554
1828	19	1.17	0.281	15.8	71.1	1.41	0.021	29.2	46.6	0.76	0.877	25.0	57.3	0.89	0.708	75	0.77	0.921
1736	17	1.18	0.279	11.8	57.4	1.70	0.002	25.1	37.7	0.80	0.800	18.0	55.1	1.11	0.276	56	0.63	0.982
1093	15	0.53	0.925	8.4	43.3	0.95	0.568	20.8	29.0	0.38	0.999	12.7	42.1	0.62	0.970	39	0.39	1.000
792	13	0.32	0.989	5.5	28.2	0.76	0.811	16.2	20.3	0.26	1.000	8.2	27.6	0.51	0.984	24	0.31	0.999
705	11	0.30	0.988	3.1	14.4	1.08	0.377	11.0	11.0	0.30	0.986	4.5	14.1	0.73	0.741	11	0.30	0.986

[a] For the F_2, F_{GH2} and Simulation tests, u_{1k} is the df for computing the mean square; h_{2k} is the numerator df for referral to tables of critical values or computing the P-value. F_R test shown on each line is the test of residuals after fitting all previous terms. Denominator df for all of the F-tests is 480. The Gollob F, F_{GH2} and Simulation tests for the AMMI analysis are from Cornelius (1993).

term by F_2 and Simulation tests are identical because they were both derived from the same pseudo random number stream. One thousand simulations were done in obtaining the u_{1k} and u_{2k} values.

For AMMI, the results for Gollob's F, F_{GH2}, and Simulation tests have been previously reported (Cornelius, 1993). The Simulation tests give the most accurate approximation to the correct significance levels (P-values) for the multiplicative components. These clearly indicate only two terms to be statistically significant if we demand strict adherence to 0.05 significance level, but P-values for the third and fourth terms are small enough to make us suspicious that these terms contain nonnull effects that are too small to be detected with high power. Since small multiplicative components are known typically to be estimated too poorly to have useful predictive value, they are probably best omitted from the model if a truncated AMMI model is to be chosen as a working model for estimation or prediction. The F_{GH2} and F_R tests, which tend to be conservative for any term that is not preceded by a term that is very large, clearly show only two significant terms. If the highly liberal Gollob tests were used as a criterion, four terms would be retained, and the apparently still more liberal F_2 criterion would retain five. Even the liberal F_2 tests show the last three terms to be negligible.

For these data, GREG analysis seems to be more definitive than AMMI analysis (which may suggest that it is a better model for these data). The F_{GH2}, Simulation and F_R tests all show three terms to be significant and the remaining five clearly nonsignificant. The liberal Gollob tests found a fourth term significant, and the still more liberal F_2 found six. This analysis clearly suggests GREG with three multiplicative terms as a "good" parsimonious model.

V. USING "SHRINKAGE" TO OBTAIN BETTER ESTIMATES

We have recently proposed "shrinkage" estimation as a method for obtaining estimates of realized performance levels of cultivars in the environments where tested that are at least as good as truncated multiplicative models fitted by least squares (Cornelius et al., 1994; Cornelius and Crossa, 1995). Shrinkage estimates are particularly appealing because they can eliminate the need for either cross validation or hypothesis testing as a criterion for model choice.

Shrinkage estimators have been proposed as improvements over more conventional estimators in a variety of statistical estimation problems. Rather well known procedures that belong to the general class of shrinkage estimators are Ridge Regression (Hoerl and Kennard, 1976) and Best Linear Unbiased Predictors (BLUP) (Henderson, 1984; Harville, 1977), both of which are equivalent to Bayes estimates under normal "priors" of the sort proposed by Lindley and Smith (1972). BLUPs have been in use as selection indices in animal breeding work for many years. Thus, the idea of shrinkage

212

estimators is neither novel nor new, but we believe our work is the first attempt to use the ideas of shrinkage estimation to obtain better estimates of multiplicative models than are provided by methods currently being used.

A. SHRINKAGE ESTIMATORS ANALOGOUS TO BLUPS

For the moment, consider the two-way random effects model

$$\bar{y}_{ij.} = \mu + \tau_i + \delta_j + (\tau\delta)_{ij} + \bar{\varepsilon}_{ij.} \tag{1}$$

Under normality and independence assumptions, and a balanced data set, empirical BLUPs of the realized performance levels of the cultivars in the sites where tested are given by

$$\bar{y}_{...} + S_\tau (\bar{y}_{i..} - \bar{y}_{...}) + S_\delta (\bar{y}_{.j.} - \bar{y}_{...}) + S_{\tau\delta} (\bar{y}_{ij.} - \bar{y}_{i..} - \bar{y}_{.j.} + \bar{y}_{...}) \tag{2}$$

where the "shrinkage factors" $S_\tau=1-1/F_\tau$, $S_\delta=1-1/F_\delta$ and $S_{\tau\delta}=1-1/F_{\tau\delta}$, where F_τ , F_δ and $F_{\tau\delta}$ are F-statistics for tests of cultivar main effects, site main effects and interaction, respectively, against the pooled error mean square, provided that these F-ratios all exceed unity (Henderson, 1984; Harville, 1977; Lindley and Smith, 1972; Feinberg, 1972). The shrinkage factors are estimates of functions of variance components, e.g., S_τ is an estimate of $(n\sigma_{\tau\delta}^2 + ne\sigma_\tau^2)/(\sigma^2 + n\sigma_{\tau\delta}^2 + ne\sigma_\tau^2)$ and $S_{\tau\delta}$ estimates $n\sigma_{\tau\delta}^2/(\sigma^2 + n\sigma_{\tau\delta}^2)$.

The assumption that the interactions in model (1) are $NID(0,\sigma_{\tau\delta}^2)$ implies that the interactions contain no transferable information, i.e., the interaction in any one cell gives no information concerning the interaction likely to occur in another cell (except that it will contribute to estimation of the unknown variance $\sigma_{\tau\delta}^2$). We submit that such is often, perhaps even usually, an incorrect assumption. In devising shrinkage estimators of multiplicative models, we sought an estimation method that would exploit the pattern recognition aspects implicit in model truncation methods, and the reduced estimation or prediction error that results from the use of BLUPs (when the underlying random effects model is appropriate, or, in the Bayesian framework, normal priors are appropriate). In other words, we seek a method that will exploit the advantages inherent in both model truncation and BLUP methodologies. In effect, we regard each λ_k^2 as if it were a variance component, and put the shrinkage estimate of λ_k equal to $S_k\hat{\lambda}_k$, where $S_k= 1-1/\ddot{F}_k$, where $\ddot{F}_k$ is the F_2-type F-statistic under the alternative hypothesis, as previously defined in describing Simulation tests. Put $S_k=0$ if $\ddot{F}_k<1$. Thus, S_k is an estimate of $n\lambda_k^2/(\ddot{u}_{1k}\sigma^2 + n\lambda_k^2)$. Note that S_k is the square of the factor that multiplies $\hat{\lambda}_k$ to obtain $\tilde{\lambda}_k$ in the scheme for Simulation tests.

B. COMPUTATION OF SHRINKAGE ESTIMATES

Most of the computation is as described for Simulation tests. We begin with Gollob's df as the value for $\ddot{u}_{1k}$. This leads to a value for $\ddot{F}_k$ from which S_k is computed. However, to obtain improved $\ddot{u}_{1k}$ values, we use the shrinkage estimate $S_k\hat{\lambda}_k$, instead of $\tilde{\lambda}_k$, as the supposed correct value of λ_k to be used in the simulation. Once the improved $\ddot{u}_{1k}$ have been obtained by simulation, these lead to new (hopefully better) values for $\ddot{F}_k$ and S_k. As with Simulation tests, the whole scheme can be iterated, but, while we have found that simulation does generally improve the shrinkage estimates, we have found little evidence of further improvement from iterating the simulation scheme. As in computing the Simulation tests, if the values of $S_k\hat{\lambda}_k$ do not give a decreasing sequence, pool as many $\hat{\lambda}_k^2$ values as necessary to obtain a nonincreasing sequence. Obtain the pooled $\ddot{F}$ statistic, from this a pooled shrinkage factor, and multiply the square root of the average value of $\hat{\lambda}_k^2$ for the pooled set by the pooled shrinkage factor. Assign this value as the shrinkage estimate of λ_k for all terms in the pooled set. (Cross validation has indicated that using $\tilde{\lambda}_k$ as the supposed correct value for λ_k in the simulation step instead of the shrinkage estimate $S_k\hat{\lambda}_k$ will not greatly reduce the efficiency of the resulting shrinkage estimates, and has the further advantage of allowing shrinkage estimates to be obtained as a "spin-off" from computing Simulation tests. Conversely, simulation results have shown cases where Simulation tests give liberal tests if shrinkage estimates $S_k\hat{\lambda}_k$ are used as "correct" values in the simulation for obtaining u_{1k} and u_{2k}. However, $\tilde{\lambda}_k$ is definitely inferior to $S_k\hat{\lambda}_k$ as an estimate of λ_k to use in a working model to estimate the realized cultivar performance levels. It seems philosophically inconsistent to use a different set of estimates as true values for the simulation than what one intends to use as predictors; thus, we prefer to use the shrinkage estimates for both purposes, but the SAS® program, which we can provide for shrinkage estimation, allows the user to use the $\tilde{\lambda}_k$ in the simulation step if such is desired.)

In practice, the above scheme is the way we obtain shrinkage estimates of multiplicative terms in any of the model forms COMM, GREG, SREG or

AMMI. We also multiply the cultivar main effects in GREG and AMMI by the shrinkage factor S_τ and the site main effects in SREG and AMMI by S_δ.[5]

C. SHRINKAGE ESTIMATION OF SHMM

As a consequence of multiplicative terms not being orthogonal to the additive intercept (shift parameter β), shrinkage estimation of SHMM is much less straightforward. A scheme for doing so (Cornelius et al., 1994; Cornelius and Crossa, 1995) is to iteratively solve the following equations:

$$\ddot{\beta} = (\bar{y}_{..} - \textstyle\sum_k S_k \hat{\bar{\alpha}}_k \hat{\bar{\gamma}}_k \sum_i \sum_j \hat{\alpha}_{ik} \hat{\gamma}_{jk} \bar{y}_{ij.}) / (1 - ge \sum_k S_k \hat{\bar{\alpha}}_k^2 \hat{\bar{\gamma}}_k^2)$$

$$\ddot{\lambda}_k = S_k \hat{\lambda}_k,$$

where $\ddot{\beta}$ and $\ddot{\lambda}_k$ denote the shrinkage estimates of β and λ_k, respectively, $\hat{\lambda}_k$, $\hat{\alpha}_{ik}$ and $\hat{\gamma}_{jk}$ are obtained from SVD of matrix Z with elements $z_{ij} = \bar{y}_{ij.} - \ddot{\beta}$, and

$$S_k = \max(1 - 1/\ddot{F}_k, 0),$$

$$\ddot{F}_k = n\hat{q}_k \hat{\lambda}_k^2 / \ddot{u}_{1k} s^2,$$

$$\hat{q}_k = (1 - ge \textstyle\sum_{m=1}^{P} \hat{\bar{\alpha}}_m^2 \hat{\bar{\gamma}}_m^2) / (1 - ge \sum_{m \neq k}^{P} \hat{\bar{\alpha}}_m^2 \hat{\bar{\gamma}}_m^2) .$$

Cornelius and Crossa (1995) give a Newton-Raphson algorithm for finding a solution for a given set of $\ddot{u}_{1k}$ values. They suggest putting $\ddot{u}_{1k} = g + e - 2k$ initially and describe a simulation strategy for finding improved $\ddot{u}_{1k}$ values,

---

[5] This begs for an explanation as to why we use S_τ and S_δ as the shrinkage factors for the main effects rather than $S'_\tau = 1 - 1/F'_\tau$ and $S'_\delta = 1 - 1/F'_\delta$ where F'_τ and F'_δ are the F-statistics for tests of main effects against the interaction. The term $S_\tau(\bar{y}_{i..} - \bar{y}_{..})$ in (2) actually derives as the sum of the BLUP of the ith cultivar main effect, $S'_\tau(\bar{y}_{i..} - \bar{y}_{..})$, and a contribution, $(S_\tau - S'_\tau)(\bar{y}_{i..} - \bar{y}_{..})$, to the BLUPs of the interactions of the ith cultivar with the environments. Thus, the ordinary least squares estimate of the main effect, $(\bar{y}_{i..} - \bar{y}_{..})$, contributes not only to the BLUP of the main effect, but also to BLUPs of the interactions. Similarly, the term $S_\delta(\bar{y}_{.j.} - \bar{y}_{..})$ is the sum of the BLUP of the main effect of the jth site and a contribution to the interactions. However, the AMMI multiplicative interaction components are constrained to be orthogonal to the main effects. We do not see a way to relax such constraints on multiplicative interactions and still have an easily computed solution, but interactions are not subject to such constraints in the BLUP methodology. If one imposes such constraints, the covariance matrix of the interactions is singular, and empirical BLUPs of cultivar and site main effects indeed become $S_\tau(\bar{y}_{i..} - \bar{y}_{..})$ and $S_\delta(\bar{y}_{.j.} - \bar{y}_{..})$. To use the usual BLUPs of main effects in conjunction with shrinkage estimates of AMMI interactions would result in the quantity $(S_\tau - S'_\tau)(\bar{y}_{i..} - \bar{y}_{..})$ and the analogous quantity for sites, $(S_\delta - S'_\delta)(\bar{y}_{.j.} - \bar{y}_{..})$, not being recoverable as contributions to estimated cell means. With $S_\tau(\bar{y}_{i..} - \bar{y}_{..})$ and $S_\delta(\bar{y}_{.j.} - \bar{y}_{..})$ used as the shrinkage estimates of additive main effects in any of the models (AMMI, GREG and SREG) which contain such effects, the contributions of ordinary least squares estimates of the main effects to estimated cell means in our shrinkage estimation strategy and under BLUP methodology are identical.

Table 3
Simulated per cell expected squared error of estimation of contributions,
$\Sigma_k \lambda_k \alpha_{ik} \gamma_{jk}$, **to cell means using shrinkage estimates with** $\ddot{u}_{1k}$ **obtained**
by simulation as compared to least squares estimates of best truncated
models and theoretically optimal shrinkage by two different optimality
criteria for $(p,q)=(8,19)$[a] **and seven sets of true** λ **values.**

True λ values	Best truncated model[b]	Shrinkage estimates[c]	λ-optimal shrinkage	Cell optimal shrinkage
10, 0,0,0,0,0,0,0	0.167	0.149	0.140	0.132
14, 6,0,0,0,0,0,0	0.342	0.266	0.269	0.245
12, 8,4,0,0,0,0,0	0.418	0.345	0.364	0.323
10,10,5,5,0,0,0,0	0.615	0.437	0.462	0.409
12,10,8,4,0,0,0,0	0.547	0.452	0.468	0.427
14, 6,4,4,2,0,0,0	0.562	0.403	0.443	0.378
5, 5,5,5,5,5,5	0.844	0.511	0.511	0.451

[a] Results shown are for an AMMI model with $(g,e)=(9,20)$ or $(20,9)$; to convert the results to other model forms with the same λ values, multiply by 45/38 for COMM with $(g,e)=(8,19)$ or $(19,8)$, by 9/8 for GREG with $(g,e)=(8,20)$ or SREG with $(g,e)=(20,8)$, or by 20/19 for GREG with $(g,e)=(19,9)$ or SREG with $(g,e)=(9,19)$.
[b] From Cornelius (1993), Cornelius et al. (1994), and Cornelius and Crossa (1995).
[c] From Cornelius and Crossa (1995) [also given by Cornelius et al. (1994) in which, due to a programming error, some of the reported results are very slightly different].

given the first set of shrinkage estimates. The SHMM shrinkage estimation scheme is mathematically messy, and it gets more messy if pooling of terms becomes necessary to insure a nonincreasing sequence of shrunken λ_k estimates, but we can provide a SAS® procedure IML™ program that seems to be highly efficient to do the computation.

D. COMPARISONS OF SHRINKAGE ESTIMATORS, TRUNCATED MODELS, AND BLUPS

For several sets of true λ_k values with $p=8$, $q=19$, simulated estimates of the expected mean squared error (over all cells) of estimation of the quantity $\Sigma_k \lambda_k \alpha_{ik} \gamma_{jk}$, this being the contribution of multiplicative terms to the mean of the (ij)[th] cell, are shown in Table 3 for the best truncated model fitted by least squares and for the shrinkage estimates. The results are for the interaction mean squared error (IMSE of Cornelius, 1993) in an AMMI model with $g=9$, $e=20$, as in the EVT16B data set, but they are also valid for COMM, GREG or SREG with the same p, q and λ_k values, if multiplied by the appropriate constant to correct for differing number of cells for the different model forms which results in $p=8$ and $q=19$. Results were obtained from 1000 simulated $p \times q$ data sets. In the analysis of each simulated data

Table 4
Root mean square predicted difference (RMSPD) from cross validation (three replicates for modeling, one replicate for validation) of truncated models estimated by least squares, shrinkage estimates ($\ddot{u}_{1k}$ obtained by simulation) and best linear unbiased predictors (BLUPs) from two examples.[a]

Model form	----------Example 1----------		----------Example 2--------	
	Best truncated model	Shrinkage estimates	Best truncated model	Shrinkage estimates
COMM	911	910	816	803
SHMM	906	903	816	801
GREG	907	906	816	803
SREG	908	908	819	808
AMMI	915	911	819	805
BLUP	----	935	----	808

[a] From Cornelius and Crossa (1995) and, except for SHMM shrinkage estimates. Cornelius et al. (1994). Example 1 is the EVT16B example (Table 1).

set, simulation for obtaining $\ddot{u}_{1k}$ values was based on 100 simulations. The shrinkage method is superior to the best truncated model in every instance. Since any method for choosing a truncated model cannot do better than to choose the best such model, these results indicate that shrinkage estimation is a much better procedure. Moreover, if such results will hold for real cultivar yield trial data, shrinkage estimates would render both cross validation and significance testing of the multiplicative terms not only unnecessary, but even of questionable relevance.

Cornelius and Crossa (1995) and Cornelius et al. (1994) cross validated shrinkage estimates and truncated models for COMM, SHMM, GREG, SREG and AMMI fitted to the EVT16B data set and to one other example (Table 4). Cross validation showed less advantage for the shrinkage estimates than was indicated by simulation results, but in the EVT16B example, shrinkage estimates and best truncated models had essentially equal predictive accuracy, and both were superior to BLUPs. In the second example, shrinkage estimates for all model forms were superior to the best truncated models. Also, for all model forms except SREG, they gave slightly more accurate prediction than BLUPs. Thus, in both examples, shrinkage estimation of multiplicative models were at least as good as the better choice of model truncation or BLUPs. Piepho (1994) found BLUPs superior in cross validation to the best AMMI model (without shrinkage) in four out of five examples (for two of the four, the best AMMI model had $p-1$ terms, and the other two were saturated). However, the margin of superiority for BLUPs was very small and possibly might be reversed if shrinkage estimators were used for AMMI. BLUPs, of course, are theoretically optimal if the assumptions for the underlying two-way random effects model are satisfied (and the appropriate variance ratios are known rather than estimated from the data), but generally not otherwise.

E. THE QUESTION OF OPTIMALITY

We have no formal proof of optimality of the proposed shrinkage estimators, but, besides the intuitively appealing argument of analogy to BLUPs, Cornelius and Crossa (1995) show that the shrinkage estimates of the multiplicative terms can, in fact, be derived as selection indices under particular definitions of empirically estimated genetic and phenotypic covariance matrices constructed from the multiplicative components.

Suppose, however, that we regard the multiplicative terms as fixed rather than as a device for defining a covariance structure for correlated interactions. Define "λ-optimal shrinkage" as estimation of λ_k by $\phi_k \hat{\lambda}_k$ where ϕ_k minimizes the expected squared error of estimation, $E(\phi_k \hat{\lambda}_k - \lambda_k)^2$, i.e., $\phi_k = \lambda_k E(\hat{\lambda}_k) / E(\hat{\lambda}_k^2) = n(\lambda_k^2 + \lambda_k w_k) / (\ddot{u}_{1k}\sigma^2 + n\lambda_k^2)$, where w_k is the bias in estimation of λ_k by $\hat{\lambda}_k$. Since λ_k is unknown, λ-optimal shrinkage cannot be computed in practice (and if λ_k was known, there would be no need for an estimate). As an estimate of ϕ_k, our S_k ignores the term involving w_k. Because w_k is generally positive, S_k is expected to underestimate ϕ_k, resulting in more shrinkage than is λ-optimal. However, λ-optimal shrinkage is not necessarily optimal for estimation of cell means. Included in Table 3 are expected mean squared errors of the contributions of the multiplicative effects to the cell means (expected IMSE for AMMI models) for λ-optimal shrinkage obtained by computer simulation. Comparison with the results for our shrinkage estimators indicates that, for estimation of cell means (or interactions), our shrinkage estimators will usually perform better than shrinkage estimators which are λ-optimal.

Define "cell-optimal" shrinkage as estimation of λ_k by $\psi_k \hat{\lambda}_k$ where ψ_k minimizes the expected per cell squared error of contributions of multiplicative effects to cell means (again, in the case of AMMI, the IMSE), viz., $E\left[\sum_i \sum_j (\sum_k \psi_k \hat{\lambda}_k \hat{\alpha}_{ik}\hat{\gamma}_{jk} - \sum_k \lambda_k \alpha_{ik}\gamma_{jk})^2\right] / ge$. The solution is $\psi_k = \sum_i \sum_j \left[\sum_{m=1}^{p} \lambda_m \alpha_{im}\gamma_{jm}\right] E\left[\hat{\lambda}_k \hat{\alpha}_{ik}\hat{\gamma}_{jk}\right] / E(\hat{\lambda}_k^2)$. It can be shown that, for a given set of λ_k values, the expected mean squared error is the same for any set of values for the α_{ik} and γ_{jk} that satisfy the orthonormality constraints. Given a set of λ_k values, simulation of cell-optimal shrinkage (Table 3) gives a theoretical lower bound for the expected per cell squared error of shrinkage estimators for multiplicative effects. Such lower bounds can be achieved only if the full set of λ_k values is known; consequently, they are not achievable in practice. The results suggest that our shrinkage estimators perform well, but also that there may be room for improvement. We are attempting to find

Table 5
Computation of shrinkage factors and shrinkage estimates of the GREG λ_k parameters for the EVT16B data.

$\hat{\lambda}_k^2$	$\hat{\lambda}_k$	$\ddot{u}_{1k}$	F_k	S_k	$S_k\hat{\lambda}_k$	$\hat{\lambda}_4$ and $\hat{\lambda}_5$ pooled
		$\ddot{u}_{1k}$ determined by counting df.				
259679430	16115	27	63.815	0.9843	15862	a
27226096	5218	25	7.226	0.8616	4496	a
8840752	2973	23	2.550	0.6079	1808	a
5402051	2324	21	1.707	0.4141	963	a
3343038	1828	19	1.167	0.1434	262	a
3014355	1736	17	1.177	0.1500	260	a
1194038	1093	15	0.528	0.0000	0	a
627737	792	13	0.320	0.0000	0	a
497079	705	11	0.300	0.0000	0	a
		$\ddot{u}_{1k}$ determined by simulation.				
259679430	16115	30.57	56.362	0.9823	15829	15829
27226096	5218	24.52	7.366	0.8643	4510	4510
8840752	2973	28.62	2.050	0.5122	1523	1523
5402051	2324	26.44	1.355	0.2622	609	609
3343038	1828	22.59	0.982	0.0000	0	148
3014355	1736	16.10	1.242	0.1949	338	148
1194038	1093	11.41	0.694	0.0000	0	0
627737	792	7.42	0.561	0.0000	0	0
497079	705	4.08	0.808	0.0000	0	0
		$\ddot{u}_{1k}$ determined by iterated simulation.				
259679430	16115	24.14	71.086	0.9859	15888	15888
27226096	5218	25.75	7.015	0.8575	4474	4474
8840752	2973	29.95	1.959	0.4894	1455	1455
5402051	2324	28.31	1.266	0.2101	488	488
3343038	1828	22.17	1.001	0.0007	1	180
3014355	1736	15.75	1.270	0.2127	369	180
1194038	1093	11.06	0.717	0.0000	0	0
627737	792	7.16	0.582	0.0000	0	0
497079	705	3.92	0.841	0.0000	0	0

[a] Fourth and fifth terms did not need to be pooled when $\ddot{u}_{1k}$ was determined by counting df.

shrinkage factors that will outperform S_k, but, at the time of this writing, none has been found that is consistently superior. The empirical and simulation evidence indicates that shrinkage estimators using S_k as the shrinkage factor may be expected to behave well in practice, and are at least as good as any other method heretofore proposed.

VI. SHRINKAGE ESTIMATION EXAMPLE

We again use the EVT16B example, for which the raw cell means are given in Table 1, and for which the variance of a cell mean $s^2/n = 150712$. Results of shrinkage estimation of AMMI and SHMM for these data are given by Cornelius and Crossa (1995). In this chapter, we will illustrate shrinkage estimation using the GREG model. We have already noted that the analyses in Table 2 suggest that GREG might be a better model choice than AMMI. Moreover, shrinkage estimation using GREG not only promises to provide estimates of the cell means that are improvements over the raw cell means, but also allows us to obtain an analysis from the shrinkage estimates that is very similar to a conventional stability analysis.

Computation of the shrinkage factors and shrinkage estimates of the λ_k, with $\ddot{u}_{1k}$ obtained by counting df (i.e., Gollob's df), by simulation and by iterated simulation, are shown in Table 5. Regardless of the method for obtaining $\ddot{u}_{1k}$, the shrinkage reduced first and second terms to about 98% and 86%, respectively, of their unshrunken values. Terms beyond the second received still more shrinkage (i.e., have smaller shrinkage factors S_k), and also were shrunk more when $\ddot{u}_{1k}$ was determined by simulation. Shrinkage factors for iterated simulation were not greatly different from the noniterated simulation. Both simulation and iterated simulation gave smaller shrinkage estimates for the fifth term than for the sixth term. Consequently, fifth and sixth terms were pooled. To illustrate, the average $\hat{\lambda}_k^2$ value for these two terms is $\hat{\lambda}_{5,6}^2 = (3343038 + 3014355)/2 = 3178696$, and $\hat{\lambda}_{5,6} = (3178696)^{1/2} = 1783$. For the simulation results, the average value of $\ddot{u}_{1k}$ for the fifth and sixth terms is $(22.59 + 16.10)/2 = 19.345$. Thus, the pooled F-statistic is $\ddot{F}_{5,6} = 3178696/(19.345)(150712) = 1.0903$, the common shrinkage factor $S_{5,6} = 1 - 1/\ddot{F}_{5,6} = 1 - 1/1.0903 = 0.0828$ and, finally, $S_{5,6}\hat{\lambda}_{5,6} = (0.0828)(1783) = 148$. Thus, the shrinkage estimates of both λ_5 and λ_6 are put equal to 148.

The proportion of the sum of squares owing to each term retained in the model is equal to the square of the shrinkage factor, i.e., S_k^2. Thus, the shrinkage estimation with $\ddot{u}_{1k}$ obtained by simulation has retained 96%, 75%, 26% and 7% of the variation due to the first through fourth terms, respectively, and less than 1% of the variation due to the pooled fifth and sixth terms. The seventh, eighth and ninth terms are discarded entirely. Of

Table 6
Shrinkage GREG estimates of cell means for the CIMMYT EVT16B maize yield trial with cultivars and sites arranged in rank order with respect to their overall means.

					Cultivar					
Site	8	2	1	7	3	9	4	5	6	Mean
19	2870	2351	2139	2976	2728	2692	2579	2719	3192	2694
10	3430	2918	2945	3049	3051	2779	2630	2792	3220	2979
7	3325	2923	2867	3382	3281	3187	3225	3422	3741	3262
1	4014	3513	3644	3453	3637	3329	3375	3637	3863	3607
8	4716	4977	5478	4355	4676	3112	2578	3153	3314	4040
2	3600	3985	3785	4372	4184	4441	4551	4351	4694	4218
18	3336	4157	3689	4776	4383	5210	5423	4833	5283	4566
20	4616	4524	4685	4417	4804	4741	5075	5372	5295	4836
17	4651	4427	4555	4268	4707	4848	5267	5492	5422	4849
5	4340	4616	4535	4574	4773	5239	5578	5398	5535	4954
9	4324	4717	4706	4739	4968	5151	5366	5395	5452	4980
15	4370	4372	4395	4560	4779	5305	5949	5956	5915	5067
3	5097	5068	5519	4952	5313	4410	4752	5480	5179	5086
4	4420	4794	4714	4790	5036	5583	5990	5837	5894	5229
11	3881	4598	4241	4989	4910	6135	6708	6090	6328	5320
16	4607	5090	5113	5118	5314	5547	5978	5961	5931	5407
14	5122	5893	5987	5667	5994	6056	6470	6489	6346	6003
6	5165	6171	6354	6187	6361	6158	6666	6822	6567	6272
13	4854	5950	5977	5926	6187	6784	7361	7104	6999	6349
12	5622	7330	7305	7119	7415	7799	8402	8166	7864	7447
Mean	4318	4619	4632	4684	4825	4925	5196	5223	5302	4858
Slope[a]	0.50	0.95	0.96	0.85	0.94	1.14	1.34	1.24	1.08	1.00
b[b]	0.52	0.97	0.98	0.86	0.95	1.12	1.31	1.23	1.06	1.00
$S_2\hat{\lambda}_2\hat{\alpha}_{i2}$	1619	1621	2501	568	1106	−1280	−1890	−798	−1198	
$S_3\hat{\lambda}_3\hat{\alpha}_{i3}$	−884	387	−60	796	143	413	−12	−708	−240	
$S_4\hat{\lambda}_4\hat{\alpha}_{i4}$	96	168	98	−357	−120	303	123	−268	−57	
$S_5\hat{\lambda}_5\hat{\alpha}_{i5}$	30	6	22	48	−84	−59	77	−37	6	
$S_6\hat{\lambda}_6\hat{\alpha}_{i6}$	83	45	−103	37	6	19	−12	−12	13	

Stability variance ($\times 10^{-3}$):

Shrinkage[c]	190	156	349	60	70	106	200	67	83	
Conv.[c]	301	117	398	167	16	225	344	170	88	

[a] Slope of regression using GREG model with shrinkage, rescaled as $\hat{\alpha}_{i1}/\bar{\hat{\alpha}}_1$ to give average regression equal to one.

[b] Slope of regression from conventional stability analysis.

[c] Stability variance estimated from shrinkage GREG estimates $= \sum_{k=2}^{6}(S_k\hat{\lambda}_k\hat{\alpha}_{ik})^2/(e-2)$;

Conv. = stability variance from conventional stability analysis.

the total variation due to multiplicative terms, shrinkage estimation retained 88% (computed as $100\Sigma S_k^2\hat{\lambda}_k^2 / \Sigma\hat{\lambda}_k^2$), with the other 12% discarded as "noise." However, only 46% of the variation due to residuals after fitting the first multiplicative term has been retained.

The cell means estimated from the shrunken GREG model are given in Table 6. The shrinkage factor for the cultivar main effects is $S_\tau = 0.9396$, giving the shrinkage estimates of the cultivar means as $\bar{y}_{..} + 0.9396(\bar{y}_{i..} - \bar{y}_{..})$ $= 0.9396\bar{y}_{i..} + 0.0604\bar{y}_{..}$. Thus, the effect is to shrink the cultivar means slightly toward the grand mean.

A. INTERPRETATION AS A STABILITY ANALYSIS

The first multiplicative term can be reparameterized for a stability analysis interpretation. If we define the environmental index $I_j = S_1\hat{\lambda}_1\bar{\hat{\alpha}}_1\hat{\gamma}_{j1}$, then the regression of the ith cultivar yields on the environmental index extracted from the shrinkage estimates is $\hat{\alpha}_{i1}/\bar{\hat{\alpha}}_1$. Like the regression coefficients in conventional stability analysis, the average of such values is unity. These values are shown as "slope" in Table 6. The values shown as "b" are the conventional stability analysis regression coefficients. There is extremely close agreement between these two sets of values and this is generally to be expected if there are large differences in productivity levels of the sites, i.e., if variation due to site main effects is much larger than the interaction. The indices $S_1\hat{\lambda}_1\bar{\hat{\alpha}}_1\hat{\gamma}_{j1}$ are shown in Table 7 along with the conventional indices $\bar{y}_{.j.} - \bar{y}_{..}$. Agreement between the two sets of indices is very close, with the correlation between them being 0.997. There are two minor reversals of rank order of sites in the two indices, namely, Sites 15 and 3, and Sites 11 and 16. The reason we get these minor discrepancies is that the GREG multiplicative model (before shrinkage) chooses the index so as to maximize the variation that can be captured by cultivar regressions on that index. (The sum of squares, on a cell mean basis, due to regression in conventional stability analysis, including the variation due to the site main effects, is 2.58906×10^8 as compared to the larger value of $\hat{\lambda}_1^2 = 2.59679 \times 10^8$.)

Also shown in Table 6 are values of $S_k\hat{\lambda}_k\hat{\alpha}_{ik}$ for $k = 2,...,6$. This value, when multiplied by $\hat{\gamma}_{jk}$, gives the contribution of the kth term to the deviation of the shrinkage GREG estimate of the $(ij)^{th}$ cell mean from the regression. The sum of squares of these contributions for the kth term, summed over sites, is $(S_k\hat{\lambda}_k\hat{\alpha}_{ik})^2$. Further summing these quantities for the second through sixth terms gives the sum of squared deviations from regression for the ith cultivar. For example, the sum of squared deviations for Cultivar 1 is $(2501)^2 + (-60)^2 + (98)^2 + (22)^2 + (-103)^2 = 6279298$. We have divided this by

Table 7
Site means, main effects and estimates of site parameters in GREG model fitted to the EVT16B data.

Site	Mean	$\hat{\gamma}_{j1}$	$S_1\hat{\lambda}_1\hat{\alpha}_1\hat{\gamma}_{j1}$	$\bar{y}_{j.} - \bar{y}_{..}$	$\hat{\gamma}_{j2}$	$\hat{\gamma}_{j3}$	$\hat{\gamma}_{j4}$	$\hat{\gamma}_{j5}$	$\hat{\gamma}_{j6}$
19	2652	−0.413	−2124	−2206	−0.158	0.121	−0.301	−0.330	0.202
10	2936	−0.367	−1887	−1922	0.036	0.016	0.169	0.155	−0.098
7	3329	−0.306	−1572	−1629	−0.096	0.011	−0.274	−0.149	−0.112
1	3613	−0.248	−1276	−1245	0.084	−0.181	0.186	0.346	0.131
8	4028	−0.192	−987	−830	0.703	0.292	0.305	−0.023	−0.199
2	4209	−0.119	−611	−649	−0.095	0.289	−0.076	0.115	0.017
18	4552	−0.042	−216	−306	−0.275	0.497	−0.066	0.164	0.148
20	4838	−0.007	−35	−20	0.032	−0.301	0.144	−0.249	0.057
17	4877	−0.002	−9	19	−0.035	−0.415	0.164	−0.094	0.176
5	4955	0.023	117	97	−0.092	−0.071	0.321	0.155	0.175
9	4971	0.025	127	113	−0.019	0.012	0.060	−0.580	−0.069
15	5049	0.050	256	191	−0.209	−0.323	−0.091	0.230	−0.254
3	5105	0.028	143	247	0.312	−0.278	−0.511	0.163	0.006
4	5230	0.078	401	372	−0.127	−0.098	0.223	−0.163	0.064
11	5307	0.109	560	449	−0.382	0.121	0.159	0.216	−0.160
16	5406	0.107	552	548	−0.019	−0.048	−0.023	0.084	0.010
14	6052	0.216	1112	1194	0.123	−0.019	0.091	0.059	0.260
6	6305	0.267	1371	1447	0.177	0.096	−0.418	0.206	−0.093
13	6332	0.294	1513	1474	−0.069	0.064	0.079	−0.169	−0.671
12	7516	0.498	2562	2658	0.112	0.215	−0.014	−0.136	0.409

$e-2 = 18$ to obtain an estimate of "stability variance" analogous to Eberhart and Russell's (1966) s_d^2. In Table 6 such values are shown as "Shrinkage" under the subhead "Stability variance"; the values shown as "Conv." are the s_d^2 values as defined by Eberhart and Russell, except that we have used the formula of Cruz Medina (1992) for the sums of squared deviations. While there is not close agreement between the two sets of values, the two cultivars with largest s_d^2 also have the largest stability variance estimated from the GREG shrinkage estimates. Apparently, based on both yield and stability, Cultivar 6 is the "winner" in this trial, since it has highest mean yield, a slope of regressions very close to one, and also has a small stability variance.

Inspection of specific values of $S_k\hat{\lambda}_k\hat{\alpha}_{ik}$ (Table 6) and $\hat{\gamma}_{jk}$ (Table 7) may pinpoint large deviations that may be of importance. For example, $\hat{\gamma}_{82} = 0.703$ indicates that $(0.703)^2$, i.e., 49%, of the variation due to the second term (after shrinkage) is due to deviations in Site 8. Moreover, since 88% of the deviations of shrinkage estimates of cell means from regression are due to the second multiplicative term, this means that 49% × 88% = 43% of all of the sum of squared deviations of the estimated cell means from regression occur in Site 8. Inspection of the raw cell means (Table 1) reveals that the generally high-yielding Cultivars 4, 5 and 6 did not perform well at Site 8, and, conversely, the generally low-yielding Cultivars 1 and 2 did well.

B. A SUGGESTED REFINEMENT FOR SHRINKING THE FIRST GREG COMPONENT

An alternative method for shrinking the first GREG component is to apply different shrinkage factors to the average regression (i.e., to the regressor variable, since rescaling will again be done to put the average regression equal to unity) and to the differences among slopes for the cultivars. The environmental index would then be $S_{1a}\hat{\lambda}_1\hat{\bar{\alpha}}_1\hat{\gamma}_{j1}$ and the slope for the ith cultivar $1+S_{1b}(\hat{\alpha}_{i1}-\hat{\bar{\alpha}}_1)/S_{1a}\hat{\bar{\alpha}}_1$. The first multiplicative component contains 27 independent parameters, $e-1=19$ (the site indices, subject to one constraint) of which are associated with the average regression, and the remaining $g-1=8$ associated with the differences in slopes. Define $\ddot{u}_{11a}=(e-1)\ddot{u}_{11}/(g+e-2)$ and $\ddot{u}_{11b}=(g-1)\ddot{u}_{11}/(g+e-2)=\ddot{u}_{11}-\ddot{u}_{11a}$. Then put $\ddot{F}_{1a}=ng\hat{\lambda}_1^2\hat{\bar{\alpha}}_1^2/\ddot{u}_{11a}s^2$, $\ddot{F}_{1b}=n\hat{\lambda}_1^2(1-g\hat{\bar{\alpha}}_1^2)/\ddot{u}_{11b}s^2$, $S_{1a}=1-1/\ddot{F}_{1a}$ and $S_{1b}=1-1/\ddot{F}_{1b}$.

For the results with $\ddot{u}_{1k}$ determined by simulation, $\ddot{u}_{11a}=(19)(30.57)/27=21.51$, and $\ddot{u}_{11b}=9.06$, $S_{1a}=0.9869$ and $S_{1b}=0.8957$. The environmental indices become 1.005 times the values in Table 7, increasing their range by 23 units. The slopes of regression become 0.55, 0.95, 0.96, 0.85, 0.95, 1.13, 1.31, 1.22 and 1.07 for the cultivars in the order they appear in Table 6. This refinement is intuitively appealing because the greater shrinkage applied to the slopes will probably increase their accuracy, particularly since the shrinkage is now toward their average value of 1.00. For this particular example, the changes in values of estimated cell means will be rather trivial. However, whether this refinement actually improves accuracy of estimation of the true slopes or of the estimates of the cell means has not yet been investigated. Also whether one can improve it by determining $\ddot{u}_{11a}$ and $\ddot{u}_{11b}$ directly in the simulation is also unknown. Note that the only parameter estimates affected by the refinement are those in the first multiplicative term.

VII. CHOOSING A MODEL FORM

The cross validation results (Table 4) for our two examples indicate that there will often be little difference among the five model forms with respect to prediction accuracy by shrinkage estimates. Consequently, many will choose a model based on their own preference for a particular form, this preference possibly based on how they wish to use the parameter estimates for interpretation. However, it would be reasonable to choose a model that, in some sense, is a "best" model for describing the particular set of data.

A. THE PRESS STATISTIC

One approach is to use the Prediction Sum of Squares (PRESS) statistic. PRESS was defined by Allen (1971) as a criterion for choosing regressor variables to be included in a multiple regression model. Specifically, PRESS is the sum of squared differences between an observation and its prediction by the regression model when the observation to be predicted is omitted from the data set to which the prediction model is fitted (thus, not allowing an observation, in any sense, to be predicting itself). For multiplicative models, we define PRESS as $\Sigma_i \Sigma_j (\bar{y}_{ij.} - \hat{y}_{[ij]})^2$, where $\hat{y}_{[ij]}$ is the prediction of $\bar{y}_{ij.}$ when $\bar{y}_{ij.}$ is omitted from the data to which the model is fitted. Evidently, PRESS primarily addresses the question of transferability of information. Presumably, the model with smallest PRESS should be the model that most successfully extracts information from the other cells that is transferable to the omitted cell. By a Newton-Raphson algorithm, the theory for which is not yet published, the FORTRAN program EIGAOV will compute PRESS for any of the five model forms and number of multiplicative terms ranging from zero to $p - 1$.

Cornelius et al. (1994) transformed PRESS into an adjusted RMSPD to make it roughly comparable to RMSPDs from data-splitting cross validation of truncated models, viz., adjusted RMSPD(PRESS) = [PRESS/ge + $(n-1)s^2/n]^{1/2}$, and reported results for the EVT16B data. The ten best PRESS models (with number of multiplicative terms shown in parentheses) and their adjusted RMSPD(PRESS) were: SHMM(2) 886, SREG(1) 892, GREG(2) 912, SHMM(1) 925, SHMM(3) 925, COMM(2) 935, GREG(1) 939, COMM(1) 942, AMMI(1) 946, AMMI(0) (i.e., additive model) 970. Badly overfitted models typically give extremely large PRESS, e.g., adjusted RMSPD(PRESS) = 30708 for AMMI(3). These results suggest that SHMM most efficiently extracts the transferable information from the EVT16B data. The conclusion is specifically with respect to the particular data set, i.e., in computed PRESS statistics from various data sets, we have not found any of the model forms to be consistently superior to the others.

B. MANDEL'S TESTS FOR CONCURRENT AND NONCONCURRENT REGRESSIONS

Analyses that partition interaction into one df due to "concurrence" of genotype (or site) regressions on site (or genotype) main effects, $g-2$ df due to "nonconcurrence" of genotype regressions, $e-2$ df due to nonconcurrence of site regressions, and $(g-2)(e-2)$ df due to the remaining interaction (Mandel, 1961) (Table 8 for the EVT16B example) provide useful diagnostics for choice of model form. The sum of squares due to concurrence is identical

Table 8
Tukey-Mandel analysis of interaction in the EVT16B data set[a].

Source of Variation	df	Mean square $\times 10^{-3}$
Concurrent regression	1	8408[b]
Nonconcurrence of site regressions	18	1170[b]
Nonconcurrence of genotype regressions	7	443[b]
Remaining interaction	126	237[b]
Pooled error	480	151

[a] From Cornelius and Crossa (1995).
[b] Significant at $P<0.01$.

to Tukey's (1949) one df for nonadditivity. The sum of squares due to nonconcurrence of genotype (or site) regressions is obtained by subtraction of the sum of squares due to concurrence from the entire sum of squares due to genotype (or site) regressions. In a balanced two-way data set, nonconcurrence of genotype regressions is orthogonal to nonconcurrence of site regressions; thus, they can be tested by simultaneous inclusion in the same model for the analysis.

The concurrence component will generally be absorbed into the first multiplicative term by SHMM, GREG or SREG fitted by least squares, but AMMI tends not to be as highly sensitive to it as the other model forms (Johnson and Hegemann, 1976; Seyedsadr and Cornelius, 1992). A significant test for concurrence (as in this example) strongly suggests SHMM as a good model form, but if accompanied by significant nonconcurrence of genotype regressions, GREG may do better. Nonconcurrence of site regressions would imply SREG as a model form. Significance of all of the regression components suggests SHMM as model form, particularly if the remaining interaction is nonsignificant, since a model that simultaneously fits all of the regression components can be reparameterized as a two-term SHMM (significance of both nonconcurrence components without significant concurrence may suggest that AMMI would do as well as SHMM; GREG or SREG, however, may also be quite satisfactory in cases where both nonconcurrence components are significant.) Nonsignificance for any of the regression components, but a significant remainder would clearly suggest AMMI as a logical model form. COMM should be a good model form in cases that suggest SHMM, but the fitted SHMM gives estimated shift ($\hat{\beta}$) close to zero. Data sets with a large number of significant multiplicative components (in any model form) are likely to present situations where the model forms will be virtually one as satisfactory as another for modeling the data.

The very large concurrence component (Table 8) suggests SHMM as a model form for the EVT16B data, which seems to agree with the implications of the PRESS statistics. However, either GREG or SREG should be nearly as satisfactory, and even AMMI quite acceptable if one, for whatever reason, prefers it as a model form for interpretation of response patterns.

VIII. UNREPLICATED TRIALS

A. TEST PROCEDURES

If the trial is unreplicated, the foregoing test procedures cannot be done for want of an estimate of the residual error σ^2 independent of the cell means (unless an appropriate estimate is available from some other trial). However, the test of Johnson and Graybill (1972) is appropriate for testing the first term in AMMI, and is readily adapted to testing the first term in GREG, SREG or COMM. Schott (1986) and Marasinghe (1985) extended the procedure to testing later terms by testing each sequentially fitted term by the Johnson-Graybill procedure as if it were the first term in a problem with p and q reduced by one for each previously fitted term. The test criterion is $\Lambda_k = SS_k/RSS_{k-1}$, where RSS_{k-1} is the residual (lack of fit) sum of squares after fitting $k-1$ terms. Large values of Λ_k constitute evidence for the alternative hypothesis. Tables of exact and/or approximate critical values have been given by Johnson and Graybill (1972) and Schuurman et al. (1973a, 1973b). Johnson and Graybill give formulas by which Λ_k can be transformed into a statistic approximately distributed as a Beta random variable for which critical values can be computed by an incomplete Beta function. However, a Beta random variable can be transformed into a statistic that has an F distribution. To test the kth term, the approximate F-test is computed as follows:

$$Q_k = [(p-k+1)\Lambda_k - 1]/(p-k),$$

$$c_1 = \frac{u_{1k} - (q-k+1)}{(q-k+1)(p-k)},$$

$$c_2 = \frac{(p-k+1)(q-k+1)u_{2k}^2 - 2u_{1k}^2}{(q-k+1)^2[(p-k+1)(q-k+1)+2](p-k)^2},$$

$$d = c_1(1-c_1) - c_2,$$

$$a = dc_1/c_2,$$

$$b = d(1-c_1)/c_2,$$

$$F_{JG/SM} = bQ_k/a(1-Q_k),$$

where u_{1k} and u_{2k} are determined under the same assumptions as for the F_{GH1} and F_{GH2} tests. The distribution of $F_{JG/SM}$ is approximated by F with df = $(2a, 2b)$.

Table 9
JG/SM and JG/YC tests of AMMI and GREG components in the EVT16B example.

Term	JG/SM				JG/YC			
	d_1	d_2	F	P	d_1	d_2	F	P
AMMI analysis								
1	41.1	188.1	4.56	<0.001	41.1	188.1	4.56	<0.001
2	34.9	145.9	1.29	0.152	35.3	153.2	1.34	0.117
3	28.7	108.7	1.17	0.276	29.4	120.1	1.26	0.189
4	22.7	76.6	1.74	0.039	23.5	89.3	1.96	0.013
5	16.9	49.6	1.60	0.102	17.7	61.2	1.88	0.036
6	11.2	27.7	1.55	0.170	11.8	36.4	1.92	0.066
7	5.6	11.1	0.10	0.994	6.1	15.9	0.12	0.992
GREG analysis								
1	46.9	224.6	21.50	<0.001	46.9	224.6	21.50	<0.001
2	40.6	179.0	4.03	<0.001	41.1	188.1	4.18	<0.001
3	34.4	138.4	1.59	0.032	35.3	153.2	1.72	0.014
4	28.4	102.8	1.28	0.187	29.4	120.1	1.44	0.088
5	22.4	72.2	0.97	0.510	23.5	89.3	1.14	0.316
6	16.6	46.5	2.03	0.030	17.7	61.2	2.51	<0.001
7	11.0	25.8	0.88	0.570	11.8	36.4	1.15	0.354
8	5.5	10.2	0.94	0.243	6.1	15.9	0.34	0.906

Operating characteristics of the JG/SM test are much like the F_{GH1} and F_{GH2} tests, which are based on the same assumptions, except it will generally have less power. Sequential tests were also proposed by Yochmowitz and Cornell (1978), who tested the kth term as though it were the first in a problem with p reduced by one for each term previously fitted, but q held at its original value. They gave another approximate method for computing P-values for a few small values of p, but an F-approximation can be computed as for the JG/SM method except that $q-k+1$ is replaced by q everywhere it appears in the formulas for c_1 and c_2. Except for the first term, the JG/YC test (as we shall denote it) tends to be liberal.

The JG/SM and JG/YC tests for AMMI and GREG analyses of the EVT16B data are shown in Table 9. The liberality of JG/YC relative to JG/SM is quite evident in the P-values obtained. Marasinghe (1985) and Schott (1986) recommended that the model be terminated with the first occurrence of a nonsignificant term. By this rule, only one term would be retained in an AMMI model. We are not convinced that this rule is sound advice. The test essentially tests the kth term against the pooling of all remaining terms. If those remaining terms include one or more which is itself nonnull, this will result in less power to detect a false null hypothesis. We are more inclined to interpret the JG/SM tests in the AMMI analysis as strong evidence that there are, in fact, four nonnull components, although perhaps suggesting that the second, third and fourth are not large.

Conversely, we believe (but have not shown) that failures in underlying assumptions, e.g., normality and homogeneity of variance, could easily lead to spuriously significant tests for one or more of the later terms. Thus, we would be cautious and skeptical about concluding on the basis of the significant test of the sixth GREG term that the GREG model indeed contains six nonnull components. On the basis of this analysis, we would truncate the GREG model with three multiplicative terms.

B. THE SEYEDSADR-CORNELIUS/SCHOTT-MARASINGHE (SC/SM) TEST FOR SHMM COMPONENTS.

A table of critical values of Λ_1 obtained by computer simulation for SHMM analysis is given by Seyedsadr and Cornelius (1993) for $p = 2(1)7(2)9(4)19$ and $q = 2(1)7(2)9(4)19, 31, 49, 99$. They also give a table of Beta distribution parameters a and b for approximating the distribution of $Q_1 = (\tilde{p}\Lambda_1 - 1)/(\tilde{p} - 1)$ as a Beta random variable, where $\tilde{p} = \min(p, q-1)$. The test is extended to testing the kth term for $k > 1$ by analogy to the Schott-Marasinghe extension of the Johnson-Graybill test to terms beyond the first in analyses of the other model forms. Specifically, the extension is to test Λ_k or $Q_k = [(\tilde{p} - k + 1)\Lambda_k - 1]/(\tilde{p} - k)$ as though it is Λ_1 or Q_1 in a problem with p and q reduced by one for each previous term tested. Like the JG/SM tests, Q_k can be transformed into an approximate F-statistic, viz., $F_{SC/SM} = bQ_k/a(1 - Q_k)$ with df $= (2a, 2b)$.

C. SHRINKAGE ESTIMATES IN UNREPLICATED TRIALS

The $\ddot{F}_k$ statistics used to obtain shrinkage factors are not required to have specified distributions, but need only to estimate $(\ddot{u}_{1k}\sigma^2 + n\lambda_k^2)/\ddot{u}_{1k}\sigma^2$ as well as we can. (For unreplicated trials, of course, $n = 1$, but we will illustrate the method using the EVT16B cell means, for which $n = 4$.) Thus, we can compute shrinkage estimates if only we can obtain a "good" estimate of σ^2/n from the cell means. A residual mean square from a truncated model should, at least in theory, provide a way to do so. Residual mean squares from fitting $0, 1, ..., p-1$ terms in AMMI and GREG analyses of the EVT16B data are shown in Table 10. When $\ddot{u}_{1k}$ for the terms fitted is obtained by counting df (i.e., number of independent parameters), the resulting residual df are $(p-k)(q-k)$ (subtract one from this value for SHMM residuals), which, according to Goodman and Haberman (1990), is correct if the k terms retained in the model are sufficiently large. Otherwise, the allocation of df to the residual is too many. For AMMI, the JG/SM tests suggest truncation either after fitting one multiplicative term or after four terms. If we choose to truncate after four terms, the mean square, 100323, is our estimate of σ^2/n. Using this estimate, we obtained a new set of $\ddot{u}_{1k}$ values by the simulation scheme used for Simulation tests (with the modification that the assumed true λ_k values for $k > 4$ were forced to zero). Residual mean squares based on the

Table 10
Residual mean squares after fitting AMMI and GREG models with various numbers of multiplicative terms to the EVT16B cell means.

No. terms fitted	JG/SM P-value[a]	$\ddot{u}_{1k}$ obtained by counting df		$\ddot{u}_{1k}$ obtained by simulation	
		df	Mean Square	df	Mean Square
		AMMI analysis			
0	------	152	410698	152.00[b]	410698
1	<0.001	126	217025	125.49	217905
2	0.152	102	175660	93.34	191963
3	0.276	80	142533	67.82	168143
4	0.039	60	100323	55.13	109182
5	0.102	42	69677	31.11	94068
6	0.170	26	46206	15.45	77751
7	0.994	12	47758	5.43	105510
		GREG analysis			
0	------	171	1811840	171.00[b]	1811840
1	<0.001	144	348230	141.07	355456
2	<0.001	119	192597	116.75	196316
3	0.032	96	146649	91.95	153108
4	0.187	75	115683	60.20	144135
5	0.510	56	95238	38.04	140209
6	0.030	39	59458	22.20	104468
7	0.570	24	46867	11.12	101146
8	0.945	11	45189	3.96	125636

[a] P-value shown is for JG/SM test of the last term fitted.
[b] The simulated df of RSS_0 was 154.50 and 171.74 for AMMI and GREG. respectively, but 152.00 and 171.00 were used because these are the known correct df for residuals after fitting the additive effects.

simulated residual df $= \Sigma_{m=k+1}^{p} \ddot{u}_{1m}$ are shown in the last column of Table 10. The estimate of σ^2/n has increased to 109182.

For GREG truncated after three terms (on the basis of JG/SM tests), the initial error estimate is 146649, which increases to 153108 in the simulation. Interestingly, in the GREG analysis where it is much more clear where the model should be truncated, the error estimate is very close to the value of s^2/n = 150712. In the AMMI analysis, where the proper truncation point was rather ambiguous, the error estimate was only about 2/3 of the pooled error estimate, but if the model had been truncated after one term, apparently the estimate would have been roughly 217000, a value substantially larger than the pooled error. There is no particular reason why the error estimate must be derived via the same model form as the form for which shrinkage estimates are desired. Thus, in this example, if shrinkage estimates were desired for AMMI (and if the pooled error estimate was not available), one might seriously consider using the error estimate obtained from the GREG analysis for computing the shrinkage factors. More generally, it may be well to obtain such estimates for each of the model forms, average those that are in close agreement, and discard any that appear to be aberrant. How well these strategies for estimating σ^2/n will perform in practice is not known.

IX. SOFTWARE

Unfortunately, we cannot presently supply a computing program that will compute all of the tests and estimators described here all in one package. A SAS® procedure IML™ program is available for computing Gollob F, F_2, F_{GH1}, F_{GH2}, F_R and Simulation tests (iterated as often as desired) for COMM, GREG, SREG or AMMI. This program will also compute estimates of E(IMSE) for truncated AMMI models, and its analogous quantity (mean squared error of contributions, $\Sigma_k \lambda_k \alpha_{ik} \gamma_{jk}$, of multiplicative terms to cell means) for truncated COMM, GREG and SREG. Another SAS® program computes shrinkage estimates for COMM, GREG, SREG and AMMI. Yet a third SAS® program computes SHMM shrinkage estimates. The SAS® programs for shrinkage estimation will compute estimates for unreplicated trials if the error estimate is computed separately and given as input. None of the SAS® programs in their present form will compute test procedures for unreplicated trials.

The FORTRAN program EIGAOV computes least squares estimates and predicted values for any of the five model forms fitted to a balanced data set (or with one missing cell) or to any specified balanced subset, computes JG/SM, Yochmowitz-Cornell, JG/YC, F_1, F_{GH1} and F_R tests, residual sums of squares within genotypes and within sites, PRESS statistics, constrained "non-crossover" one-term SHMM solutions (Crossa et al., 1995a, 1995b), Tukey-Mandel analyses (Table 8) (including the regression coefficients), and decomposition of residuals from Tukey-Mandel models into multiplicative components. [The Mandel genotypes regression model is the same as the conventional stability analysis model of Finlay and Wilkinson (1963) and Eberhart and Russell (1966).] None of the tests except the F_R test is known to be valid if dimensions of the problem exceed those for which values of u_{1k} and u_{2k} are given by Mandel (1971), i.e., $p \leq 19$, $q \leq 99$. F_1 tests are done precisely only for the first three terms in COMM, GREG, SREG and AMMI, with each term beyond the third tested as though it is the third in a problem with p sequentially reduced by one for each such term sequentially tested. For SHMM, the F_1 test is precisely done for the first five terms, with subsequent terms tested as though each is the fifth in a problem with reduced p. EIGAOV has been successfully run on IBM® mainframe, VAX® and UNIX® systems. Output is formatted for wide carriage printers (132-character record length). At the time of this writing, development of a version which will run on a personal computer has not been completed.

Persons desiring the SAS® programs should send a diskette or an e-mail address to the first author along with their request. Inquiries concerning authorized use of the FORTRAN program EIGAOV should also be directed to the first author.

REFERENCES

Allen, D.M. 1971. Mean square error of prediction as a criterion for selecting variables. *Technometrics* 13:469-475.

Boik, R.J. 1986. Testing the rank of a matrix with applications to the analysis of interaction in ANOVA. *Journal of the American Statistical Association* 81:243-248.

Bradu, D., and K.R. Gabriel. 1978. The biplot as a diagnostic tool for models of two-way tables. *Technometrics* 20:47-68.

Cornelius, P.L. 1980. Functions approximating Mandel's tables for the means and standard deviations of the first three roots of a Wishart matrix. *Technometrics* 22:613-616.

Cornelius, P.L. 1993. Statistical tests and retention of terms in the additive main effects and multiplicative interaction model for cultivar trials. *Crop Science* 33:1186-1193.

Cornelius, P.L., and J. Crossa. 1995. Shrinkage estimators of multiplicative models for crop cultivar trials. Technical Report No. 352, University of Kentucky Department of Statistics, Lexington, Kentucky.

Cornelius, P.L., J. Crossa and M.S. Seyedsadr. 1994. Tests and estimators of multiplicative models for variety trials. *Proceedings of the 1993 Kansas State University Conference on Applied Statistics in Agriculture*. Manhattan, Kansas.

Cornelius, P.L., M. Seyedsadr and J. Crossa. 1992. Using the shifted multiplicative model to search for "separability" in crop cultivar trials. *Theoretical and Applied Genetics* 84:161-172.

Cornelius, P.L., D.A. Van Sanford and M.S. Seyedsadr. 1993. Clustering cultivars into groups without rank-change interactions. *Crop Science* 33:1193-1200.

Crossa, J., and P.L. Cornelius. 1993. Recent developments in multiplicative models for cultivar trials. In: D.R. Buxton et al. (Eds.) *International Crop Science I*. Crop Science Society of America, Madison, Wisconsin.

Crossa, J., P.L. Cornelius, K. Sayre and J.I. Ortiz-Monasterio R. 1995a. A shifted multiplicative model fusion method for grouping environments without cultivar rank change. *Crop Science* 35:54-62.

Crossa, J., P.L. Cornelius and M.S. Seyedsadr. 1995b. Using the shifted multiplicative model cluster methods for crossover genotype-by-environment interaction. Chapter 7, this volume.

Crossa, J., P.L. Cornelius, M. Seyedsadr and P. Byrne. 1993. A shifted multiplicative model cluster analysis for grouping environments without genotypic rank change. *Theoretical and Applied Genetics* 85:577-586.

Cruz Medina, R. 1992. Some exact conditional tests for the multiplicative model to explain genotype-environment interaction. *Heredity* 69:128-132.

Eberhart, S.A., and W.A. Russell. 1966. Stability parameters for comparing varieties. *Crop Science* 6:36-40.

Feinberg, G.E. 1972. Discussion on D.V. Lindley and A.F.M. Smith, Bayes estimates for the linear model. *Journal of the Royal Statistical Society*, Series B, 34:30-32.

Finlay, K.W., and G.N. Wilkinson. 1963. The analysis of adaptation in a plant breeding programme. *Australian Journal of Agricultural Research* 14:742754.

Fisher, R.A., and W.A. MacKenzie. 1923. Studies in variation II: The manurial response in different potato varieties. *Journal of Agricultural Science* 13:311-320.

Gabriel, K.R. 1978. Least squares approximation of matrices by additive and multiplicative models. *Journal of the Royal Statistical Society*, Series B, 40:186-196.

Gauch, H.G. Jr. 1988. Model selection and validation for yield trials with interaction. *Biometrics* 44:705-715.

Gauch, H.G. Jr. and R.W. Zobel. 1988. Predictive and postdictive success of statistical analyses of yield trials. *Theoretical and Applied Genetics* 76:1-10.

Gollob, H.F. 1968. A statistical model which combines features of factor analytic and analysis of variance techniques. *Psychometrika* 33:73-115.

Goodman, L.A., and S.J. Haberman. 1990. The analysis of nonadditivity in two-way analysis of variance. *Journal of the American Statistical Association* 85:139-145.

Harville, D.A. 1977. Maximum likelihood approaches to variance component estimation and to related problems. *Journal of the American Statistical Association* 72:320-338.

Henderson, C.R. 1984. *Applications of Mixed Models in Animal Breeding.* University of Guelph. Guelph, Ontario, Canada.

Hoerl, A.E., and R.W. Kennard. 1970. Ridge regression: Biased estimation for nonorthogonal problems. *Technometrics* 12:55-67.

Hogg, R.V., and A.T. Craig. 1978. *Introduction to mathematical statistics.* Fourth edition. Macmillan. New York.

Johnson, D.E. 1976. Some new multiple comparison procedures for the two-way AOV model with interaction. *Biometrics* 32:929-934.

Johnson, D.E., and F.A. Graybill. 1972. An analysis of a two-way model with interaction and no replication. *Journal of the American Statistical Association* 67:862-868.

Johnson, D.E., and V. Hegemann. 1976. The power to two tests for nonadditivity. *Journal of the American Statistical Association* 71:945-948.

Lindley, D.V., and A.F.M. Smith. 1972. Bayes estimates for the linear model. *Journal of the Royal Statistical Society*, Series B, 34:1-41.

Mandel, J. 1961. Non-additivity in two-way analysis of variance. *Journal of the American Statistical Association* 56:878-888.

Mandel, J. 1969. The partitioning of interaction in analysis of variance. *Journal of Research of the National Bureau of Standards*, Series B, 73:309-328.

Mandel, J. 1971. A new analysis of variance model for non-additive data. *Technometrics* 13:1-18.

Marasinghe, M.G. 1985. Asymptotic tests and Monte Carlo studies associated with the multiplicative interaction model. *Communications in Statistics, A. Theory and Methods.* 14:2219-2231.

Piepho, H.P. 1994. Best linear unbiased prediction (BLUP) for regional yield trials: A comparison to additive main effects and multiplicative interaction (AMMI) analysis. *Theoretical and Applied Genetics* 89:647-654.

Piepho, H.P. 1995. Robustness of statistical tests for multiplicative terms in the additive main effects and multiplicative interaction model for cultivar trials. *Theoretical and Applied Genetics* 90:438-443.

Schott, J.R. 1986. A note on the critical values in stepwise tests for multiplicative components of interaction. *Communications in Statistics, A. Theory and Methods* 15:1561-1570.

Schuurman, F.J., P.R. Krishnaiah and A.K. Chattopadhyay. 1973a. On the distributions of the ratios of the extreme roots to the trace of the Wishart matrix. *Journal of Multivariate Analysis* 3:445-453.

Schuurman, F.J., P.R. Krishnaiah and A.K. Chattopadhyay. 1973b. Tables for the distributions of the ratios of the extreme roots to the trace of Wishart matrix. Aerospace Research Laboratories Technical Report 73-0010, United States Air Force.

Seyedsadr, M., and P.L. Cornelius. 1991. Functions approximating the expectations and standard deviations of sequential sums of squares in the shifted multiplicative model for a two-way table. Technical Report No. 322, University of Kentucky Department of Statistics, Lexington, Kentucky.

Seyedsadr, M., and P.L. Cornelius. 1992. Shifted multiplicative models for nonadditive two-way tables. *Communications in Statistics. B. Simulation and Computation* 21:807-822.

Seyedsadr, M.S., and P.L. Cornelius. 1993. Hypothesis testing for components of the shifted multiplicative model for a nonadditive two-way table. *Communications in Statistics, B. Simulation and Computation* 22:1065-1078.

Seyedsadr, S.M. 1987. Statistical and computational procedures for estimation and hypothesis testing with respect to the shifted multiplicative model. Ph.D. dissertation, University of Kentucky, Lexington, Kentucky.

Tukey, J.W. 1949. One degree of freedom for non-additivity. *Biometrics* 5:232-242.

Williams, E.R., and J.T. Wood. 1993. Testing the significance of genotype-environment interaction. *Australian Journal of Statistics* 35:359-362.

Yochmowitz, M.G., and R.G. Cornell. 1978. Stepwise tests for multiplicative components of interaction. *Technometrics* 20:79-84.

Zobel, R.W., M.J. Wright and H.G. Gauch, Jr. 1988. Statistical analysis of a yield trial. *Agronomy Journal* 80:388-393.

APPENDIX

Following are the approximating functions for u_{1k} and u_{2k} for F_{GH2} (or F_{GH1}) tests for COMM, GREG, SREG and AMMI adapted from Cornelius (1980) and, for SHMM, adapted from Seyedsadr and Cornelius (1991). Put $r = q - k + 1$, $c = p - k + 1$, $h = r/c$, $m_1 = \log_e r$ and $m_2 = \log_e c$, where q and p are as defined in Section III, and $\log_e$ is the natural logarithm. Note that the right-hand sides of the following equations will depend on k through the dependence of r, c, m_1 and m_2 on k; also that $r \geq c$ (since $q \geq p$).

A. FOR COMM, GREG, SREG OR AMMI

$u_{1k} = -1.5262 + 1.060602(r + c) + 0.01623859rc$

$\quad + (0.1765587 \times 10^{-5})\, r^2c^2 - (0.669151 \times 10^{-4})(r^2c + rc^2)$

$\quad + 0.39497(m_1 + m_2) + 0.3817419(m_1^2 m_2 + m_1 m_2^2)$

$\quad - 0.0935053(m_1^3 + m_2^3) + 0.0782977(m_1^3 m_2 + m_1 m_2^3)$

$\quad - 0.0858762\, m_1^2 m_2^2 ,$

$u_{2k} = 0.065167(r + c) - (0.251751 \times 10^{-6})(r^3c + rc^3)$

$\quad + (1.075584 \times 10^{-6})\, r^2c^2 - 0.26275\, \log_e(r^2 + c^2)$

$\quad + 1.854934\, \log_e(r^2c + rc^2) - 1.64568(m_1 + m_2)$

$\quad - 0.0061384(m_1^3 m_2 + m_1 m_2^3) .$

These formulas are not valid if $r > 99$, $c > 19$, or $c < 2$. If $c = 1$, put $u_{1k} = r$ and $u_{2k} = (2r)^{1/2}$.

B. FOR SHMM

$u_{1k} = -1.50554 + 1.0941560(r + c) + 0.01990084rc$

$\quad - (4.822005 \times 10^{-6})r^2c^2 + 0.3521341 m_1 m_2(m_1 + 2)$

$\quad + 0.06170702(m_1^2 m_2^2) - 1.54776(e^{-r} + e^{-c})$

$\quad + (1.785228 \times 10^{-8})(r^3c^2 + r^2c^3) - 0.0934906(|r - c|) ,$

$u_{2k} = 1.730323(m_1 + m_2) - 0.3855876(m_1^2 + m_2^2) + 0.0807067(m_1^3 + m_2^3)$

$\quad + 0.1399678(h + h^{-1}) - (3.04894 \times 10^{-3})(h^2 + h^{-2})$

$\quad + (4.74447 \times 10^{-7})(h^4 + h^{-4}) + 0.0370850(|h - h^{-1}|) .$

These formulas are not valid if $r > 99$, $c > 19$, or $c < 2$. If $c = 1$, put $u_{1k} = r-1$ and $u_{2k} = [2(r-1)]^{1/2}$. Also, if $r = c = 2$, put $u_{1k} = 3$ and $u_{2k} = 6^{1/2} = 2.44949$.

Chapter 9

NONPARAMETRIC ANALYSIS OF
GENOTYPE x ENVIRONMENT INTERACTIONS BY RANKS

Manfred Hühn[1]

I. INTRODUCTION

The statistical analysis of genotype x environment interactions is of an outstanding interest in the field of applied statistics as well as for the analysis of experiments in plant breeding and crop production. We refer to the review articles by Freeman (1973), Hill (1975), Skrøppa (1984), Cox (1984), Freeman (1985, 1990), Westcott (1986), and Crossa (1990). Quite different statistical methods have been proposed for estimation and partitioning of genotype x environment interactions (variance components, regression methods, multivariate analyses, cluster techniques).

The analysis of genotype x environment interactions is closely linked with the quantitative estimation of phenotypic stability of genotypes over environments. For this topic, we refer to the review articles by Utz (1972), Freeman (1973), Wricke and Weber (1980), Lin et al. (1986), and Becker and Léon (1988).

When genotype x environment interactions are present, the effects of genotypes and environments are statistically nonadditive, which means that differences between genotypes depend on the environment. Existing genotype x environment interactions may, but must not necessarily, lead to different rank orders of genotypes in different environments.

In many practical situations, the experimenter is not interested in a knowledge of the numerical amount of genotype x environment interactions per se, but is only interested in the existence (or non-existence) of genotype x environment interactions insofar as they lead to different orderings of genotypes in different environments. This concept of genotype x environment interaction is closely related to the concept of selection in plant breeding. The breeder is mainly interested in rank orders of genotypes in different environments and in changes of these rank orders. He/she is interested in questions such as whether the best genotype in one environment is also the best in other environments, which means that relative characterizations and comparisons of the genotypes (orderings) are often more important than absolute characterizations and comparisons. Therefore, it is an obvious idea to use rank information for a quantitative description of these relationships.

---

[1] Institute of Crop Science and Plant Breeding, University of Kiel, Olshausenstrasse 40, D-24118 Kiel, Germany.

236

For two genotypes A and B, and two environments X and Y, the basic types of relationships between genotype x environment interactions and changes of rank orders are demonstrated schematically in Figure1.

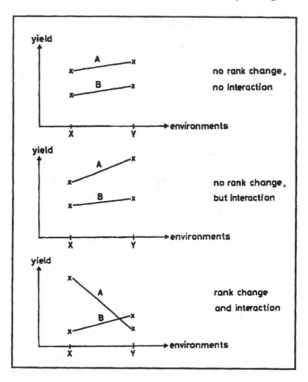

Figure 1 Genotype x environment interactions and changes of rank orders - different type of relationships (for two environments X and Y and two genotypes A and B) (modified from Wricke, 1965).

Some authors introduced the terms qualitative interactions (crossover interactions) and quantitative interactions (noncrossover interactions). For noncrossover interactions, the true treatment differences vary in magnitude but not in direction, whereas for crossover interactions, the direction of true treatment differences varies. Although these terms and the corresponding tests of significance for these effects have been developed in the field of medicine, they can be appropriately applied to questions concerning genotype x environment interactions in crop improvement.

From a breeder's point of view, interaction is tolerable as long as it does not affect the rank orders. So the question arises, under which circumstances does interaction become rank-interaction (Haldane, 1946; Baker, 1988, 1990).

Some interesting statistical test procedures have been published by Azzalini and Cox (1984), Berger (1984), Gail and Simon (1985), and Zelterman (1990). For applications with agricultural crops, see, for example,

Baker (1988, 1990), and Virk and Mangat (1991). Another interesting statistical approach for a test of interaction under order restrictions (identical rankings of genotypes in different environments) has been developed by Denis (1979, 1982). Closely connected with these concepts of crossover interactions versus noncrossover interactions is the mathematical concept of separability versus non-separability introduced by Gregorius and Namkoong (1986).

The shifted multiplicative model by Seyedsadr and Cornelius (1992) was originally developed for analyzing nonadditivity in a two-way classification. It provides an interesting statistical tool for the investigation of separability versus non-separability (Cornelius et al., 1992; Crossa et al., 1993).

Although all these statistical procedures are concerned with rank orders and changes of rankings, they will not be further discussed in this paper since they are all based on the original absolute data. But here we are only concerned with analyses based on the ranks of genotypes in different environments.

For data sets with more than two genotypes and more than two environments, the genotype x environment interactions are commonly calculated by analysis of variance techniques leading to an estimated variance component for genotype x environment interactions. For a two-way table with $\underline{l}$ genotypes (rows) and $\underline{m}$ environments (columns), the relationships between the numerical amount of the variance component of genotype x environment interactions and the rank changes of the genotypes in different environments are of particular interest in the field of practical applications.

With regard to these problems, many prominent statements have been published. One should be quoted here: "Procedure for estimation of genotype x environment interaction variances has been outlined and numerous estimates have actually been obtained and reported. In general, however, these have been interpreted with reference to genotype-environment interaction as a source of nongenetic variance among selection units of one kind or another.

In contrast, quantitative genetics has not shown how such estimates can logically be employed in decisions concerning target populations of environments. For example, I do not know how large a component of genotype-environment interaction variance can be, as a fraction or multiple of genetic variance, when there is no variation among environments in the rank order values of genotypes (or family groups of genotypes) nor am I aware that this has been discussed in the literature" (Comstock, 1977).

The purpose of this paper is to present some contributions to this topic or, more precisely, to discuss some approaches for the analysis of genotype x environment interactions by using the ranks of genotypes in different environments. Nonparametric statistics for genotype x environment interactions based on ranks provide a useful alternative to parametric approaches currently used, which are based on absolute data. Some essential

238

advantages of nonparametric statistics compared to parametric ones are: reduction or even avoidance of the bias caused by outliers, no assumptions are needed about the distribution of the analyzed values, homogeneity of variances, and additivity (linearity) of effects are not necessary requirements. Statistics based on ranks and rank-orders are often easy to use and interpret.

If some of the necessary assumptions are violated, the validity of the inferences obtained from the standard statistical techniques, for example, analysis of variance, may be questionable or lost. Compared with this fact, the results of nonparametric estimation and testing procedures, which are based on ranks, must be quite reliable.

Most of the following nonparametric approaches are not at all new. But many of them have been published in periodicals that are not commonly available to plant breeders and other scientists. Therefore, many of these attractive statistical tools may not have received proper attention by a broader audience.

The following investigations are divided into three separate parts:
1. Nonparametric estimation and testing of genotype x environment interactions by ranks.
2. Application of Spearman's rank-correlation and Kendall's coefficient of concordance a) to quantify genotype-environment relationships and b) to group environments or genotypes.
3. Genotype-specific measures for a quantitative estimation of phenotypic stability.

These topics will be discussed from the application point of view without going into any details of nonparametric statistics.

II. THEORETICAL INVESTIGATIONS

A. INTRODUCTORY COMMENTS

All following investigations are based on a two-way table with l rows = genotypes and m columns = environments. The value of genotype i in environment j in replication k is denoted by x_{ijk} (i=1,2,..,l; j=1,2,..,m; k=1,2,..,n) and we apply the well-known dot notation:

$$x_{ij.} = \sum_{k=1}^{n} x_{ijk}, \bar{x}_{ij.} = \frac{1}{n} \sum_{k=1}^{n} x_{ijk} \quad \text{etc.}$$

For this two-way classfication, we use the well-known model

$$x_{ijk} = \mu + \alpha_i + \beta_j + (\alpha\beta)_{ij} + e_{ijk} \tag{1}$$

where

μ = overall mean
α_i = effect of genotype i

β_j = effect of environment j

$(\alpha\beta)_{ij}$ = interaction between genotype i and environment j and

e_{ijk} = experimental error.

These terms are usually estimated by

$$\hat{\mu} = \bar{x}_{...}, \hat{\alpha}_i = \bar{x}_{i..} - \bar{x}_{...}, \hat{\beta}_j = \bar{x}_{.j.} - \bar{x}_{...}, \left(\alpha\beta \right)_{ij} = \bar{x}_{ij.} - \bar{x}_{i..} - \bar{x}_{.j.} + \bar{x}_{...}$$

and $\hat{e}_{ijk} = x_{ijk} - \bar{x}_{ij.}$.

If we shift from the individual values x_{ijk} in (1) to means, one obtains

$$\bar{x}_{ij.} = \mu + \alpha_i + \beta_j + (\alpha\beta)_{ij} + \bar{e}_{ij.}. \tag{2}$$

The mean error $\bar{e}_{ij.}$ is often assumed to be zero. If the analysis of genotype x environment interactions should be based on means, we denote (for abbreviation) $\bar{x}_{ij.}$ by x_{ij} and (2) reduces to

$$x_{ij} = \mu + \alpha_i + \beta_j + \underbrace{\left[(\alpha\beta)_{ij} + \bar{e}_{ij.} \right]}_{\gamma_{ij}} = \mu + \alpha_i + \beta_j + \gamma_{ij}. \tag{3}$$

If the genotypes are ranked separately within each environment j, giving the lowest value a rank of 1 and the highest value a rank of l, the environmental effects have no influence on these ranks. Let r_{ij} be the rank of $x_{ij} \cdot r_{ij}$ gives the rank of $\left[\alpha_i + (\alpha\beta)_{ij} + \bar{e}_{ij.} \right]$. Using this ranking procedure, therefore, one applies a confounding and simultaneous evaluation of genotypic effects and interaction effects (and the mean error). But, for many applications, such a confounding must be considered as a considerable disadvantage. One important example for this situation is the estimation of the phenotypic stability of genotypes over environments where the breeder often intends to measure yield stability independent from the yield level (discussed later).

An adjustment of the phenotypic values x_{ijk} with regard to the genotypic effects can be realized by the transformation:

$$x_{ijk}^* = x_{ijk} - \left(\bar{x}_{i..} - \bar{x}_{...} \right). \tag{4}$$

Ranking the transformed values $\overline{x_{ij}^*}$ (denoted by x_{ij}^*) separately within each environment j leads to the ranks r_{ij}^* which now reflect the ranks of $\left[(\alpha\beta)_{ij} + \bar{e}_{ij}.\right]$. Using this approach, interaction effects (and the mean error) only are utilized.

The rank orders of the genotypes, which are obtained by using the original data x_{ij} or by using the transformed values x_{ij}^*, may be quite different (see: numerical example).

B. NONPARAMETRIC ESTIMATION AND TESTING OF GENOTYPE x ENVIRONMENT INTERACTIONS BY RANKS

In this section, four nonparametric approaches are presented without giving any theoretical discussion of the mathematical and statistical facts and relationships from nonparametric statistics. No proofs and no theoretical comparison of the efficiency of these nonparametric procedures are included. Here, we only mention the computational performance of the different approaches and describe the necessary steps for a numerical calculation of the test statistics for testing the existence of genotype x environment interactions. For all statistical discussions, we refer to the relevant literature cited.

Approaches 1-3 (Bredenkamp, Hildebrand, Kubinger) are based on the usual statistical concept of an interaction, that is: deviations from additivity in the linear model. But, the 4th approach (van der Laan and de Kroon) uses a modified concept of interaction (rank-interaction), which can be defined and described as follows: rank-interaction (discordance) exists if the rankings of the genotypes are not identical in different environments. If these rankings are identical, rank-interaction is said not to exist (concordance) (de Kroon and van der Laan, 1981; de Kroon, 1986).

Not all the interaction effects in the usual sense (deviations from additivity in the linear model) are of such a magnitude as to cause changes of the rankings of genotypes in different environments (Figure 1). In this concept of rank-interaction, the common interactions are only utilized insofar as they lead to different rankings of the genotypes. This is, of course, a concept with particular relevance for the breeder who is interested in selection.

C. BREDENKAMP-APPROACH
i)   Transformation of the original data
$x_{ijk}(i = 1,2,\ldots,l; j = 1,2,\ldots,m; k = 1,\ldots,n)$ into ranks
$R_{ijk} : x_{ijk} \rightarrow R_{ijk}.$

ii) The test statistic for a test of genotype x environment interaction

$$\frac{12lm}{N^2(N+1)}\sum_{i=1}^{l}\sum_{j=1}^{m}\left(R_{ij.}^2 - \left\{1/m^2\right\}R_{i..}^2 - \left\{1/l^2\right\}R_{.j.}^2\right) + 3(N+1) \qquad (5)$$

is approximately chi-squared distributed with (l-1)(m-1) degrees of freedom and N=nlm (Bredenkamp, 1974; Lienert, 1978; Erdfelder and Bredenkamp, 1984; Bortz et al., 1990).

D. HILDEBRAND-PROCEDURE

i) Transformation of the original data x_{ijk} (i=1,2,..,l; j=1,2,..,m; k=1,2,..,n) into x_{ijk}^* by $x_{ijk} \rightarrow x_{ijk}^* = x_{ijk} - \bar{x}_{i..} - \bar{x}_{.j.} + 2\bar{x}_{...}$ (6)

ii) Transformation of the x_{ijk}^* into ranks $R_{ijk}: x_{ijk}^* \rightarrow R_{ijk}$.

iii) The test statistic for a test of genotype x environment interaction

$$\frac{12}{lm(N+1)} \sum_{i=1}^{l} \sum_{j=1}^{m} (\bar{R}_{ij.} - \bar{R}_{i..} - \bar{R}_{.j.} + \bar{R}_{...})^2 \qquad (7)$$

is approximately chi-squared distributed with (l-1)(m-1) degrees of freedom and N=nlm. (Hildebrand, 1980; Kubinger, 1986; Bortz et al., 1990).

E. KUBINGER-TEST

i) Transformation of the original data x_{ijk} (i=1,2,..,l; j=1,2,..,m; k=1,2,..,n) into ranks

$R_{ijk}: x_{ijk} \rightarrow R_{ijk}$.

ii) Transformation of the R_{ijk} into R_{ijk}^* by

$$R_{ijk} \rightarrow R_{ijk}^* = R_{ijk} - \bar{R}_{i..} - \bar{R}_{.j.} \qquad (8)$$

iii) Transformation of the R_{ijk}^* into ranks $R_{ijk}^{**} : R_{ijk}^* \rightarrow R_{ijk}^{**}$.

iv) The test statistic for a test of genotype x environment interaction

$$\frac{12}{lm(N+1)} \sum_{i=1}^{l} \sum_{j=1}^{m} (\bar{R}_{ij.}^{**} - \bar{R}_{i..}^{**} - \bar{R}_{.j.}^{**} + \bar{R}_{...}^{**})^2 \qquad (9)$$

is approximately chi-squared distributed with (l-1)(m-1) degrees of freedom and N=nlm. (Kubinger, 1986; Bortz et al., 1990).

F. RANK-INTERACTION BY VAN DER LAAN AND DE KROON

i) Separately for each j: Transformation of the original data x_{ijk} (i=1,2,..,l; j=1,2,..,m; k=1,2,..,n) into ranks $R_{ijk}: x_{ijk} \rightarrow R_{ijk}$.

ii) The test statistic for a test of genotype x environment rank-interaction

$$\frac{12}{n^2 l(nl+1)}\left(\sum_{i=1}^{l}\sum_{j=1}^{m}R_{ij.}^2 - \frac{1}{m}\sum_{i=1}^{l}R_{i..}^2\right) \tag{10}$$

is approximately chi-squared distributed with (l-1)(m-1) degrees of freedom. (de Kroon and van der Laan, 1981; de Kroon, 1986; van der Laan, 1987).

Since this approach is based on a modified definition of genotype x environment interaction, the null-hypothesis (H_0) and the alternative hypothesis (H_1) will be expressed more explicitly than those for the previous approaches 1-3 that are based on the usual concept of genotype x environment interaction.

$$H_0: \alpha_1+(\alpha\beta)_{1j}=\alpha_2+(\alpha\beta)_{2j}= ..... = \alpha_l+(\alpha\beta)_{lj} \tag{11}$$

$$(j=1,2,..,m)$$

H_1: 'H_0 does not hold'. $\qquad\qquad$ (12)

H_1 implies that for at least one j, at least one of the equalities stated under H_0 is violated.

On ranking within each environment j, the ranks in cell (i,j) are denoted by R_{ij1}, R_{ij2}, ..., R_{ijn}.

In de Kroon and van der Laan (1981), it was shown that the test statistic T

$$T = \frac{12}{nl(nl+1)}\sum_{i=1}^{l}\sum_{j=1}^{m}\sum_{k=1}^{n}(\overline{R}_{ij.}-\overline{R}_{.j.})^2 \tag{13}$$

can be partitioned into two terms T_1 and T_2 with

$$T_1 = \frac{12}{nl(nl+1)}\sum_{i=1}^{l}\sum_{j=1}^{m}\sum_{k=1}^{n}(\overline{R}_{i..}-\overline{R}_{...})^2 \tag{14}$$

$$T_2 = T - T_1 \tag{15}$$

T_1 is sensitive to differences between the α_i, while T_2 is sensitive to differences between the $(\alpha\beta)_{ij}$. More precisely, T_1 is sensitive to concordance between the different environments, with respect to the ordering of $\alpha_1+(\alpha\beta)_{1j},.....,\alpha_l+(\alpha\beta)_{lj}$, and T_2 is sensitive to discordance between the environments.

The authors proved that for the common number n of observations per cell tending to infinity, T_1 and T_2 are asymptotically independent under H_0. The asymptotic distributions under H_0 are derived. For T, a chi-squared distribution with m(l-1) degrees of freedom, for T_1, a chi-squared distribution with 1-1 degrees of freedom, and for T_2 a chi-squared distribution with $(l-1)(m-1)$ degrees of freedom.

The exact distribution under H_0 of the test statistic T was tabulated for some values of 1, m, and n (de Kroon and van der Laan, 1981). Furthermore, the authors give some recommendations on how to use the three test statistics T, T_1, and T_2 dependent on the alternative of the null hypothesis which is considered important. This passage from the original publication will be quoted here: "Obviously T serves as an omnibus test, whereas T_1 will be used if one is mainly concerned with differences between the αi and T_2, if one aims at proving the presence of rank-interaction. The three test statistics have one-sided critical regions: only large values lead to rejection of the null hypothesis. One can also use both statistics T_1 and T_2 simultaneously, each at confidence level 1/2 α, rejecting H_0 if at least one of the outcomes of T_1 or T_2 is larger than the corresponding critical value. In this way, it is possible to carry out an omnibus test with the possibility of detecting simultaneously which component of H_0 is not true. This procedure is somewhat conservative, but the difference between α and $1-(1-1/2\ \alpha)^2$, being equal to 1/4 α^2. is for small values of α negligible ($\alpha=0.05 \rightarrow 1/4^2\alpha = 0.0006$). The combined confidence level $1-(1-1/2\ \alpha)^2$, for T_1 and T_2 is calculated as if T_1 and T_2 are independent, which is obviously not true. However, as we mentioned before, it can be proved that under H_0, the statistics T_1 and T_2 are asymptotically independent, as n tends to infinity. In agreement with the concepts of concordance and discordance, the power of the test procedure by the statistic T_2 depends on the presence of differences between the αi's which are tested by means of the statistic T_1. Otherwise stated: the power of T_1 will become smaller as the amount of rank-interaction (discordance) becomes larger, whereas the power of T_2 will become smaller as concordance becomes larger. This statement is a consequence of the obvious fact that T_1 and T_2 are not independent under the alternative. After all, this is not amazing, since as soon as there is perfect concordance, there cannot be discordance and if discordance is as large as possible, there cannot be concordance" (de Kroon and van der Laan, 1981).

Based on extensive data sets from German official registration trials (1985-1989) with several agricultural crops [winter oilseed rape (Brassica napus), faba bean (Vicia faba), oat (Avena sativa), fodder beet (Beta vulgaris), sugar beet (Beta vulgaris)] the four nonparametric procedures (Bredenkamp, Hildebrand, Kubinger, van der Laan-de Kroon) and a 'classical' analysis of variance had been studied and compared recently. A comparison of the methods can be easily carried out by comparing the numerical values of the test statistics (chi-squared values for identical numbers of degrees of freedom for the four nonparametric procedures and F-values for the analysis of variance which can be approximately expressed as chi-squared values for the same number of degrees of freedom).

For each of the factors, cultivars, locations, and interactions, one obtains the relationships: .

{analysis of variance}$\geq${Hildebrand}$\cong${Kubinger}>{van der Laan-de Kroon} >{Bredenkamp}. This empirically obtained result is in line with the theoretical expectations.

If the assumptions for parametric methods cannot be accepted, the van der Laan-de Kroon method is recommended for the crossover interaction concept, while the Hildebrand or Kubinger method should be applied for the usual interaction concept. Both procedures (Hildebrand, Kubinger) are approximately equivalent. The Bredenkamp method cannot be recommended. These results are yet unpublished (Hühn and Léon, 1995).

G. APPLICATION OF SPEARMAN'S RANK-CORRELATION AND KENDALL'S COEFFICIENT OF CONCORDANCE TO QUANTIFY GENOTYPE-ENVIRONMENT RELATIONSHIPS AND TO GROUP ENVIRONMENTS OR GENOTYPES

From a selection point of view, the breeder is much more interested in rankings and changes of rankings than in an estimation of interaction effects per se. Two environments with quite different yielding capacities but similar rankings of the tested genotypes must be considered (in this selection point of view) to be more similar among each other than two environments with similar yielding capacities but different rankings of the tested genotypes. Grouping of environments that are similar in this sense must be of particular interest for the establishment of test environments in practical breeding work as well as in designing official registration trials. The more similar the two environments, the more similar the rankings of the tested genotypes. That means the "distance" between these two environments can be quantitatively expressed by Spearman's rank correlation coefficient between the rankings of the genotypes in these two environments. Based on this very simple measure of distance between environments, numerous techniques of cluster analysis can be applied to group the environments with respect to this measure of similarity.

Three well-known and frequently used clustering procedures will be mentioned here:
1. average-linkage-algorithm, where the similarity of an individual with a cluster is defined by the mean similarity to all cluster members.
2. complete-linkage-algorithm, where the similarity of an individual with a cluster is defined by the similarity with the most dissimilar member of the cluster.
3. single-linkage-algorithm, where the similarity of an individual with a cluster is defined by the similarity with the most similar member of the cluster.

But, for an analysis of genotype x environment interactions and for a classification of environments, the algorithm no. 3 seems inappropriate. The most suitable method is the "complete-linkage-algorithm" since the similarity of a certain environment with a group of environments should be determined by the lowest rank correlation coefficient.

The necessary next step must be the generalization from two to more environments. The similarity resp. dissimilarity of the rankings of the tested genotypes in several environments can be quantitatively expressed by Kendall's parameter of concordance W. W is defined as the ratio of the observed variance of the rank sums and the maximum variance of the rank sums. With the exception of a certain standardization, W has the meaning of the mean of all Spearman rank correlation coefficients for all possible pairwise comparisons of the rankings of the genotypes in different environments (Kendall, 1962; Gibbons, 1985). If one denotes the Spearman rank correlation coefficient between environments j and j' with $\rho_{jj'}$ the following relation between W and the $\rho_{jj'}$'s is well-known:

$$\bar{\rho} = \frac{\sum\limits_{j<j'} \rho_{jj'}}{\binom{m}{2}} = \frac{mW-1}{m-1}. \tag{16}$$

(16) can be rewritten as

$$W = \bar{\rho} + \frac{1-\bar{\rho}}{m}. \tag{17}$$

From a breeder's point of view, it is desirable to have a maximum association (W=1) between the rankings in different environments. W=1 implies identical rank orders of genotypes in each environment, which means $\rho_{jj'} = 1$ for each pair j and j' and, therefore, $\bar{\rho} = 1$.

If the ranks of genotypes are assigned at random within each environment, then all of the expected individual Spearman coefficients and, hence, their expected mean too are zero ($\rho_{jj'} = 0$ for each pair j and j' with $\bar{\rho} = 0$). This case corresponds to an expected W of 1/m. For an increasing number of environments (m → ∞) this value tends to W=0.

For large W, one might expect a high chance that the ranking will be similar to or even the same as in other environments if this set of genotypes is transferred to another environment (not too different from the ones used for computing W). Such a high predictability can be obscured, of course, by two sources: genotype-environment interaction and experimental error.

For applications, it is of particular interest to know whether the concordance between the rankings significantly differs from one completely due to chance.

Several statistical tests of significance for W are available. Based on Fisher's z-distribution

$$z = \frac{1}{2}\ln\frac{(m-1)W}{1-W} \tag{18}$$

with $v_1 = (l-1) - \dfrac{2}{m}$ and $v_2 = (m-1)(l-1) - \dfrac{2(m-1)}{m}$ degrees of freedom.

For a sufficiently large number of genotypes, an asymptotic test based on the chi-squared distribution can be applied:

$$\chi^2 = m(l-1)W \tag{19}$$

with l-1 degrees of freedom (Kendall, 1962; Sprent, 1989). For the handling of existing ties, we refer to Kendall (1962) and Lienert (1978).

If W is significant, there are differences in the genetic effects α_i. Note that m(l-1)W is the statistic of the Friedman-test. If W is significant, the Friedman-test for equality of all genotypic means will also be significant (Gibbons, 1985). If W is nonsignificant, this may have two different causes. Either there are no differences in the genetic effects, or the γ_{ij} effects, i.e., the experimental error and genotype-environment-interaction, are so large in relation to the genetic effects that differences in the α_i's are obscured. The latter reason is more probable in practical applications. The Friedman-test was originally developed to test the equality of a location parameter (no treatment differences, i.e., equal genotypic means). It can be simultaneously used as a test for consistency of rankings (Sprent, 1989).

In a recent publication (Hühn et al., 1993) relationships between parametric approaches (variance components) and nonparametric approaches (rank orders) had been investigated: on the parametric side, the l x m x n data set can be characterized by the variance components, i.e., genotypic, environmental, interaction, and error variances. On the nonparametric side, the same data set can be described by the similarity respective dissimilarity of the rank orders of genotypes in different environments. Relationships between the similarity of the rank orders and these variance components or functions of them would be of particular interest in the point of view of Comstock's previously cited statement. Furthermore, such investigations can help to clarify the nature of rank-interaction.

For the extensive crossclassified data sets (with replications) from German official registration trials (see, preceding section 'Nonparametric estimation and testing of genotype x environment interactions by ranks') the variance components σ_α^2 (genotypes), σ_β^2 (environments), $\sigma_{\alpha\beta}^2$ (interaction), and σ_ε^2 (error) were estimated using the linear model in (1) by applying elementary statistical theory (for a mixed effects model with genotypes = fixed and all other factors = random). If, however, we only use the cell means, interaction and error cannot be separated and the estimated variance components are: genotypes, environments, and residual (= interaction plus mean error).

Relationships between the similarity of the rank orders (measured by Kendall's W) and functions of the diverse variance components had been developed in Hühn et al. (1993). W had been shown to be approximately equal to the ratio of genotypic variance and sum of genotypic variance plus residual variance:

$$W \cong \frac{\sigma_\alpha^2}{\sigma_\alpha^2 + \sigma_{\alpha\beta}^2 + \sigma_\varepsilon^2/n} \qquad \text{(with replications)}$$

$$W \cong \frac{\sigma_\alpha^2}{\sigma_\alpha^2 + \sigma_\gamma^2} \qquad \text{(without replications)} \qquad (20)$$

where $\sigma_\gamma^2 = \sigma_{\alpha\beta}^2 + \sigma_\varepsilon^2/n$.

This expression (20) gives a quite clear and even simple relationship between the nonparametric coefficient of concordance W and the parametric ratio $\sigma_\alpha^2/(\sigma_\alpha^2 + \sigma_\gamma^2)$. This result answers Comstock's question and comment given before. Furthermore, it provides a clear and meaningful interpretation of W, since (20) can be rewritten as:

$$W \cong \frac{1}{1 + \sigma_\gamma^2/\sigma_\alpha^2}. \qquad (21)$$

W depends on only one parameter: the ratio $\sigma_\gamma^2/\sigma_\alpha^2$ of both variance components. This result (21) can be easily explained by considering Figures 2a and b for two genotypes and two environments.

In both figures the effects γ_{ij} (i=1,2; j=1,2) are the same, but the differences between the two genotypic effects are different (large in Figure 2a, small in Figure 2b). In Fig. 2a, the interaction does not cause a rank change because it is small in relation to the difference between the genotypic effects. More precisely, the rank-interaction in Fig. 2b (in environment 1) occurs because the absolute value of $(\gamma_{11}-\gamma_{21})$ is larger than that of $(\alpha_2-\alpha_1)$:

$$x_{21} - x_{11} = (\mu + \alpha_2 + \beta_1 + \gamma_{21}) - (\mu + \alpha_1 + \beta_1 + \gamma_{11})$$
$$= (\alpha_2 - \alpha_1) + (\gamma_{21} - \gamma_{11}) > 0 \text{ in Fig. 2a} \qquad (22)$$
$$< 0 \text{ in Fig. 2b}$$

and this leads to

$$(\alpha_2 - \alpha_1) > (\gamma_{11} - \gamma_{21}) \text{ in Fig. 2a}$$
$$(\alpha_2 - \alpha_1) < (\gamma_{11} - \gamma_{21}) \text{ in Fig. 2b.} \qquad (23)$$

In terms of variances, σ_γ^2 is the same in both figures, while σ_α^2 is larger in Fig. 2a than in Fig. 2b. These considerations make plausible that rank-interaction is more likely to occur the smaller σ_α^2 is relative to σ_γ^2, or the larger σ_γ^2 is relative to σ_α^2. The similarity respective dissimilarity of the rank orders of genotypes in different environments will, therefore, be determined by the ratio $\sigma_\gamma^2 \, / \, \sigma_\alpha^2$. This conclusion is in perfect accordance with the approximation (21).

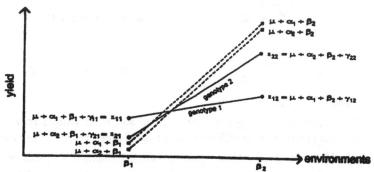

Figure 2a. Schematic representation for 2 genotypes and 2 environments (without rank change).

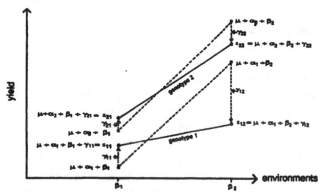

Figure 2b. Schematic representation for 2 genotypes and 2 environments (with rank change).

The approximation (20) provides an interesting further interpretation of Kendall's W; by (20) one obtains:

$$\frac{1-W}{W} = \frac{measure \ of \ discordance}{measure \ of \ concordance} \equiv \frac{\sigma_\gamma^2}{\sigma_\alpha^2}. \tag{24}$$

That means the discordance (1-W) expressed in units of t he concordance (W) equals the ratio of the two variance components σ_γ^2 and σ_α^2.

For $\sigma_\gamma^2 = 0$ it follows that W=1 (perfect concordance). $\sigma_\gamma^2 = 0$ is equivalent to $\sigma_{\alpha\beta}^2 = \sigma_\varepsilon^2 = 0$. For zero variance components for interaction and for error, one obtains identical rank orders of genotypes in different environments (W=1), i.e., maximum association. Relationship (20), therefore, is in complete agreement with the expectations.

Approximation (20) provides a useful condition for a consideration of consistency among rankings. It links the nonparametric view (rank orders - W) and the parametric characterization (variance components - σ_α^2 and σ_Y^2) of the two-way layout.

For some applications it may be of interest to know whether the concordance between the rankings significantly differs from one completely due to chance. The null hypothesis (H_o) and the alternative hypothesis (H_1) of this test are H_o: the rankings are effectively random and H_1: there is evidence of concordance in rankings. For sufficiently large numbers of genotypes and environments, an asymptotic test based on the chi-squared distribution can be applied as has been given in (19). Combination with (20) gives the following condition for significant concordance in rankings:

$$\sigma_\alpha^2 \geq \lambda_1 \cdot \sigma_Y^2 \qquad\qquad (25)$$

where $\lambda_1 = \chi_{l-1}^2 \ / \left[m(l-1) - \chi_{l-1}^2 \right]$ *with* χ_{l-1}^2 = value of the chi-squared distribution with l-1 degrees of freedom (and any given error probability). That means in a two-way classification a significant consistency of rankings occurs if the genotypic variance exceeds a certain multiple λ_1 of the residual variance σ_Y^2. A better approximation than (19) had been suggested by Iman and Davenport (1980): (m-1)W/(1-W) has approximately an F-distribution with l-1 and (m-1)(l-1) degrees of freedom (Sprent, 1989). Combination of $F_{l-1,(m-1)(l-1)}$=(m-1)W/(1-W) with (20) leads to the following condition for significant concordance in rankings:

$$\sigma_\alpha^2 \geq \lambda_2 \cdot \sigma_Y^2 \qquad\qquad (26)$$

with $\lambda_2 = F \ / \ (m-1)$.

Numerical values of λ_1 *and* λ_2 are presented in Tables 1a and 1b (for some error probabilities, numbers $\underline{l}$ of genotypes and numbers $\underline{m}$ of environments). For all cases one obtains $\lambda_1 > \lambda_2$. The differences between λ_1 *and* λ_2 are largest for small l, small m and small error probabilities (Tables 1a and 1b).

The null hypothesis of all the aforementioned tests of significance (H_o = all the rankings are effectively random) implies that the similarity between the rankings in any two environments is non-significant. That means that the 'true' Spearman rank correlation coefficients are zero for all possible pairs of environments. H_o is rejected if the genotypic variance exceeds a certain fraction $\left(\lambda_1 \ or \ \lambda_2 \right)$ of the residual variance. Rejection of H_o, however, implies the existence of at least one pair of environments j and j'

Table 1a. Numerical values of λ_1 for some error probabilities, numbers l of genotypes and numbers m of environments.

	error probability								
	10 %			5 %			1 %		
m \\ l	5	10	15	5	10	15	5	10	15
5	0.637	0.242	0.150	0.903	0.312	0.188	1.975	0.497	0.285
10	0.485	0.196	0.123	0.603	0.232	0.144	0.929	0.318	0.192
15	0.431	0.178	0.112	0.512	0.204	0.128	0.714	0.263	0.162
20	0.402	0.168	0.106	0.465	0.189	0.119	0.616	0.236	0.146
25	0.383	0.161	0.102	0.436	0.179	0.113	0.559	0.219	0.136
30	0.370	0.156	0.099	0.416	0.173	0.109	0.520	0.207	0.129
35	0.360	0.153	0.097	0.401	0.167	0.106	0.493	0.198	0.124
40	0.352	0.150	0.095	0.389	0.163	0.103	0.471	0.191	0.120

Table 1b. Numerical values of λ_2 for some error probabilities, numbers l of genotypes and numbers m of environments.

l	error probability 10%			5%			1%		
m	5	10	15	5	10	15	5	10	15
5	0.584	0.235	0.147	0.752	0.293	0.182	1.194	0.433	0.263
10	0.453	0.191	0.121	0.539	0.222	0.140	0.737	0.293	0.183
15	0.407	0.174	0.110	0.469	0.197	0.125	0.605	0.248	0.156
20	0.382	0.164	0.105	0.432	0.184	0.117	0.539	0.224	0.142
25	0.366	0.158	0.101	0.408	0.175	0.111	0.498	0.209	0.132
30	0.355	0.154	0.098	0.392	0.168	0.107	0.470	0.199	0.126
35	0.346	0.150	0.096	0.379	0.164	0.104	0.449	0.191	0.121
40	0.339	0.148	0.094	0.370	0.160	0.102	0.433	0.185	0.118

252

with a Spearman rank correlation coefficient different from zero: $\rho_{jj'} \neq 0$. That means either $\rho_{jj'} = 1$ or there is a rank interaction. For a more detailed discussion of these results, see Hühn and Piepho (1994).

The grouping and categorization of environments with similar rankings of the tested genotypes can be carried out by quite different procedures:

1. Application of Spearman's rank correlation coefficient as a measure of distance between two environments.
2. Application of Kendall's coefficient of concordance W as a measure of similarity of a group of environments.
3. Using different strategies of clustering from the "classical" cluster analysis (for example: average-linkage-method, complete-linkage-method, etc.) (see: Abou-El-Fittouh et al., 1969; Lin and Butler, 1990).

An obvious idea of grouping would be the requirement of the two criteria

a) the size of the groups should be as large as possible and
b) the rank correlation coefficients within each group for all possible pairs of environments are not statistically significantly different from each other.

For a certain given number h with 1<h<m those h environments can be derived (by using the W-parameter) where the h rankings of the tested genotypes are as much similar as possible amongst each other (= largest W-value). For each of the $\binom{m}{h}$ possible groupings of h environments, W is calculated and the group with the maximum value of W is selected (see: numerical example). This procedure may be of some interest if in variety and registration trials the number of test environments must be reduced and restricted to a given number h.

The previous procedures have been introduced for a classification of the environments. But, of course, they can be applied in the same way for a classification of the genotypes. But, these topics will not be further followed up in this paper.

H. GENOTYPE-SPECIFIC MEASURES FOR A QUANTITATIVE ESTIMATION OF PHENOTYPIC STABILITY

In the previous nonparametric procedures for a test of the existence or non-existence of genotype x environment interactions, the interaction effects between **all** the tested genotypes and **all** the included environments are utilized. But, in the field of practical applications, this global measure of interactions and the statistical inferences based on this approach are often of minor importance. The breeder, for example, is much more interested in quantitative characterizations of the individual genotypes and of the individual environments.

These considerations are closely connected with the problem of an estimation of phenotypic stability. The term "phenotypic stability" characterizes the ability of a genotype to react to varying environmental conditions with "stable" responses. There are quite different concepts and definitions of "stability." For each of the proposed concepts of stability, numerous stability parameters have been developed to measure the stability quantitatively (Utz, 1972; Freeman, 1973; Wricke and Weber, 1980; Lin et al., 1986; Becker and Léon, 1988). Some years ago, several rank-based measures of stability were proposed (Hühn, 1979). These are based on the ranks of genotypes in each environment as described previously. A genotype i is stable over environments if its ranks are similar over environments. It has maximum stability if its ranks are the same over environments. These rank stability measures from Hühn (1979) define stability in the sense of homeostasis or the ability of a genotype to stabilize itself in different environments. Only this stability concept will be applied in this contribution. Each statistic that measures the similarity or dissimilarity of the ranks for each row = genotype can be used as an appropriate stability parameter.

All the following investigations are based on the means $\bar{x}_{ij.} \equiv x_{ij}$ of the genotypes in different environments (see previous discussion) with r_{ij} = rank of genotype i in environment j if the genotypes are separately ranked within each environment.

Two of the parameters proposed by Hühn (1979) have been recently reviewed by new investigations (Nassar and Hühn, 1987; Hühn and Nassar, 1989 and 1991; Hühn, 1990; Piepho, 1992):

$$S_i^{(1)} = \frac{\sum\limits_{j<j'}\left|r_{ij} - r_{ij'}\right|}{\binom{m}{2}} = \frac{2}{m(m-1)}\sum_{j=1}^{m-1}\sum_{j'=j+1}^{m}\left|r_{ij} - r_{ij'}\right|. \qquad (27)$$

$S_i^{(1)}$ is the mean of the absolute rank differences of a genotype i over the m environments (= Gini's mean difference).

$$S_i^{(2)} = \frac{1}{m-1}\sum_{j=1}^{m}(r_{ij} - \bar{r}_{i.})^2 \quad with \quad \bar{r}_{i.} = \frac{1}{m}\sum_{j=1}^{m}r_{ij}. \qquad (28)$$

$S_i^{(2)}$ is the common variance among the ranks over the m environments. The $\bar{r}_{i.}$ can be interpreted to be the expectation of each r_{ij} under the hypothesis of maximum stability (= equal ranks). $S_i^{(2)}$, therefore, has the meaning of the deviation of the observed ranks r_{ij} from maximum stability.

For a genotype i with maximum stability one obtains $S_i^{(1)} = S_i^{(2)} = 0$. Both stability parameters $S_i^{(1)}$ = mean absolute rank difference and $S_i^{(2)}$ = variance of the ranks for ranks of a genotype in different environments have been frequently used as rank-based measures in the field of practical applications, for example: Hühn, 1979; Thomson and Cunningham, 1979; Kleinschmit, 1983; Skrøppa, 1984; Böhm and Schuster,

1985; Hühn and Léon, 1985; Léon, 1985, 1986; Clair and Kleinschmit, 1986; Magnussen and Yeatman, 1986; Nachit and Ketata, 1986; Rognli, 1986; Becker, 1987; Wanyancha and Morgenstern, 1987; Weber and Wricke, 1987; Becker and Léon, 1988; Léon and Becker, 1988; Hühn, 1990; Krenzer et al., 1992; Piepho and Lotito, 1992.

Environmental effects have no influence on stability as defined in (27) and (28). However, differences among genotypes would have an effect on the $S_i^{(1)}$ and $S_i^{(2)}$ stability measures and may lead to differences in stability among genotypes when in fact there is no genotype-environment interaction. To avoid this ambiguity, one may correct the x_{ij} values for the genotypic effects by applying the transformation (4). The stability measures $S_i^{(1)}$ and $S_i^{(2)}$ should be computed now using the ranks based on the transformed values (4) if one intends to measure phenotypic yield stability independent from the yield level. This computational procedure is further strengthened by the fact that the most commonly used statistical model considers genotypic effects as fixed effects while all other effects are defined to be random (Freeman, 1973; Piepho, 1992).

The null-hypothesis of no genotype-environment interaction effects implies "All genotypes are equally stable" (with maximum stability). For this situation, approximate but easily applied statistical tests of significance based on the normal distribution were developed by Nassar and Hühn (1987) and Hühn and Nassar (1989, 1991).
The statistic

$$Z_i^{(m)} = \frac{\left[S_i^{(m)} - E\{ S_i^{(m)} \} \right]^2}{\mathrm{var}\{ S_i^{(m)} \}} \;,\; m = 1,2 \tag{29}$$

has an approximate chi-square distribution with 1 degree of freedom and the statistic

$$S^{(m)} = \sum_{i=1}^{l} Z_i^{(m)} \;,\; m = 1,2 \tag{30}$$

may be approximated by a chi-square distribution with 1 degrees of freedom with $E\{ S_i^{(m)} \}$ = expectation (= mean) of $S_i^{(m)}$ and $\mathrm{var}\{ S_i^{(m)} \}$ = variance of $S_i^{(m)}$.

The test statistic $S^{(m)}$ can be used to decide whether or not there are significant differences in stability between genotypes. To test the stability of a single genotype i, the test statistic $Z_i^{(m)}$ can be applied where the null-hypothesis that the mean stability for the genotype i is $E\{ S_i^{(m)} \}$ is tested against the alternative that the mean stability deviates from this expectation.

If the null hypothesis is rejected, the genotype i may be stable $\left[i.e., S_i^{(m)} < E\{S_i^{(m)}\}\right]$ or unstable $\left[i.e., S_i^{(m)} > E\{S_i^{(m)}\}\right]$. Further aspects of these tests of significance were discussed in Hühn (1990), Hühn and Nassar (1991), Krenzer et al. (1992) and Piepho (1992). If the chi-squared test by (30) indicates significant differences in stability between all tested genotypes, one may look for stability differences among certain genotypes using standard procedures for multiple comparisons among the observed $S_i^{(m)}$-values. Here, the number of degrees of freedom should be infinity since the variance of $S_i^{(m)}$ is known and is not being estimated.

To carry out these tests of significance, the explicit formulae for the means $E\{S_i^{(m)}\}$ and variances $\text{var}\{S_i^{(m)}\}$ under the null hypothesis are needed. They have been derived and explained in Nassar and Hühn (1987) and Hühn and Nassar (1989):

$$E\{S_i^{(1)}\} = \frac{l^2 - 1}{3\ l} \tag{31}$$

$$E\{S_i^{(2)}\} = \frac{l^2 - 1}{12} \tag{32}$$

$$\text{var}\{S_i^{(1)}\} = \frac{(l^2 - 1)\left[(l^2 - 4)(m + 3) + 30\right]}{45\ l^2 m(m - 1)} \tag{33}$$

$$\text{var}\{S_i^{(2)}\} = \frac{(l^2 - 1)}{36m}\left[\frac{l^2 - 4}{5} = +\frac{l^2 - 1}{2(m - 1)}\right]. \tag{34}$$

By applying (29)-(34), the tests of significance can be easily carried out. Extensive investigations (mainly by simulation studies) on the statistical properties (power, robustness etc.) of these tests of significance have been carried out by Piepho (1992). They revealed, for example, quite favorable robustness characteristics of the rank-based measure $S_i^{(1)}$. The previous rankbased tests were only approximately valid. The approximation generally improved with the number of environments (Piepho, 1992).

Finally, an interesting analogy between Wricke's ecovalence (Wricke, 1962) and the sum of squares of genotype-environment interactions on the parametric side and $S_i^{(2)}$ and Kendall's W on the nonparametric side will be presented.

By simple calculations one obtains:

$$\frac{ml(l^2 - 1)}{12(m - 1)}(1 - W) = \sum_{i=1}^{l} S_i^{(2)}. \tag{35}$$

In (35), the constant $ml(l^2-1)/12/(m-1)$ can be regarded as a standardizing factor. From a breeder's point of view, the ideal situation corresponds to W=1 or 1-W=0, i.e., maximum concordance or minimum discordance. This is equivalent to: $S_i^{(2)} = 0$ for each i, i=1,2,...,l, i.e., equal ranks in each environment (for each genotype).

From (35) we see that, apart from a standardization, $S_i^{(2)}$ is a genotype's contribution to the discordance (1-W) in the data set. By analogy, Wricke's ecovalence T_i is a genotype's contribution to the genotype-environment interaction sum of squares (SSGE):

$$SSGE = \sum_{i=1}^{l} T_i \tag{36}$$

where $T_i = \sum_{j=1}^{m} \left(x_{ij} - \bar{x}_{i.} - \bar{x}_{.j} + \bar{x}_{..} \right)^2$. Maximum stability occurs for $T_i = 0$.

The breeder's objective is to minimize both the ecovalences and the $S_i^{(2)}$'s, i.e., to minimize the interaction sum of squares and to maximize Kendall's W (minimize discordance).

The conditions (25) and (26) for significant concordance in rankings, of course, can also be expressed for $\sum_{i=1}^{l} S_i^{(2)}$:

$$\sum_{i=1}^{l} S_i^{(2)} \le \frac{lm(l^2-1)}{12(m-1)(1+\lambda_1)} \tag{37}$$

$$\sum_{i=1}^{l} S_i^{(2)} \le \frac{lm(l^2-1)}{12(m-1)(1+\lambda_2)}. \tag{38}$$

These upper bounds for the sum of the stability parameters of all the tested genotypes may be of some interest if one analyzes phenotypic stability by the parameter $S_i^{(2)}$. These relationships have also been discussed in Hühn and Piepho (1994).

For practical applications in crop production and plant breeding, a combination and simultaneous consideration of the measurements for yield and for stability in only one parameter is of particular interest and importance.

Some procedures have been published dealing with this problem (construction of an index, diverse parameters based on the deviations from the maximum yield in each environment, etc.), but only very few of these approaches are based on ranks.

The parameter

$$\sum_{j=1}^{m} \frac{\left| r_{ij} - \bar{r}_{i.} \right|}{\bar{r}_{i.}} \tag{39}$$

has been proposed by Hühn (1979) and applied by Léon (1985, 1986), Becker and Léon (1988), Hühn (1990), Kang (1990), and Kang and Pham (1991). It gives the sum of the absolute deviations of the ranks r_{ij} from their mean $\bar{r}_{i.}$, where these deviations are expressed in $\bar{r}_{i.}$ -units. This parameter realizes a confounding and simultaneous evaluation of yield stability and yield since the numerator measures stability (= variability of the ranks r_{ij}), while the denominator reflects yield level (= mean of the ranks r_{ij}).

An additional, but only slightly modified rank-based stability parameter

$$\sum_{j=1}^{m} \frac{\left(r_{ij} - \bar{r}_{i.} \right)^2}{\bar{r}_{i.}} \tag{40}$$

was proposed by Hühn (1979) and applied by Léon (1985, 1986), Becker and Léon (1988), Kang (1990), and Kang and Pham (1991). Both rank-based measures (39) and (40), however, are conceptually quite similar. These parameters (39) and (40) express stability in units of yield. If one intends such a simultaneous consideration and confounding of yield and stability by application of (39) or (40), the transformation (4) of the original data x_{ij}, of course, cannot be applied in the calculations of the denominator $\bar{r}_{i.}$ in (39) and in (40), since hereby the effect of genotype i would be eliminated from the data. But, the denominator in (39) and (40) must reflect the yield level of genotype i. For the application of (39) and (40), therefore, two computational procedures are available:
1. Numerator and denominator are both calculated with the original untransformed data x_{ij}.
2. Numerator is calculated with the transformed data x_{ij}^{*}, while the denominator is based on the original untransformed data x_{ij}.

Another very simple method based on ranks for combining yield and stability has been proposed by Schuster and Zschoche (1981) and by Kang (1988). Ranks were assigned for mean yield (highest yield = lowest rank of 1) as well as for stability variance parameters (Kang) or ecovalence (Schuster and Zschoche) (lowest variance or ecovalence = lowest rank of 1) and both ranks were summed (Kang) or multiplied (Schuster and Zschoche). The lowest rank-sum or rank-product would be the most desirable. But, no statistical tests of significance of these procedures have been provided. Kang and Pham (1991) have compared several methods of simultaneous selection for yield and stability (see, also: Francis and Kannenberg, 1978; Gravois et al., 1990).

Furthermore, for each procedure dealing with simultaneous evaluation of yield and stability, the appropriate weighting of both characteristics is an unsolved problem. Kang and Pham (1991) discussed indices derived from equal weights for stability variance and yield, and from 2, 3, 4, and 5 times more weight for yield than stability variance.

In a recent publication (Nassar et al., 1994), we proposed three rank-based measures that combine stability and performance and investigated, by simulation, their type I error and power for small samples. The measures are:

$$\sum_{j<j'} \frac{|r_{ij} - r_{ij'}|}{\bar{r}_{i.}}. \tag{41}$$

The rank r_{ij} in the numerator of eq. (41) represents the rank of the transformed (= adjusted) phenotypic value $x_{ij}^* = x_{ij} - (\bar{x}_{i.} - \bar{x}_{..})$ while the ranks in the denominator are for the uncorrected phenotypic values, x_{ij}.

Other rank-based measures, combining stability and performance, are:

$$\ln\left[\sum_{j=1}^{m}(r_{ij}-0)^2 / m\right] \tag{42}$$

where ln denotes the natural logarithm and

$$\sum_{j=1}^{m}|r_{ij} - 0| / m \tag{43}$$

where 0 is an optimum for performance with regard to a certain quantitative trait. In the case of yield, the optimum may be considered to be 1, the highest rank among l genotypes under consideration.

The measures (41), (42) and (43) have an asymptotic normal distribution (Nassar et al., 1994) and, therefore, one may use the standard analysis of variance and the F statistic to test for differences among genotypes with regard to stability and performance. Multiple comparisons among the means, such as Duncan's multiple range test or the Newman-Keuls test, may be applied if the F test is significant.

A simulation study was run to determine the applicability of the F test (its type I error) for small samples and the power of each measure. Comparisons of the observed with expected type I error values, based on the F test in the one way analysis of variance, for testing the null hypothesis that all l genotypes have equal means with regard to a given stability-performance measure, lead to the following main results:

1. For l equal or less than 20, the observed type I error rate is larger than its expected value.
2. Fairly good agreement between observed and expected values was seen for larger l values.
3. Although the F test is liberal, it may still be used for l as small as 10 when the type I error ≤ 0.05 without any serious error.
4. Studies on power show that (41) is, in general, substantially less powerful than the other measures.
5. In general, power increased with an increase in the genotypic or interaction variance and decreased with an increase in the error variance.
6. Comparison of power for l=10 and l=30 revealed that the power increased (for measures (42) and (43)) with an increase in the number of genotypes.
7. One may conclude that the measures (42) and (43) are similar with regard to type I error. However, the measure (43) is preferable since it has a slightly higher power and does not require transformation of the data.
8. For developing cultivars that perform well over a range of environments, the breeder may select among genotypes that have low values for the measure (43). Progress from selection may be predicted as usual based on normal theory and on the heritability of the measure.

For a detailed discussion (with extensive numerical results on type I error and power) see Nassar et al. (1994). Furthermore, we think that the development of rank-based counterparts of the concept of risk aversion (Barah et al., 1981) and of Eskridge's safety-first selection indices (Eskridge, 1990 a,b, 1991 a,b,c, 1992; Helms, 1993) might be elaborated to efficient procedures for simultaneous consideration of yield and stability. These ideas, however, have not yet been formulated and worked out in the literature.

I. NUMERICAL EXAMPLE

The application of all the previous theoretical investigations is demonstrated below for a data set with l=6, m=4, and n=4: Four winter-rapeseed (*Brassica napus* L.) cultivars (Jupiter [A], Skrzeszowicki [B], Jet Neuf [C] and Planet [D]) were crossed in a diallel design with reciprocals. For this study, the reciprocals were mixed. Varieties A, C, and D originated from Sweden, France, and Germany, respectively, and were bred as lineal varieties (Frauen, 1983, pers. comm.), while cultivar B originated from Poland and was supposed to be an open-pollinated variety. The resulting 6 F_1's were tested in 1983/84 at four locations (Göttingen, Hohenlieth, Hohenschulen, and

Heidmoor, all situated in the northern part of Germany). Each experiment was arranged in a randomized complete block design with four replications. Here, we only use the values of the trait "seed yield" expressed in dt/ha. This data set of l=6 genotypes (F_1's), m=4 environments (locations) and n=4 replications (blocks) is presented in Table 2.

Table 2. Experimental results for the trait "seed yield" (dt/ha) of 6 F_1's at 4 locations in 4 replications.

Locations

genotypes	1		2		3		4	
A x C	28.90	27.80	30.60	32.10	33.43	37.22	34.07	21.96
	29.60	25.10	32.00	30.85	37.39	37.51	31.88	31.24
B x D	30.60	22.80	35.65	32.45	41.03	30.81	32.36	32.86
	28.00	26.40	29.80	32.80	29.84	32.48	32.86	30.72
A x B	23.80	24.00	43.05	31.05	37.08	29.03	27.92	28.70
	23.90	28.50	34.00	38.30	34.99	30.94	35.20	32.56
B x C	24.90	28.20	31.80	39.85	37.27	34.66	30.07	29.72
	29.70	26.60	40.90	36.60	35.06	35.95	33.98	30.87
A x D	29.40	28.20	37.95	42.70	37.30	36.07	31.70	29.79
	24.20	31.50	38.85	44.00	33.95	32.26	32.97	30.99
C x D	26.80	26.30	31.80	41.00	41.94	37.15	24.05	31.52
	30.30	32.40	42.05	35.65	28.57	37.38	35.10	34.68

These data are a small and selected part of an extensive data set obtained by Léon (1985) for 26 genotypes, 2 years, and 4 locations. For further information on experimental design, agronomic practices, measurement of the trait "seed yield", etc., see Léon (1985). In this study, these data are used for demonstration purposes only. No discussion of certain genotypes or certain locations will be made. Statistical significances are indicated by: *, **, and n.s. = significance at 0.05, 0.01 level of error probability, and no significance, respectively.

All parameters that have been introduced previously will be estimated numerically and all the procedures will be applied to this data set. We, therefore, proceed from the standard two-way classification model (1). The estimates of the terms μ, α_i, β_j are:

$$\hat{\alpha}_1 = -1.05$$
$$\hat{\alpha}_2 = -1.06$$
$$\hat{\alpha}_3 = -0.97 \qquad \hat{\beta}_1 = -4.99$$
$$\hat{\alpha}_4 = 0.48 \qquad \hat{\beta}_2 = 3.67$$
$$\hat{\alpha}_5 = 1.46 \qquad \hat{\beta}_3 = 2.57$$
$$\hat{\alpha}_6 = 1.14 \qquad \hat{\beta}_4 = -1.25$$
$$\mu = 32.40$$

The estimates of the interaction terms $(\alpha\beta)_{ij}$ are summarized in Table 3. This table of interaction terms shows no clear pattern.

Table 3. Estimates of the interaction terms $(\alpha\beta)_{ij}$ from model (1) for 6 genotypes and 4 locations.

locations

genotypes	1	2	3	4
A x C	1.49	-3.64	2.47	-0.32
B x D	0.60	-2.34	-0.37	2.11
A x B	-1.40	1.49	-1.00	0.90
B x C	-0.54	0.73	0.28	-0.48
A x D	-0.55	3.34	-1.54	-1.26
C x D	0.40	0.41	0.15	-0.96

Next, we proceed from the individual values x_{ijk} to the means $\overline{X}_{ij.}$ (denoted by x_{ij}). These x_{ij} and, additionally, their ranks within each environment are presented in Table 4.

Table 4. Means and ranks of 6 genotypes at 4 locations (untransformed data).

locations

genotypes	1		2		3		4	
	value	rank	value	rank	value	rank	value	rank
A x C	27.85	4	31.39	1	36.39	6	29.79	1
B x D	26.95	2	32.67	2	33.54	2	32.20	6
A x B	25.05	1	36.60	3	33.01	1	31.09	2
B x C	27.35	3	37.29	4	35.73	4	31.16	3
A x D	28.32	5	40.87	6	34.89	3	31.36	5
C x D	28.95	6	37.62	5	36.26	5	31.34	4

Application of the transformation (4) leads to the X_{ijk}^*'s. The corresponding means $\overline{X_{ij.}^*}$ (denoted by X_{ij}^*) and, additionally, their ranks within each environment are given in Table 5.

Table 5. Means and ranks of 6 genotypes at 4 locations (transformed data).

genotypes	locations							
	1		2		3		4	
	value	rank	value	rank	value	renk	value	rank
A x C	28.90	6	32.44	1	37.44	6	30.84	4
B x D	28.01	5	33.73	2	34.60	3	33.26	6
A x B	26.01	1	37.56	5	33.97	2	32.05	5
B x C	26.87	3	36.81	4	35.25	5	30.68	3
A x D	26.86	2	39.41	6	33.43	1	29.90	1
C x D	27.81	4	36.48	3	35.12	4	30.20	2

The rank orders of the genotypes in both cases (original data x_{ij} and transformed data $\overset{*}{X}_{ij}$) can be quite different with reasonably low Spearman rank correlation coefficients r_s: $r_s = 0.26$ (location 1), $r_s = 0.77$ (location 2 and location 3) and $r_s = -0.09$ (location 4). The results of the "classical" parametric analysis of variance are presented in Table 6.

Table 6. Results of the "classical" parametric analysis of variance for the winter-rapeseed seed yield data (for a fixed effect model).

source of variation	degrees of freedom	sum of squares	mean squares	F-ratio	signi- ficance
genotypes	5	109.21	21.84	1.89	n.s.
locations	3	1198.42	399.47	34.50	**
interaction	15	141.66	9.44	0.82	n.s.
error	72	834.02	11.58		

The interaction between genotypes and locations and the main effect "genotypes" are nonsignificant while the differences between locations are highly significant. The results of the four nonparametric approaches (Bredenkamp, Hildebrand, Kubinger, van der Laan) for an estimation and testing of genotype x environment interactions are summarized in Table 7. (The test statistic which has been applied in the test of van der Laan is T_2.)

Table 7. Results of the 4 nonparametric approaches for the testing of genotype x environment interactions.

approach	numerical value of the test statistic	degrees of freedom	signi- ficance
Bredenkamp	$x^2 = 8.53$	15	n.s.
Hildebrand	$x^2 = 23.28$	15	n.s.
Kubinger	$x^2 = 22.28$	15	n.s.
van der Laan	$x^2 = 14.94$	15	n.s.

The four nonparametric procedures lead to the same conclusion of nonsignificant genotype x environment interactions but with quite different numerical values of the test statistic χ^2. The ordering of these numerical values is:

$\{Hildebrand(=23.28)\} \cong$

$\{Kubinger\ (=22.28)\}>\{van\ der\ Laan\ (=14.94)\}>\{Bredenkamp(=8.53)\}$. This result agrees with the conclusion obtained from the "classical" parametric analysis of variance in Table 6. For this data set, the nonparametric approaches and the well-known parametric procedure do not lead to differing inferences. But, in numerous calculations with other data sets, the conclusions obtained with both approaches (parametric and nonparametric) were different.

For reasons of completeness the three test statistics T, T_1, and T_2 from the nonparametric procedure of van der Laan and de Kroon shall be given: T = 21.61, T_1 = 6.67, and T_2 = 14.94. They are chi-squared distributed with 20, 5 and 15 degrees of freedom, respectively. None of these test statistics is significant.

Spearman's rank correlation coefficients for all pairs of environments are given in Table 8 for the original data x_{ij} (right upper part of the correlation matrix) and for the transformed values $\overset{*}{X}_{ij}$ (left lower part of the correlation matrix).

Table 8. Spearman's rank correlation coefficients for all pairs of environments for the original data xij (right upper part) and for the transformed values $\overset{*}{X}_{ij}$ (left lower part).

	1	2	3	4
1	—	0.54	0.71	0.14
2	-0.94	—	-0.09	0.43
3	0.71	-0.77	—	-0.37
4	0.26	-0.49	0.14	—

264

Based on these numerical values, Kendall's coefficients of concordance W can be easily calculated for each grouping of h environments with $2 < h \leq 4$. The results are presented in Table 9. (The tests of significance for W have been carried out by (19)).

Table 9. Kendall's coelffcients of concordance for different groups of environments (untransformed and transformed data).

h	group of envir.	original data x_a	transformed values $\overset{\cdot}{\chi}ij$
4	(1,2,3,4)	W=0.42	W=0.11
	(1,2,3)	W=0.59	W=0.11
	(1,2,4)	W=0.58	W=0.07
3	(1,3,4)	W=0.44	W=0.58
	(2,3,4)	W=0.33	W=0.09

The differences between W-values for both cases (untransformed and transformed data) are considerable, indicating the strong effect of the transformation (4). All W-values are statistically nonsignificant. The grouping with the largest W-value is the combination of environments 1, 2, and 3 (untransformed data) and the combination of environments 1, 3, and 4 (transformed data). The results of two different procedures of cluster analysis (average linkage approach and complete linkage algorithm) are compiled in Table 10.

Table 10. Results of the cluster-analyses.

untransformed data x_u	
complete linkage	average linkage
(1,3)	(1,3)
(2,4)	(2.4)
((1,3),(2,4))	((1,3), (2,4))
(1,3)	(1,3)
((1,3), 4)	((1,3), 4)
(((1,3), 4), 2)	(((1,3), 4), 2)
compete linkage	average linkage

transformed data $\overset{\cdot}{X}ij$

For the untransformed data as well as for the transformed data, both procedures of cluster analysis (complete linkage and average linkage) lead to the same groupings of environments. Finally, the stability parameters $S_i^{(1)}$ and $S_i^{(2)}$ for a quantitative estimation of the phenotypic stability of each individual genotype will be discussed. The numerical results are given in Table 11. All values of $Z_i^{(1)}$ and $Z_i^{(2)}$ (untransformed and transformed) as well as $S^{(1)}$ and $S^{(2)}$ (untransformed and transformed) are nonsignificant.

Table 11. Numerical values of the stability parameters $S_i^{(1)}$ and $S_i^{(2)}$ (and the corresponding test statistics $Z_i^{(1)}$ and $Z_i^{(2)}$)for both data sets (untransformed and transformed).

genotypes	untransformed data				transformed data			
	$S_i^{(1)}$	$Z_i^{(1)}$	$S_i^{(2)}$	$Z_i^{(2)}$	$S_i^{(1)}$	$Z_i^{(1)}$	$S_i^{(2)}$	$Z_i^{(2)}$
A x C	3.00	2.44	6.00	3.20	2.83	1.72	5.58	2.39
B x D	2.00	0.01	4.00	0.39	2.33	0.33	3.33	0.06
A x B	1.17	1.31	0.92	1.34	2.50	0.68	4.25	0.60
B x C	0.67	3.55	0.33	2.25	1.17	1.31	0.92	1.34
A x D	1.50	0.43	1.58	0.60	2.67	1.15	5.67	2.55
C x D	1.00	1.95	0.67	1.70	1.17	1.31	0.92	1.34

$$S^{(1)}=9.69 \qquad S^{(2)}=9.48 \qquad S^{(1)}=6.50 \qquad S^{(2)}=8.28$$

Using the numerical mean squares from Table 6 for a mixed effects model (genotypes = fixed and all other factors = random), one obtains the following estimates for the variance components:

$$\sigma_\alpha^2 \ (genotypes) \qquad = 0.77$$

$$\sigma_\beta^2 \ (environments) \qquad = 16.16$$

$$\sigma_{\alpha\beta}^2 \ (interaction) \qquad = 0$$

$$\sigma_\epsilon^2 \ (error) \qquad = 11.58$$

(The variance component for interaction has been defined to be zero although its numerical estimated value is slightly negative.)

Application of the theoretical approximation (20) gives W=0.21, while the data-based observed value is W=0.42 (Table 9). The agreement between observed and expected values, therefore, is fairly good. The proportionality factors $\lambda 1$ and $\lambda 2$ from eqs. (25) and (26) are: $\lambda 1=1.24$ or 3.07 for an error probability of 5% or 1%, respectively, and $\lambda 2=0.97$ or 1.52 for an error probability of 5% or 1%, respectively. That means that no significant concordance in rankings can be obtained by application of eqs. (25) or (26).
For the sum of all stability parameters $S_i^{(2)}$ one obtains:

$$\sum_{i=1}^{6} S_i^{(2)} = 13.50 \ \text{(for untransformed data) and} \ \sum_{i=1}^{6} S_i^{(2)} = 20.67 \ \text{(for}$$

transformed data). The upper bounds from eqs. (37) and (38) for significant concordance in rankings are presented in Table 12.

Table 12. Numerical values of the upper bounds $\sum_{i=1}^{6} S_i^{(2)}$ for significant concordance in rankings (by eqs. (37) and (38)).

	error probability	
	5%	1%
eq. (37)	10.42	5.73
eq. (38)	11.84	9.26

Since $\sum_{i=1}^{6} S_i^{(2)}$ is larger than these upper bounds (for untransformed as well as for transformed data), no significant concordance in rankings is obtained.

Finally, the numerical values for the combined measures of yield and stability (eqs. 39-43) are presented in Table 13. The differences between the two computational procedures I and II can be small (AxC, BxD, BxC, CxD) or large (AxB, AxD). But, in this context these numerical values will not be discussed further, since the data set is only for demonstration purposes.

Table 13. Numerical values of the combined measures of yield and stability (eqs. 39-43) for the six genotypes, where the two computational procedures are denoted by I and II, respectively (I=Numerator and denominator are both based on untransformed data xij, II=Numerator based on transformed data $\overset{\bullet}{X}_{ij}$ and denominator based on untransformed data xij).

genotypes		combined measures of yield and stability				
		eq. (39)	eq. (40)	eq. (41)	eq. (42)	eq. (43)
A x C	I	2.67	6.00	6.00	2.60	3.00
	II	2.33	5.58	5.67	—	—
B x D	I	2.00	4.00	4.00	2.48	3.00
	II	2.00	3.33	4.67	—	—
A x B	I	1.71	1.57	4.00	2.93	4.25
	II	4.00	7.28	8.57	—	—
B x C	I	0.57	0.29	1.14	1.87	2.50
	II	0.86	0.78	2.00	—	—
A x D	I	0.74	1.00	1.89	1.03	1.25
	II	1.47	3.58	3.37	—	—
C x D	I	0.40	0.40	1.20	0.41	1.00
	II	0.60	0.55	1.40	—	—

III. CONCLUSIONS AND RECOMMENDATION

The main results of the previous theoretical investigations can be summarized as follows:

a) The "classical" parametric approaches for an analysis of genotype x environment interactions are based on several assumptions (normality, homogeneity of variances, additivity (linearity) of effects). If some of these assumptions are violated, the validity of the inferences obtained from the standard statistical techniques may be questionable or lost.

In the field of practical applications, all the necessary assumptions for an effective "classical" analysis will be hardly simultaneously fulfilled. If there exists some doubt on the validity of these requirements, we, therefore, propose the application of nonparametric approaches for an analysis of genotype x environment interactions.

b) Several nonparametric techniques for an analysis of genotype x environment interactions are available from nonparametric statistics theory. In our view, the procedure proposed by van der Laan and de Kroon seems to be the most appropriate one, particularly for applications in plant breeding (selection). This approach uses a modified concept of interaction (rank-interaction), where the common interaction terms are only utilized insofar as they lead to different rankings of the genotypes in different environments. Such a concept must, of course, be of particular relevance for the plant breeder who is interested in rankings and selection.

Furthermore, exact critical values of the test statistics have been tabulated for some values of l=number of genotypes, m=number of environments and n=number of replications in each genotype-environment combination. For sufficiently large numbers n efficient approximate tests of significance based on the chi-squared distribution are available.

c) An attractive statistical tool for the comparison and grouping of environments (or genotypes) can be easily carried out: Two environments are considered more similar if they produce more similar rankings of the tested genotypes. The similarity between two environments, that means the "distance" between these two environments, can be quantitatively expressed by Spearman's rank correlation coefficient between the rankings of the genotypes in these two environments. Based on this simple measure of distance between environments, numerous techniques of cluster analysis can be applied to group the environments with respect to this measure of similarity.

An appropriate generalization from two up to an arbitrary number of environments can be easily carried out by applying Kendall's coefficient of concordance. This quite simple quantitative measure for similarity of rankings of the tested genotypes in several environments has received little attention in the literature.

d) In the last decades, the quantitative estimation of the phenotypic stability of genotypes has attracted a still increasing interest in plant breeding and crop production. The trait "stability" has been explicitly included in diverse breeding activities and breeding decisions.

For practical applications we prefer the use of $S_i^{(1)}$. This stability parameter is very easy to compute and allows a clear and relevant interpretation (mean absolute rank difference between the environments).

e) The two stability rank orders of the genotypes which will be obtained by using the uncorrected data x_{ij} and by using the corrected values $\overset{*}{x}_{ij}$ are often considerably different. The rank correlations are medium or low. Thus, we may conclude: differences among genotypes would have an effect on the proposed rank-based measures $S_i^{(1)}$ and $S_i^{(2)}$ and may lead to differences in stability among genotypes when, in fact, there is no genotype-environment interaction. Therefore, we propose the use of the corrected data $\overset{*}{x}_{ij}$ if one intends to estimate phenotypic stability independent from genotypic effects.

f) The four nonparametric approaches (Bredenkamp, Hildebrand, Kubinger, van der Laan), which have been used in this paper for an analysis of genotype x environment interactions, can also be applied for an analysis of the main effects "genotypes" and "environments." The explicit formulae of the test statistics can be quoted from the original publications of Bredenkamp, Hildebrand, Kubinger and van der Laan. But, furthermore, these formulae for both analyses (main effects and interaction) have been compiled in Hühn (1988a and 1988b) and Hühn and Léon (1995).

g) Further research is needed for the development of an efficient simultaneous evaluation of yield and stability. A few recent publications, however, offer some quite interesting and viable approaches and perspectives (Kang and Pham, 1991; Nassar et al., 1994).

REFERENCES

Abou-El-Fittouh, H.A., J.O.Rawlings, and P.A.Miller. 1969. Classification of environments to control genotype by environment interactions with an application to cotton. *Crop Science* 9: 135-140.

Azzalini, A., and D.R.Cox. 1984. Two new tests associated with analysis of variance. *Journal of Royal Statistics Society B.* 46: 335-343.

Baker, R.J. 1988. Tests for crossover genotype-environmental interactions. *Canadian Journal of Plant Science.* 68: 405-410.

Baker, R.J. 1990. Crossover genotype-environmental interaction in spring wheat. In: Kang, M.S. (ed.) *Genotype-by-environment interaction and plant breeding.* Louisiana State University Agricultural Center, Baton Rouge, La., pp 42-51.

Barah, B.C., H.P.Binswanger, B.S.Rana, and N.G.P.Rao. 1981. The use of risk aversion in plant breeding; concept and application. *Euphytica* 30: 451-458.

Becker, H.C. 1987. *Möglichkeiten zur züchterischen Verbesserung der Ertragssicherheit bei Getreide.* Habil Schrift Univ., Hohenheim.

Becker, H.C., and J.Léon. 1988. Stability analysis in plant breeding. *Plant Breeding* 101: 1-23.

Berger, R.L. 1984. Testing for the same ordering in several groups of means. In: T.J.Santner, and A.C.Tamhane. *Design of experiments. Ranking and selection.* Marcel Dekker, Inc. New York, pp 241-249.

Böhm, H., and W.Schuster. 1985. Untersuchungen zur Leistungsstabilität des Kornertrages von Mais (Zea mays L.). *Journal of Agronomy and Crop Science* 154: 222-231.

Bortz, J., G.A.Lienert, and K.Boehnke. 1990. *Verteilungsfreie Methoden in der Biostatistik.* Springer Berlin-Heidelberg-New York.

Bredenkamp, J. 1974. Nonparametrische Prüfung von Wechselwirkungen. *Psychologische Beiträge* 16: 398-416.

Clair, J.B.S., and J.Kleinschmit. 1986. Genotype-environment interaction and stability in ten-year height growth of Norway spruce clones (Picea abies Karst.). *Silvae Genetica* 35: 177-186.
Comstock, R.E. 1977. Quantitative genetics and the design of breeding programs. In: *Proceedings of the Int.Conf.on Quantitative Genetics August 16-21, 1976*. E.Pollak, O.Kempthorne, and T.B.Bailey, Jr. (eds.). The Iowa State University Press, Ames, pp 705-718.
Cornelius, P.L., M.Seyedsadr, and J.Crossa. 1992. Using the shifted multiplicative model to search for "separability" in crop cultivars trials. *Theoretical Applied Genetics* 84: 161-172.
Cox, D.R. 1984. Interaction. *Int.Statist.Rev.* 52: 1-31.
Crossa, J. 1990. Statistical analyses of multilocation trials. *Advances in Agronomy*. 44: 55-85.
Crossa, J., P.L.Cornelius, M.Seyedsadr, and P.Byrne. 1993. A shifted multiplicative model cluster analysis for grouping environments without genotypic rank change. *Theoretical Applied Genetics* 85: 577-586.
Denis, J.B. 1979. Sous-modèles interactifs respectant des contraintes d'ordre. *Biométrie-Praximétrie* 19: 49-58.
Denis, J.B. 1982. Test de l'interaction sous contraintes d'ordre. *Biométrie-Praximétrie* 22: 29-45.
Erdfelder, E., and J.Bredenkamp. 1984. Kritik mehrfaktorieller Rangvarianzanalysen. *Psychologische Beiträge* 26: 263-282.
Eskridge, K.M. 1990a. Safety-first models useful for selecting stable cultivars. In: Kang, M.S. (ed.) *Genotype-by-environment interaction and plant breeding*. Louisiana State University Agricultural Center, Baton Rouge, La., pp 151-168.
Eskridge, K.M. 1990b. Selection of stable cultivars using a safety-first rule. *Crop Science* 30: 369-374.
Eskridge, K.M., and B.E.Johnson. 1991a. Expected utility maximization and selection of stable plant cultivars. *Theoretical Applied Genetics* 81: 825-832.
Eskridge, K.M. 1991b. Screening cultivars for yield stability to limit the probability of disaster. *Maydica* 36: 275-282.
Eskridge, K.M., P.F.Byrne, and J. Crossa. 1991c. Selection of stable varieties by minimizing the probability of disaster. *Field Crops Research* 27: 169-181.
Eskridge, K.M., and R.F.Mumm. 1992. Choosing plant cultivars based on the probability of outperforming a check. *Theoretical Applied Genetics* 84: 494-500.
Francis, T.R., and L.W.Kannenberg. 1978. Yield stability studies in short-season maize. I. A descriptive method for grouping genotypes. *Canadian Journal of Plant Science* 58: 1029-1034.
Freeman, G.H. 1973. Statistical methods for the analysis of genotype-environment interactions. *Heredity* 31: 339-354.
Freeman, G.H. 1985. The analysis and interpretation of interactions. *Journal of Applied Statistics* 12: 3-10.
Freeman, G.H. 1990. Modern statistical methods for analyzing genotype x environment interactions. In: Kang, M.S. (ed.) *Genotype-by-environment interaction and plant breeding*. Louisiana State University Agricultural Center, Baton Rouge, La., pp 118-125.
Gail, M., and R.Simon. 1985. Testing for qualitative interactions between treatment effects and patient subsets. *Biometrics* 41: 361-372.
Gibbons, J.D. 1985. Nonparametric statistical inference (2nd ed.). *Statistics: textbooks and monographs* 65, Marcel Dekker, Inc. New York, Basel.
Gravois, K.A., K.A.K.Moldenhauer, and P.C.Rohman. 1990. Genotype-by-environment interaction for rice yield and identification of stable, high-yielding genotypes. In: Kang, M.S. (ed.) *Genotype-by-environment interaction and plant breeding*. Louisiana State University Agricultural Center, Baton Rouge, La., pp 181-188.
Gregorius, H.R., and G.Namkoong. 1986. Joint analysis of genotypic and environmental effects. *Theoretical Applied Genetics* 72: 413-422.
Haldane, J.B.S. 1946. The interaction of nature and nurture. *Ann. Eugenics* 13: 197-205.
Helms, T.C. 1993. Selection for yield and stability among oat lines. *Crop Science* 33: 423-426.
Hildebrand, H. 1980. Asymptotisch verteilungsfreie Rangtests in linearen Modellen. *Med.Inform.Stat.* 17: 344-349.
Hill, J. 1975. Genotype-environment interactions - a challenge for plant breeding. *Journal of Agricultural Science* 85: 477-493.
Hühn, M. 1979. Beiträge zur Erfassung der phänotypischen Stabilität. I. Vorschlag einiger auf Ranginformationen beruhenden Stabilitätsparameter. *EDV in Medizin und Biologie* 10: 112-117.
Hühn, M., and J.Léon. 1985. Genotype x environment interactions and phenotypic stability of Brassica napus. *Z.Pflanzenzüchtg.* 95: 135-146.
Hühn, M. 1988a. Einige neuere Ansätze zur nichtparametrischen Analyse von Genotyp-Umwelt-Wechselwirkungen. *Mitt.Ges.Pflanzenbauwiss.* 1: 162-166.
Hühn, M. 1988b. Einige neuere Ansätze zur nichtparametrischen Analyse von Haupteffekten für "Genotypen" und "Umwelten". *Mitt.Ges.Pflanzenbauwiss.* 1: 167-170.
Hühn, M., and R.Nassar. 1989. On tests of significance for nonparametric measures of phenotypic stability. *Biometrics* 45: 997-1000.
Hühn, M. 1990. Nonparametric measures of phenotypic stability. Part 1: Theory and Part 2: Applications. *Euphytica* 47:189-201.

270

Hühn, M., and R.Nassar. 1991. Phenotypic stability of genotypes over environments: On tests of significance for two nonparametric measures. *Biometrics* 47: 1196-1197.

Hühn, M., S.Lotito, and H.P.Piepho. 1993. Relationships between genotype x environment interactions and rank orders for a set of genotypes tested in different environments. *Theoretical Applied Genetics* 86: 943-950.

Hühn, M., and H.P.Piepho. 1994. Relationships between Kendall's coefficient of concordance and a nonparametric measure of phenotypic stability with implications for the consistency in rankings as affected by variance components. *Biometrical Journal* 36: 719-727.

Hühn, M., and J.Léon. 1995. Nonparametric analysis of cultivar performance trials - Experimental results and comparison of different procedures based on ranks. *Agron. Journal* (in press).

Iman, R.L., and J.M.Davenport. 1980. Approximations of the critical region of the Friedman statistics. *Communications in Statistics* A 9: 571-595.

Kang, M.S. 1988. A rank-sum method for selecting high-yielding, stable corn genotypes. *Cereal Research Communications* 16: 113-115.

Kang, M.S. 1990. Understanding and utilization of genotype-by-environment interaction in plant breeding. In: Kang, M.S. (ed.) *Genotype-by-environment interaction and plant breeding*. Louisiana State University Agricultural Center, Baton Rouge, La., pp 52-68.

Kang, M.S., and H.N.Pham. 1991. Simultaneous selection for high yielding and stable crop genotypes. *Agron. Journal* 83: 161-165.

Kendall, M.G. 1962. *Rank correlation methods*. 3. ed., Griffin London.

Kleinschmit, J. 1983. Concepts and experiences in clonal plantations of conifers. Proc. 19th Meeting Can Tree Impr Assoc Part 2: *Symp on Clonal Forestry: Its impact on tree improvement and our future forests*. Toronto (Ontario) 1983; Zsuffa, Rauter, and Yeatman (eds.) pp 26-56.

Krenzer, E.G., Jr., J.D.Thompson, and B.F.Carver. 1992. Partitioning of genotype x environment interactions of winter wheat forage yield. *Crop Science* 32: 1143-1147.

Kroon, J.P.M.de, and P.van der Laan. 1981. Distribution-free test procedures in two-way layouts; a concept of rank-interaction. *Statistica Neerlandica* 35: 189-213.

Kroon, J.P.M.de. 1986. Rank Interaction. *Encyclopedia of Statistical Sciences* 7: 592-594.

Kubinger, K.D. 1986. A note on nonparametric tests for the interaction in two-way layouts. *Biometrical Journal* 28: 67-72.

Laan, P.van der 1987. Extensive tables with exact critical values of a distribution-free test for rank interaction in a two-way layout. *Biuletyn Oceny Odmian* 12: 195-202.

Léon, J. 1985. *Beiträge zur Erfassung der phänotypischen Stabilität unter besonderer Berücksichtigung unterschiedlicher Heterogenitäts- und Heterozygotiegrade sowie einer zusammenfassenden Beurteilung von Ertragshöhe und Ertragssicherheit*. Dissertation Universität Kiel.

Léon, J. 1986. Methods of simultaneous estimation of yield and yield stability. In: Biometrics in Plant Breeding. Proc.6th Meeting EUCARPIA-section *"Biometrics in Plant Breeding"*, Birmingham, UK, pp 299-308.

Léon, J., and H.C.Becker. 1988. Repeatability of some statistical measures of phenotypic stability - Correlations between single year results and multi years results. *Plant Breeding* 100: 137-142.

Lienert, G.A. 1978. *Verteilungsfreie Methoden in der Biostatistik. Band II* (2. Aufl.). Verlag A. Hain, Meisenheim am Glan.

Lin, C.S., M.R.Binns, and L.P.Lefkovitch. 1986. Stability analysis: Where do we stand? *Crop Science* 26: 894-900.

Lin, C.S., and G.Butler. 1990. Cluster analyses for analyzing two-way classification data. *Agron. Journal* 82: 344-348.

Magnussen, S., and C.W.Yeatman. 1986. Accelerated testing of Jack pine progenies: a case study. In: *Proc. Conf. on Breeding Theory/Progeny Testing/Seed Orchards*. IUFRO/North Carolina State Univ., Williamsburg, pp 107-121.

Nachit, M.M., and H.Ketata. 1986. Breeding strategy for improving durum wheat in mediterranean rainfed areas. *Proc. Int.Wheat Conf., Rabat* (Marocco).

Nassar, R., and M.Hühn. 1987. Studies on estimation of phenotypic stability: Tests of significance for nonparametric measures of phenotypic stability. *Biometrics* 43: 45-53.

Nassar, R., J.Léon, and M.Hühn. 1994. Tests of significance for combined measures of plant stability and performance. *Biometrical Journal* 36: 109-123.

Piepho, H.P. 1992. *Vergleichende Untersuchungen der statistischen Eigenschaften verschiedener Stabilitätsmasse mit Anwendungen auf Hafer, Winterraps, Ackerbohnen sowie Futter- und Zuckerrüben*. Dissertation Universität Kiel.

Piepho, H.P., and S.Lotito. 1992. Rank correlation among parametric and nonparametric measures of phenotypic stability. *Euphytica* 64: 221-225.

Rognli, O.A. 1986. Genotype x environment interaction for seed production in timothy (Phleum pratense L.). In: *Biometrics in Plant Breeding*. Proc.6th Meeting EUCARPIA-section "Biometrics in Plant Breeding", Birmingham, UK: 282-298.

Schuster, W., and K.H.Zschoche. 1981. Wie ertragsstabil sind unsere Rapssorten? *DLG-Mitteilungen* 12/1981: 670-671.

Seyedsadr, M., and P.L.Cornelius. 1992. Shifted multiplicative models for nonadditive two-way tables. *Comm Stat B Simul Comp* 21: 807-832.
Skrøppa, T. 1984. A critical evaluation of methods available to estimate the genotype x environment interaction. *Studia Forestalia Suecica* 166: 3-14.
Sprent, P. 1989. *Applied nonparametric statistical methods*. Chapman and Hall, London New York.
Thomson, N.J., and R.B.Cunningham. 1979. Genotype x environment interactions and evaluation of cotton cultivars. *Australian Journal of Agricultural Research* 30: 105-112.
Utz, H.F. 1972. Die Zerlegung der Genotyp x Umwelt-Interaktionen. *EDV in Medizin und Biologie* 3: 52-59.
Virk, D.S., and B.K.Mangat. 1991. Detection of cross over genotype x environment interactions in pearl millet. *Euphytica* 52: 193-199.
Wanyancha, J.M., and E.K.Morgenstern. 1987. Genetic variation in response to nitrogen fertilizer levels in tamarack families. *Canadian Journal of Forest Research* 17:1246-1250.
Weber, W.E., and G.Wricke. 1987. Wie zuverlässig sind Schätzungen von Parametern der phänotypischen Stabilität? *Vortr.Pflanzenzüchtg.* 12: 120-133.
Westcott, B. 1986. Some methods of analysing genotype-environment interaction. *Heredity* 56: 243-253.
Wricke, G. 1962. Über eine Methode zur Erfassung der ökologischen Streubreite in Feldversuchen. *Zeitschr.f.Pflanzenzüchtg.* 47: 92-96.
Wricke, G. 1965. Die Erfassung der Wechselwirkung zwischen Genotyp und Umwelt bei quantitativen Eigenschaften. *Z.Pflanzenzüchtg.* 53: 266-343.
Wricke, G., and W.E.Weber. 1980. Erweiterte Analyse von Wechselwirkungen in Versuchsserien. In: W.Köpcke & K.Überla (eds.), *Biometrie - heute und morgen*. Springer Verlag Berlin-Heidelberg-New York, pp 87-95.
Zelterman, D. 1990. On tests for qualitative interactions. *Statistical Probability Letters* 10: 59-63.

Chapter 10

ANALYSIS OF MULTIPLE ENVIRONMENT TRIALS USING THE PROBABILITY OF OUTPERFORMING A CHECK*

K. M. Eskridge[1]

I. INTRODUCTION

An important consideration in plant breeding is the development of new germplasm that consistently outperforms accepted cultivars over a broad range of environments. In recent years, plant breeders have placed increased emphasis on "head-to-head" or pairwise entry comparisons as a means of comparing test entries to checks across a wide range of customers' environments. Trends have been toward more sites with simpler designs at individual sites and the use of easily understood decision methods that integrate data across many different tests and provide an understanding of the environmental and management factors that contribute to an entry's instability (Bradley et al., 1988).

In response to these trends in breeding programs, pairwise mean comparison methods have been proposed to compare entries with checks (Jones, 1988; Bradley et al.,1988). These approaches compare entries with checks using methods that make statements about 'true' mean trait values over a population of environments. Such pairwise mean comparisons maximize the number of locations in the comparisons, avoid the problem of imbalance that occurs when differing sets of entries are included in different tests, and are useful for making comparisons in the presence of genotype-environment (GE) interaction. However, pairwise mean comparisons may not provide information most relevant to plant breeders.

A major concern of most plant breeders is the identification of test entries that have a high <u>probability</u> of outperforming the check in environments where the check is normally grown. Pairwise mean comparison methods may provide little insight regarding this probability nor do these methods aid with understanding environmental and management factors that contribute to an entry's instability. Pairwise test entry evaluation could be enhanced by developing precise decision making tools that quantify the probability that a test entry outperforms a check, that may be used with multiple traits and can be used to identify environmental and management factors that contribute to an entry's instability.

One group of decision making models that is easy to understand and may be used to quantify the chances that an entry will outperform the check is based on the assumption of **safety-first** behavior (Eskridge, 1990a; 1990b; 1991; Eskridge et al., 1991). Safety-first decision models may be used to quantify the probability that a new entry outperforms the check cultivar in environments where the check cultivar is well suited. More concisely, the decision maker is primarily concerned with:

$$P(X_i - X_c > 0) \qquad [1]$$

[1]Department of Biometry, University of Nebraska, Lincoln, NE 68583-0712

where X_i and X_c are responses of the ith entry and the check cultivar respectively, and P(Z) denotes the probability of event Z occurring (Eskridge and Mumm, 1992). Eq. [1] is defined as the reliability of the ith test entry where a reliable entry has a high probability of outperforming the check. Reliability is a commonly used concept in machine life testing and quality engineering (Nelson, 1982).

Reliability of test entries (Eq. [1]) can be useful to breeders in identifying superior entries for several reasons. The reliability of an entry is an easily understood measure of riskiness. Choice of an entry with a reliability near 0.5 is risky since it will fail to outperform the check in 50% of the environments. Another test entry with a reliability of 0.9 is much less risky since it would be expected to fall short of the check in only 10% of the environments. Also, reliability is directly related to several commonly used measures of stability (Eskridge and Mumm, 1992; Lin et al., 1986). These relationships allow one to understand why some entries might be considered unstable but still highly reliable and thus preferred over more stable entries. Furthermore, reliability explicitly incorporates genotype-environment interaction since the test-check differences are evaluated across environments.

Reliability may be easily extended to consider many correlated traits simultaneously. Multiple trait reliability is the probability of the test entry out-performing the check for all traits under consideration. Multiple trait reliability is easy to compute and interpret. In contrast, interpretation of multiple trait data using common stability methods is difficult since traits are frequently correlated and nonnormally distributed (Lin et al., 1986; Eskridge et al., 1994).

Reliability as expressed in eq. [1] is not useful for graphically describing how test entries respond relative to a check across environments. However, a simple generalization of eq. [1] by replacing 0 with d and computing probabilities for all possible values of d gives a reliability *function* which may be used to analyze across environment variability (Eskridge et al., 1993). Graphing the reliability function provides a simple way of evaluating across environment performance relative to a check. The location, slope and shape of a reliability function may be used to assess a test entry's stability, to determine how test entries perform relative to a check across a range of environments and to identify environments where test entries have performance problems.

The objectives of this paper are to (i) demonstrate different methods of estimating both single trait and multiple trait reliability based on field trial information, (ii) show how reliability is related to other commonly used stability parameters, (iii) evaluate the usefulness of reliability as a genetic parameter by analyzing its repeatability relative to other measures of stability and (iv) to show how to analyze across-environment variability using reliability functions.

II. RELIABILITY METHODS AND APPLICATIONS

A. ESTIMATING THE PROBABILITY OF OUTPERFORMING A CHECK

If the probability of outperforming a check, or reliability, is to be useful in aiding the breeder with identifying superior entries, it is necessary to use field trial information to estimate and test hypotheses regarding these reliabilities. Reliabilities may be estimated and tested using at least three different approaches.

1. Normally Distributed Differences.

If test-check differences are normally distributed, then the probability that the ith test entry outperforms the check may be estimated using standard normal tables. Define the test-check difference to be the difference $D_i = X_i - X_c$ where X_i and X_c are responses for the ith test entry and the check respectively. The probability that the test entry outperforms the check is the probability that $D_i > 0$. This probability is estimated by using a standard normal table to determine the probability of a standard normal random variable falling above the negative ratio of the mean difference to the standard deviation of the difference ($-\bar{D}_i/s_{Di}$) (Eskridge and Mumm, 1992; Nelson, 1982). These estimates are computed for all test entries with the assumption that the best test entry is the one with the highest estimated probability of outperforming the check. Also, approximate $100(1-\alpha)\%$ confidence intervals for the reliability of a test entry may be estimated (Nelson, 1982). Additionally, when the check and test entries under consideration are common in some set of environments, the Wald test may be used to test equality of the reliabilities, and contrasts among reliabilities, for several test entries simultaneously (see Appendix).

2. Binomial Approach.

The assumption of normality may not be justified with some traits. An alternative approach to estimating reliability that does not require normality is to use the sample proportion of environments where the test entry outperforms the check (Eskridge and Mumm, 1992). For example, assume there are n environments where both the test and check entries are present and there are X environments where the test-check difference is greater than 0. Then X has a binomial distribution and X/n is an unbiased estimate of the reliability of the test entry. Confidence intervals for proportions (Nelson, 1982) may be used to obtain interval estimates of the reliability of a test entry. Additionally, when the check and test entries under consideration are all present in some set of environments, Cochran's Q test (Cochran, 1950) may be used to test equality of reliabilities, and contrasts among reliabilities, for several test entries simultaneously. The idea of this test is similar to an F test for treatment in a randomized complete block experiment with environments as blocks, test entries as treatments and where the responses for each environment-entry cell is 0 if $D_i \leq 0$ and 1 if $D_i > 0$ (Winer, 1971).

3. Nonparametric Approach Based On Residuals.

The binomial approach to estimating reliability is easy to understand and it does not require that the trait values follow any special probability distribution (e.g. normality). However, a major limitation of using the binomial approach is that precise estimates may require a large number of environments.

An alternative nonparametric approach is based on the residuals from a linear model (Eskridge et al. (1994) and Appendix). This approach can be considerably more precise than the binomial approach since all data in the trial are used to estimate the reliability for every test entry. The reliability for an entry using the binomial approach is computed only using the observations from that entry and the check. Assume the probability distributions of the test-check differences for the test entries differ only by their means and possibly their variances. Then the residual deviations for all entries will have identical probability distributions after either (1) standardizing the residuals to have standard deviation (SD) of 1 when variances differ among entries or (2) leaving the residuals unstandardized when trait variances do not differ. When variances are heterogeneous across entries, standardized residuals are multiplied by the SD for the entry of interest and the entry mean is added to the product to obtain a pseudo-observation. When residual variances are equal across entries, an entry's mean is added to the unstandardized residuals to produce pseudo-observations. Variances of the residuals should be tested for equality across genotypes using a test that is not sensitive to nonnormality such as the Levene test modified by replacing means with medians (Conover et al., 1981).

The pseudo-observations for a test entry are estimated test-check differences that could be expected to result if all non-check plots in the trial had been planted to this entry. The proportion of all pseudo-observations greater than zero is the estimated probability of outperforming the check. This approach can be considerably more precise than the binomial approach. For example, when the true reliability is 0.85, there are 5 entries in the trial and the pseudo-observations are uncorrelated within an environment; approximately only 5 environments are needed for the estimate to be within 0.15 of the true value, while the binomial approach requires 23 environments (see Appendix).

Confidence intervals for reliabilities using the nonparametric approach are simply estimated for the gth test entry by

$$\hat{p}_g \pm Z(1-\alpha/2)\sqrt{V(\hat{p}_g)}$$

where $\hat{p}_g$ is the nonparametric reliability estimate, $Z(1-\alpha/2)$ is the $1-\alpha/2$ quantile from a standard normal random variable and

$$V(\hat{p}_g) = (1/n)^2 [\sum_j^n (\hat{p}_j^g)^2 - (1/n)(\sum_j^n \hat{p}_j^g)^2]$$

where $\hat{p}_j^g$ is the proportion of entry g pseudo-observations > 0 based on residuals from location j (see Appendix).

4. Application.

Data from an on-farm dry bean trial will be used to demonstrate these three methods of estimating reliability (Nuland and Eskridge, 1992). [For other examples of using of these reliability methods, see Eskridge and Mumm (1992) and Eskridge et al. (1994)]. The trial involved 4 varieties (BillZ, Fiesta, Othello, and UI114=check) grown at 17 different farms throughout western Nebraska and eastern Colorado in 1991.

Yield reliabilities (Table 1) for all three methods give quite similar values indicating that responses are approximately normally distributed. Regarding yield, all three methods show that BillZ is most reliable having about a 70% chance of outperforming UI114. Fiesta is least likely to outperform UI114 in yield with all three reliability methods giving an estimated reliability of slightly over 0.50. Seed size reliabilities for all three methods indicate that Fiesta is the most reliable by a large margin even though it is the least stable based on the size of the slope coefficient. Fiesta's larger yield offsets the larger slope and this is reflected in the larger reliability value. Both BillZ and Othello have less than a 50% chance of outperforming UI114 regarding seed size.

B. PROBABILITY OF SEVERAL TRAITS OUTPERFORMING A CHECK

Although identification of entries that are reliable regarding a single trait is important, breeding decisions are rarely based on a single trait alone. Development of superior entries requires evaluation of many different traits that are often intercorrelated and that may or may not be normally distributed. Breeders often identify and use checks to establish minimum values of acceptability for important traits. Germplasm is then selected which is likely to exceed the check for multiple traits across diverse environments.

Single trait reliability methods are easily extended to the multiple trait situation. Multiple trait or multivariate reliability allows the breeder to base selection decisions on the probability of obtaining acceptable performance for all traits under consideration when tested over multiple environments. An entry with high probability of all traits falling above a check would be more reliable, and thus preferred, over those with lower probability values. A single resulting multivariate probability value can reflect the overall selection value of an entry and is analogous to using a multiple trait selection index.

An entry's probability of all p traits under consideration being larger than the trait values for the check is estimated based on test-check differences. For example, if X_{ig} is the response of ith trait for the gth entry, X_{ic} is the ith trait's response for the check, and $D_{ig} = X_{ig}\text{-}X_{ic}$, then the probability of acceptable performance for the gth entry, is

$$P(D_{1g} > 0, D_{2g} > 0, \ldots , D_{pg} > 0) \qquad [2]$$

where $P(Z)$ denotes the probability of event Z.

Table 1
Yield and seed size means, mean differences, standard deviations of differences, slope coefficients, normal (RN), binomial (R) and nonparametric (RNP) reliabilities of yield for 4 varieties from a dry bean trial 17 grown in locations where UI114 is the check variety.

	Mean	Mean Difference	Std. dev of Difference	Slope	RN[+]	R[++]	RNP
			Yield				
BillZ	2048	155.20	296.9	1.057	0.699	0.765	0.725
Othello	1964	71.06	289.9	1.252	0.597	0.588	0.647
Fiesta	1901	8.06	296.8	0.837	0.511	0.529	0.569
UI114	1893	0	-	0.854	-	-	-
			Seed Size				
Variety	Mean	Mean Difference	Std. dev Difference	Slope	RN[+]	R[++]	RNP
Fiesta	1.2442	0.099	0.083	1.151	0.885	0.941	0.922
Othello	1.1341	-0.011	0.087	1.016	0.450	0.412	0.373
BillZ	1.0978	-0.047	0.056	0.935	0.197	0.235	0.275
UI114	1.1452	0	-	0.898	-	-	-

[+] Wald test indicates significant difference among yield reliabilities at $p = 0.102$ and significant difference among seed size reliabilities at $p = 0.0001$.

[++] Cochrans' Q test indicates significant difference among yield reliabilities at $p = 0.157$ and significant difference among seed size reliabilities at $p = 0.0001$.

1. Multivariate Normal Approach.

If the 1xp vector of traits' test-check differences ($\underline{D}_g$) has a multivariate normal distribution, then probabilities in [2] may be estimated using a FORTRAN subroutine (Schervish, 1984; 1985) that computes standard multivariate normal probabilities. Estimates of the mean vector and the covariance matrix of the test-check differences are used for each entry. Estimates are then used to standardize each trait and [2] is estimated by computing the probability that each of the traits (i=1,...,p) is above the standardized bounds, $-\bar{D}_{ig}/s_{Dig}$.

2. Multiple Trait Binomial Approach.

For many traits, the assumption of multivariate normality may be unjustified which limits the usefulness of the multivariate normal approach. The single trait binomial approach, which does not require normality, may be easily extended to the multiple trait case. Multiple trait reliability in [2] is estimated by the proportion of environments where the test entry outperforms the check for all p traits. Confidence intervals and tests based on Cochran's Q apply as in the single trait case.

3. Multiple Trait Nonparametric Residual Approach.

The multiple trait binomial estimates may be imprecise even with a large number of environments. The univariate nonparametric residual approach is easily modified to apply to the multiple trait case. This approach can give more precise estimates since all observations from a trial are used to estimate reliability for each entry. Pseudo-observations are estimated for all p traits for each entry. The proportion of all pseudo-observations with all p traits outperforming the check is used as the estimated reliability for each entry (see Appendix). Confidence intervals are obtained as in the single trait case.

4. Interpretation Using Single Trait Reliabilities.

Single trait or univariate reliabilities for each of the p traits may be used to aid with identifying individual traits that contribute to small multivariate reliability values. In situations where all traits are statistically independent, the product of the univariate probabilities is the multivariate probability. Many traits of economic importance are not independent. Nevertheless, the size of the univariate probabilities can be used as a rough way of identifying which individual traits are likely causing small multivariate reliability values for certain entries.

5. Application.

Multiple trait reliability methods will be demonstrated with data from a multiple environment study on wheat quality traits (Peterson, et al., 1992; Eskridge et al., 1994). Five flour quality traits were measured on each of eighteen hard red winter wheat entries grown in fourteen environments over two years and seven locations. Thirteen entries were released varieties and five were experimental lines. Scout 66 was used as the check. Flour quality was evaluated for protein concentration

280

(%) by macro-Kjeldahl; mixing time (min) and mixing tolerance (0-9) using 10 gram National Manufacturing mixograph; SDS sedimentation volume using the standard method modified to test 2 gram flour samples; and kernel hardness (1-9) using microscopic evaluation of individual crushed grains. Mixing time and tolerance were quality measures of dough development. SDS sedimentation provided an estimate of protein quality and is a reflection of loaf volume potential.

Multiple trait reliability values for the normal, binomial and nonparametric residual approaches gave broadly similar results when compared to Scout 66 (Table 2). Karl, Centurk 78 and Bennet had the highest probabilities of all quality traits outperforming Scout 66. All the remaining entries had less than a 16% chance of meeting or exceeding the quality of the check (with the exception of Lancota which had a binomial reliability estimate of 0.214). At least five of the entries had less than a 2% chance, depending on the reliability method used. The binomial estimates had less precision and thus deviated somewhat from the other two approaches.

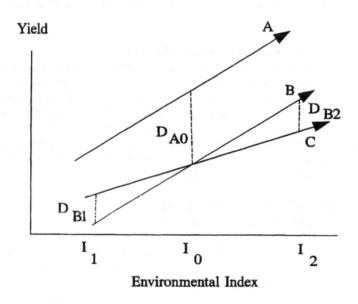

Figure 1. Stability based on regression analysis for three hypothetical entries.

Table 2

Single-trait reliabilities of protein concentration, mixing tolerance, SDS sedimentation and kernel hardness values and 3 different multi-trait reliability estimates for 17 genotypes grown over 14 environments with Scout 66 as the check.

Genotype	Flour protein	Mixing time	Mixing tolerance	Sedimentation	Kernel hardness	Multi-trait		
						Non-parametric	Normal	Binomial
Centurk 78	0.605	0.975	0.874	0.723	0.744	0.311	0.346	0.429
Karl	1.00	0.958	0.866	0.832	0.639	0.445	0.395	0.429
Bennet	0.945	0.819	0.803	0.576	0.782	0.294	0.340	0.214
Cody	0.517	0.987	0.735	0.311	0.727	0.139	0.064	0.071
Lancota	0.979	0.445	0.563	0.618	0.887	0.168	0.110	0.214
TAM107	0.261	0.739	0.647	0.466	0.744	0.092	0.031	0.000
Century	0.429	0.882	0.475	0.500	0.588	0.097	0.063	0.143
N87U102	1.00	0.798	0.769	0.143	0.769	0.071	0.084	0.143
TX86V1110	0.975	0.206	0.538	0.887	0.782	0.063	0.082	0.071
Arkan	0.958	0.773	0.311	0.655	0.597	0.080	0.106	0.071
Colt	0.874	0.597	0.429	0.655	0.483	0.088	0.099	0.071
Arapahoe	0.706	0.874	0.403	0.176	0.555	0.038	0.015	0.000
Redland	0.130	0.975	0.387	0.777	0.223	0.013	0.006	0.000
OK83396	0.941	0.269	0.071	0.143	0.727	0.000	0.007	0.000
N87U110	1.00	0.063	0.046	0.550	0.559	0.000	0.002	0.000
Siouxland	0.542	0.790	0.105	0.063	0.559	0.004	0.002	0.000
N87U113	1.00	0.450	0.055	0.055	0.727	0.000	0.004	0.000

The entries' single trait reliabilities relative to Scout 66 based on the nonparametric residual approach roughly indicated the relative contributions of individual traits to the overall reliability. Redland showed relatively low reliability compared to Scout 66 for flour protein, mixing tolerance and kernel hardness. OK83396, N87U110, N87U113, N87U102 and TX86V1110 each had high probabilities of meeting or exceeding flour protein (>0.92) and kernel hardness (>0.54) levels of Scout 66, but each had lower probabilities for mixing characteristics and/or SDS sedimentation values that reduced overall reliability for all traits and overall acceptability of these lines.

C. RELATIONSHIP BETWEEN RELIABILITY AND OTHER STABILITY MEASURES

Single trait reliability can be shown to be directly related to several commonly used stability measures when test-check differences are normally distributed.

The relationship between reliability and the Finlay and Wilkinson (1963) slope coefficient (β_i) can be shown pictorially. Assume there are three entries (A, B, and C (check)) with responses to the environmental index and no deviations from regression as shown in Figure 1. If A and B have slopes of 1 and C < 1, then from Figure 1, A would be preferred over B because it has identical stability (slope) to B but larger yield. This will also be reflected in the reliability as indicated in Figure 2. By computing the differences between A and B and the check C and plotting these differences against the index, the regression line for C is rotated to the x axis and the other lines follow. The distribution of these differences over environments can be thought of as being represented by normal distributions as in Figure 2. If these differences are normally distributed over environments, the reliability of a test entry outperforming the check is the area under the normal curve above the x axis (Figure 2). Thus, entry A has nearly all the area of its normal curve above 0, while B only has about half. Therefore the reliability is close to 100% for A but only about 50% for B, so A is preferred as with the regression approach.

The relationship between reliability and other stability parameters can also be demonstrated algebraically. When test check differences are normally distributed, reliability is $P(Z > -\mu_{di}/\sigma_{di})$ where μ_{di} is the true mean of the test-check for test entry i and σ_{di} is true standard deviation of these differences (Eskridge and Mumm, 1992). Equations for σ_{di}^2 expressed in terms of stability parameters may be substituted into $P(Z > -\mu_{di}/\sigma_{di})$ to demonstrate how reliability explicitly weighs the importance of the difference in performance (μ_{di}) relative to stability. In general, for a given positive difference in mean performance, stability parameters that result in larger values of σ_{di}^2 will reduce the reliability of the test entry.

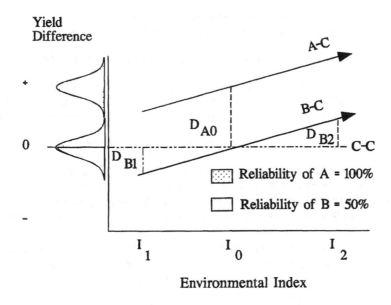

Figure 2. Reliability based on the difference from the check ($Y_t - Y_c$) for two hypothetical test entries.

Finlay and Wilkinson's regression coefficient (β_i) (1963), Eberhart and Russell's deviation mean square ($\sigma_{\delta i}^2$) (1966) and Shukla's stability variance (σ_i^2) (1972) are functionally related to the variance of the test entry check differences ($\sigma_{d_{ti}}^2$) in the following ways (see Appendix):

Shukla: $\quad\quad\quad \sigma_{d_{ti}}^2 = \sigma_i^2 + \sigma_c^2$

Eberhart-Russell: $\quad \sigma_{d_{ti}}^2 = (\beta_i - \beta_c)^2 \sigma_I^2 + \sigma_{\delta i}^2 + \sigma_{\delta c}^2$

Finlay-Wilkinson: $\quad \sigma_{d i}^2 = (\beta_i - \beta_c)^2 \sigma_I^2$

where any parameter with subscript c is that parameter for the check and $\sigma_I^2 =$ variance of the environmental index. By using these different definitions of stability in expressing $\sigma_{d_i}^2$ it can be seen how reliability is related to these particular

stability parameters when the mean difference (μ_{d_i}) is positive.

For Shukla's model, holding constant the mean difference (μ_{d_i}) and other terms,

as Shukla's variance for the test entry (σ_i^2) becomes larger, the ratio μ_{d_i}/σ_{d_i}

becomes smaller and the reliability expressed as $P(Z > -\mu_{d_i}/\sigma_{d_i})$ is reduced. This result is reasonable since the larger Shukla's variance the less stable and more unreliable the entry. For Eberhart and Russell's deviation mean

square ($\sigma_{d_i}^2$) holding other terms constant, an increase in the ith entry's $\sigma_{d_i}^2$ also

reduces the reliability of the entry. Again, this result is reasonable since larger mean square deviations imply less stability and, thus, less reliability. Similarly, for Finlay and Wilkinson's approach, holding all other terms constant, the larger the absolute difference between slope coefficients of the test entry and the check ($\beta_i - \beta_c$), the smaller the reliability.

The above reasoning is useful to see how reliability is theoretically related to these stability parameters. However, holding μ_{d_i} and 'other terms' constant as the stability parameters of interest are varied is not possible in application. Eskridge and Mumm (1992) empirically evaluated the relationships that the normal (RN_i) and binomial reliability (R_i) estimates had with three stability

measures (σ_i^2, β, $\sigma_{\delta_i}^2$) and the mean yield. Kendall's Tau Rank correlations

across entries in CIMMYT experimental variety trials (EVTs) were used to assess the relationships between these various statistics (Table 3).

Reliabilities were strongly correlated with the mean yield ($r > 0.70$) in all EVTs indicating that varieties with large means tend to have larger reliabilities. R_i and RN_i were somewhat positively correlated ($r > 0.35$) with b_i in EVTs 14B, 16A and 16B. This result appeared to be caused by b_i being somewhat correlated with the mean ($r > 0.46$) which, in turn, was highly correlated with reliability. Deviation

mean squares ($S_{\delta_i}^2$) and Shukla's stability variance were poorly correlated ($r <$

0.38) with both reliability estimates (R_i, RN_i) in five of six EVTs. These poor correlations indicated that both deviation mean squares and Shukla's stability variance had an insignificant impact on reliabilities. In EVT 16A, deviation mean squares and Shukla's variance were positively correlated ($r > 0.63$) with both reliabilities. For this EVT, large correlations were likely caused by two distinct groups of varieties, one group with large means, large deviation mean squares and

Table 3
Rank correlations between mean yield (Mean), binomial and normal reliability (R_i and RN_i), regression coefficient (b_i), mean square deviation $(S_{\delta i}^2)$ and stability variance $(\hat{q}_i^2)$ for 1988 CIMMYT EVTs.

EVT	Measure	R_i	RN_i	b_i	$S_{\delta i}^2$	$\hat{q}_i^2$
12	Mean	0.743	0.944	0.333	0.111	-0.067
12	R_i	1.000	0.743	0.114	0.111	0.057
12	RN_i		1.000	0.167	0.000	-0.056
13	Mean	0.805	0.890	0.010	0.162	0.086
13	R_i	1.000	0.805	-0.112	0.022	0.045
13	RN_i		1.000	-0.121	0.011	0.033
14A	Mean	0.731	0.883	0.103	-0.118	-0.088
14A	R_i	1.000	0.696	-0.017	-0.087	-0.070
14A	RN_i		1.000	0.000	-0.100	-0.083
14B	Mean	0.711	0.961	0.661	0.322	0.275
14B	R_i	1.000	0.670	0.573	0.380	0.325
14B	RN_i		1.000	0.647	0.294	0.255
16A	Mean	0.702	0.758	0.562	0.524	0.505
16A	R_i	1.000	0.794	0.449	0.633	0.633
16A	RN_i		1.000	0.648	0.736	0.714
16B	Mean	0.853	0.905	0.467	0.233	0.200
16B	R_i	1.000	0.834	0.441	0.167	0.147
16B	RN_i	0.834	1.000	0.352	0.086	0.067

large values of Shukla's stability statistic, while the other group had relatively small values of means, deviation mean squares and Shukla's variances.

286

Table 4

Rank correlations between two randomly separated sets of environments based on the mean yield, binomial and normal reliabilities (R_i and RN_i), regression coefficient (b_i) mean square deviation ($S_{\delta i}^2$) and stability variance ($\hat{q}^2$) for four 1988 CIMMYT EVTS when environments were randomly separated in six different runs.

EVT	Number of varieties	Run	Mean yield$_i$	R_i	RN_i	b_i	$S_{\delta i}^2$	$\hat{q}^2$
12	10	1	0.511	0.561	0.222	0.689	0.022	0.156
		2	0.467	0.159	0.333	0.156	0.289	0.200
		3	0.244	0.223	0.222	0.600	0.067	0.200
		4	0.511	0.381	0.556	0.200	-0.022	0.244
		5	0.289	0.571	0.333	0.600	0.200	0.244
		6	0.022	-0.125	-0.222	0.511	-0.111	0.200
		mean	0.341	0.295	0.241	0.459	0.074	0.207
13	15	1	0.752	0.485	0.604	-0.181	0.219	0.086
		2	0.752	0.281	0.560	0.105	-0.162	-0.238
		3	0.810	0.417	0.604	0.429	-0.181	-0.200
		4	0.771	0.330	0.406	0.048	-0.505	-0.467
		5	0.790	0.694	0.626	0.048	-0.219	-0.200
		6	0.638	0.339	0.451	0.295	0.101	0.010
		mean	0.752	0.424	0.542	0.124	-0.140	-0.168
14A	17	1	0.588	0.356	0.500	0.103	-0.029	-0.015
		2	0.338	0.423	0.283	0.308	-0.103	-0.103
		3	0.206	0.056	0.283	0.029	-0.059	-0.074
		4	0.323	0.212	0.267	0.191	-0.059	0.000
		5	0.529	0.247	0.350	0.103	-0.176	-0.147
		6	0.559	0.351	0.550	0.088	0.015	0.073
		mean	0.424	0.274	0.372	0.137	-0.068	-0.044
14B	19	1	0.485	0.528	0.516	0.427	-0.017	0.053
		2	0.497	0.171	0.516	0.474	0.252	0.287
		3	0.532	0.472	0.386	0.321	0.123	0.146
		4	0.450	0.446	0.464	0.275	0.111	0.181
		5	0.567	0.424	0.647	0.263	0.216	0.158
		6	0.661	0.504	0.608	0.649	0.356	0.427
		mean	0.532	0.424	0.523	0.401	0.173	0.209

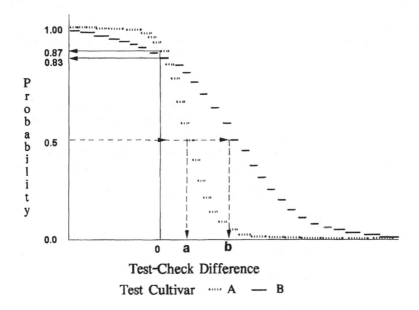

Figure 3. Reliability function of test-check differences for two hypothetical test entries illustrating reliabilities for test entries A and B (0.87, 0.83) and their medians of test-check differences (a and b).

D. COMPARING THE REPEATABILITY OF RELIABILITIES WITH OTHER MEASURES

For the reliability of an entry to be useful to the plant breeder, it is necessary that this value actually be representative of the genetic characteristics of the entries under consideration. If a parameter estimate is truly a measure of the genetic features of an entry, then the ranking of a group of entries based on the estimated parameter should be fairly consistent between any two different sets of environments randomly sampled from the population of environments under consideration.

Eskridge and Mumm (1992) assessed the repeatability of single-trait reliability as a measure of genetic characteristics compared to other measures. For each of four CIMMYT experimental variety trials (EVT), environments were randomly separated into two sets and calculations were made for each set. This process was repeated six times (runs) to ensure dependable results. Mean yields, reliabilities

(R_i, RN_i), Finlay and Wilkinson's (1963) regression coefficients (b_i), Eberhart and Russell's (1966) mean square deviations $(S_{\delta i}^2)$, and Shukla's (1972) stability variances $(\hat{q}^2)$ were estimated for each set of environments and each run. For each statistic, rank correlations across varieties were then computed. Large rank correlations indicated that an estimated parameter consistently ranked varieties over the two sets of environments, thus meaning it was repeatable and a useful measure of genetic features of a variety.

The sizes of rank correlations differed substantially depending on the EVT and statistic (Table 4). For all EVTs, the rank correlations for both reliabilities $(R_i$ and $RN_i)$ were somewhat lower than rank correlation for the mean yield, meaning that both reliability measures were not as repeatable as the mean. However, in most runs, both reliabilities had rank correlations that were larger than rank correlations of the joint regressions coefficient (b_i), and substantially larger than rank correlations of the mean square deviation $(S_{\delta i}^2)$ and the stability variance $(\hat{\sigma}_i^2)$.

This result demonstrated that reliabilities, although not as repeatable as the mean, had a considerably better repeatability than several of the most commonly used stability statistics. In addition, the repeatabilities of Eberhart and Russell's mean square deviation $(S_{\delta i}^2)$ and Shukla's stability variance $(\hat{q}^2)$ were generally quite small $(r < 0.2)$ indicating that neither appeared to measure the genetic features of an entry.

E. EVALUATING ACROSS ENVIRONMENT VARIABILITY USING RELIABILITY FUNCTIONS

Reliability is easy to understand, emulates farmers' and plant breeders' perceptions of what is important and can be used with unbalanced data sets that are likely to arise with multiple location testing. However, reliability is somewhat limited since (1) it is not useful in graphically comparing the responses of different test entries regarding their similarity to the check over environments, (2) it does not aid the plant breeder with identifying environmental factors that may cause a test entry's performance to fall below the check and (3) two test entries may have nearly identical probabilities of outperforming the check but one may be clearly superior.

One approach that circumvents these limitations of reliability is by estimating and graphing the reliability function of each test entry (Eskridge et al., 1993). This function gives the probabilities that a test entry outperforms the check by more than an amount d for all possible values of d. For any particular trait, the reliability function for the ith test entry is defined as

$$R_i(d) = P(X_i - X_c > d) \qquad [3]$$

where P(Z) represents probability of event Z, X_i and X_c are the responses of the trait of interest for the test entry and the check respectively and d is any difference. When $d = 0$, $R_i(d)$ is the reliability of the test entry.

Reliability functions are easy to understand, do not require balanced data sets and the locations, slopes and shapes of the functions may be used to describe and compare the across environment performance of test entries relative to a check. In addition, the functions are directly related to the stability of the test entry and the check.

1. Estimation, Graphing And Testing Reliability Functions.

If reliability functions of test-check differences are to be useful, it is necessary to estimate these functions based on field trial information. One approach is to obtain the fraction of environments in the trial where the test- check difference is greater than d and use this value as an estimate of $R_i(d)$. For example, if there are n environments where both the test and check entries' trait values are available, and there are X environments where the test-check difference is greater than d, then X/n is an unbiased estimate of $R_i(d)$. Computing these estimates for all possible values of d results in an estimated reliability function. Confidence bounds may also be computed for the reliability function for any value of d (Nelson, 1982).

Graphs of the estimated reliability functions are plotted as step functions (Figure 3). These graphs may be used to describe how test-check differences vary over environments. Reliabilities may also be simply obtained from the graph by identifying the value of the function when the test-check difference is 0.

If the decision maker is interested in statistically testing equality of the reliability functions of two test entries, the Hollander bivariate symmetry test may be used assuming the two test entries and the check are common in some set of representative environments (Eskridge et al., 1993; Hollander and Wolfe, 1973; Kotz and Johnson, 1983).

2. Location, Slope And Shape Of Reliability Functions.

The more the reliability function is shifted to the right, the larger the mean difference and the more likely the test entry will outperform the check cultivar, assuming slope and shape are unchanged.

A steep reliability function indicates the test entry responds similarly to the check over environments. Steep functions result when test-check differences for all environments are nearly constant. A gradually sloped function reflects more variability in performance differences and less similarity in response relative to the check. In Figure 3, entry A has a smaller median difference (median of A = a; median of B = b) but has a response more similar to the check than entry B. Also, since entry B has a more gradual slope, it has a larger fraction of environments where it outperformed the check by a larger margin than what would be expected if B simply had a larger mean performance.

In addition, the slope of the reliability function for a test entry is directly related to the stability parameters (β, $\sigma_{\delta_i}^2$ and σ_i^2) of the test and check entries. For any particular trait, if the test-check differences for the ith test entry are normally distributed with mean μ_{d_i} and standard deviation σ_{d_i}, then the reliability function is $R_i(d)=P[Z>(d-\mu_{d_i})/\sigma_{d_i}]$ where Z is a standard normal random variable and d is any value. The slope of the reliability function is mathematically related to the stability parameters β, $\sigma_{\delta_i}^2$ and σ_i^2. This relationship may be shown using the same reasoning that was used to show how reliability was related to these parameters. For two test entries with the same μ_{di}, the entry with the larger $\sigma_{d_i}^2$ as caused by a larger value of $|\beta_i-\beta_c|$ or $\sigma_{\delta_i}^2$ or σ_i^2 , will have a reliability function with a more gradual slope. Thus a more gradual slope will indicate a less stable test entry. In Figure 3, entry B is less stable than A.

The shape of the reliability function may also be useful in understanding what environmental factors either contribute or detract from the test entry's response as compared to the check. For example, in Figure 3, entry B has a larger chance of falling far below the check as indicated by its longer tail on the left below entry A. Identification of environmental and management conditions that contributed to these poor responses relative to the check can aid with understanding where the entry has problems.

Figure 3 also demonstrates an important limitation of basing selection solely on reliability. The two test entries have nearly identical reliabilities, but most decision makers would consider entry B to be clearly superior. Entry B outperforms the check by a larger amount than A as indicated by most of its reliability function being to the right of entry A.

3. Application.

To demonstrate the approach, data were taken from on-farm corn strip test yield trials grown cooperatively by Pioneer Hi-bred International and farmers in Iowa, Illinois and Nebraska over three years (Eskridge et al., 1993). The set of hybrids grown varied from location to location. Plots were chosen where the 4 hybrids used in the analysis were grown together. Hybrid 4 was generally well-adapted to the environments in this study and was used as the check. Estimated reliability functions for the three test hybrids were used to graphically describe how the differences varied over environments, identify environments where hybrids performed poorly and estimate the reliabilities of the test hybrids.

Means, mean differences from the check and their standard deviations, stability statistics and reliabilities were obtained for each of the four corn hybrids (Table 5). The hybrids' reliability functions illustrated how the test-check differences varied over environments (Figure 4). The Hollander bivariate symmetry test failed to reject the hypothesis of coincidence of reliability functions of hybrids 1 and 3 ($p\approx0.40$) but the reliability functions of both hybrids 1 and 3 differed from hybrid

2 (p<0.001). The most striking feature of the three reliability functions was that hybrid 2 was shifted to the right and had a more gradual slope as compared to the other two test hybrids. The shift to the right indicated hybrid 2 had higher mean performance.

Hybrid 2's more gradual slope indicated that it was less stable than either of the other two test hybrids. This larger instability was reflected through larger values of $|b_i-b_c|$, S_{bi}^2 and σ_i^2.

In environments where hybrid 2's yield fell more than 0.5 ton ha^{-1} below the check, its performance dropped off faster than would have been expected as indicated by the 'notch' in the left tail of hybrid 2's reliability function. Identification of environmental conditions which contributed to hybrid 2's poor performance relative to the check would have been useful in determining where the hybrid had performance problems. However, given the available information

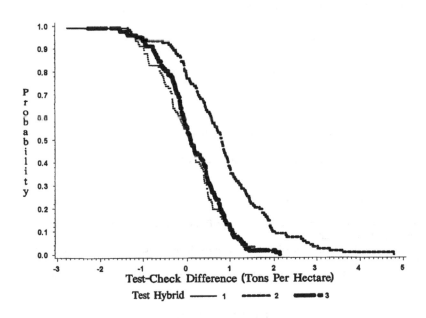

Figure 4. Reliability functions of test-check differences for three Pioneer corn hybrids based on data from on-farm trials conducted in years 1988 through 1990.

on these locations, it was not possible to identify why hybrid 2 performed so poorly relative to the check in these environments and the notch was considered to be caused by unexplained variation associated with strips in the plots.

If selections among these hybrids were to be based on yield alone, the reliability functions demonstrated how hybrid 2 should be preferred to hybrids 1 and 3 by all decision makers since it had a larger probability of outperforming the check by more than d bushels for all values of d. This superior performance was indicated by hybrid 2's reliability function never falling to the left of either hybrid 1 or 3.

Table 5
Means, standard deviations of yields, yield differences and reliabilities of
four corn hybrids over 122 common environments.

Hybrid	Yield[+] Mean	Test-Check Differences[+] Mean	Test-Check Differences[+] Std Dev.	Stability Statistics[*] b_1	Stability Statistics[*] $S_{\delta i}^2$	Stability Statistics[*] q_i^2	$R_i(0)^{++}$
1[A]	9.90	0.04	0.79	0.93	0.20	.023	0.53[a]
2[B]	10.69	0.83	1.01	1.10	0.47	0.81	0.77[b]
3[A]	9.99	0.13	0.75	0.97	0.18	0.17	0.56[a]
4(Check)	9.86	.	.	1.00	0.29	0.37	.

[+] Metric tons ha^{-1}
[*] b_1 - Finlay and Wilkinson's regression coefficient (1963),

$S_{\delta_1}^2$ -Eberhart and Russell's deviation mean square (1966),

q_i^2 -Shulka's stability variance (1972)

[++]The estimated chance that the test hybrid outperforms the check in a future environment using the binomial reliability estimator.
A,B Hybrid reliability functions with common letters do not differ ($\alpha = 0.05$) based on Hollander bivariate symmetry test.
a,b Hybrid reliabilities with common letters do not differ ($\alpha = 0.05$) based on Cochran's Q test (Cochran, 1950).

III. DISCUSSION

The usefulness of reliability functions of test-check differences in identifying superior entries is based on several assumptions. First, decision makers using the approach are assumed to be primarily interested in the performance of test entries relative to a check. If a decision maker values entries using other approaches such as an index of economic worth, the index should be used directly. Second, the approach is based on the assumption that the trial information is obtained from environments that are representative of those environments where the check is well-adapted.   If the check is planted outside its range of adaptability, its performance may fall well below a test entry whereas in its range of adaptability, it may outperform the test entry. Thus, comparing a test entry with a check planted in environments to which it is not adapted may give reliability functions that are biased against the check.

There are several advantages to using reliability in comparing and identifying superior test entries. The approach is conceptually straightforward and is easily understood by many different types of decision makers, be they breeders, growers or management.   Moreover, because the method is pairwise in nature it can maximize the number of environments for a comparison between a test entry and a check. Any location or year where the test entry and check are grown within a reasonable proximity of one another will provide additional information on the reliability.

Also, reliability is functionally related to several commonly used stability statistics $(b_i, S_{\delta_i}^2, \hat{\sigma}_i^2)$. However reliability estimates are more repeatable than these stability statistics and thus are better measures of genetic characteristics of entries. In addition, these stability statistics $(b_i, S_{\delta_i}^2, \hat{\sigma}_i^2)$ can only be used as relative measures since each depends on the particular set of entries being evaluated (Lin et al.,1986). Reliability of a test entry has a broader inference base than $b_i, S_{\delta_i}^2,$ or $\hat{\sigma}_i^2$ because it only depends on the check and the particular test entry being considered and does not depend on other test entries in the trial.

Another advantage of reliability is that it is an index that explicitly weighs the importance of the difference in performance relative to stability. This property relieves the plant breeder from having to make decisions about how to weigh the importance of performance to stability when making final selections.   Viewed as an index that explicitly combines both performance and stability through the ratio $\bar{D}_i/s_{D_i}$, reliability compares favorably with other indices that combine performance and stability. Reliability has fewer assumptions and is easier to understand than the expected utility stability indices proposed by Barah et al. (1981) and Eskridge and Johnson (1991). Reliability does not require special 'disaster' parameters as do safety-first stability indices (Eskridge, 1990; Eskridge et al., 1991; Eskridge, 1991). Reliability is based on reasonable behavioral assumptions about how plant breeders make selections and has a better theoretical basis than ad hoc yield-stability indices such as those discussed in Kang and Pham (1991). In addition, reliability can be used to obtain a complete ranking of entries under test in contrast to stochastic dominance as used by Menz (1980) to categorize wheat varieties as risk-efficient or risk-inefficient.

Reliability methods are easily applied to the multiple trait case where numerous correlated traits are to be evaluated simultaneously. A single resulting multiple-trait reliability value can reflect the overall selection value of a genotype and is analogous to using a multiple trait selection index. Single-trait reliabilities can then be used to help identify limitations of traits of the test entries and show the impact of individual traits on overall reliability.

Reliability estimates do not require that the trait(s) under consideration have any specific type of probability distribution. Both the binomial approach and the nonparametric residual approach give unbiased estimates of reliabilities without assuming normality, although the nonparametric method can produce more precise reliability estimates. Confidence intervals and tests on reliabilities may also be conducted.

Reliability functions provide a basis for graphically comparing test entries over a range of environments. The location, slope and shape of the functions give a general indication of how the differences are distributed over environments. Steeper slopes indicate less variability and more similarity in response relative to the check. Functions shifted more to the right imply larger performance differences compared to the check while the shapes of reliability functions may be useful in identifying environments where a entry has performance problems relative to the check. Additional information on these environments could then be used to identify environmental and management factors that contribute to such problems. Reliability functions are related to several commonly used stability statistics. Also, coincidence of the functions of test entries may be statistically tested. There are several consequences that should be considered when using reliability to identify superior plant entries. Since the approach compares test entries with a common check, the choice of a check can have a major impact on the estimated reliability functions. In situations where comparisons with several checks are desired, it may be necessary to compute different reliabilities using different checks. Also, reliability can be a useful aid to plant breeders when evaluating entries in the presence of GE interaction, but breeders should not use the approach in lieu of understanding the biological nature of these interactions.

IV. APPENDIX

A. TESTING EQUALITY OF RELIABILITIES FOR k TEST ENTRIES BASED ON NORMALLY DISTRIBUTED DIFFERENCES AND THE WALD TEST

Assume test trial data are available for the check and k test entries in n environments. Let

$$d_{ij} = Y_{ij} - Y_{cj}$$

be the difference between the response of the ith test entry and that of the check in the jth environment. Let the 1xk vector of differences of the k test entries for the jth environment $\mathbf{d}_j = (d_{1j}\ d_{2j}...d_{kj})$ be one of n independent samples from a multivariate normal distribution with mean vector $\boldsymbol{\mu} = (\mu_{d1}\ \mu_{d2}\ ...\ \mu_{dk})$ and a kxk covariance matrix Σ. Σ has σ_{di}^2 as the ith diagonal element and σ_{dij} as the i,jth off diagonal element. The marginal distribution for the ith entry then has a univariate normal distribution with mean μ_{di} and variance σ_{di}^2 (Anderson, 1984). Further, define

$$p_i = P(d_{ij} > 0) \text{ and note } p_i = 1 - \Phi(-\mu_{di}/\sigma_{di})$$

where $\Phi(.)$ is the standard normal distribution function. Then the hypothesis of interest is

$$H_0: p_1 = p_2 = ... = p_k .$$

To test H_0 using Wald's test (Nelson, 1982), define k-1 constraint functions:

$$h(\mu_{d1},...,\mu_{dk}, \sigma_{d1}^2,...,\sigma_{dk}^2) = p_1 - p_i = 0, \; i = 2,...,k.$$

Obtain a 2k x (k-1) matrix of partial derivatives:

$$H(\theta) = \{\delta h_i/\delta \theta_j\}; \; 1=2,...,k \text{ and } j=1,...,2k$$

where θ_j is the appropriate parameter μ_{di} or σ_{di}^2 The 2k x 2k covariance matrix of the maximum likelihood estimators

$$\hat{\theta} = (\hat{\mu}_{d1},...,\hat{\mu}_{dk}, \hat{\sigma}_{d1}^2,...,\hat{\sigma}_{d2}^2)$$

is defined as $\Sigma(\hat{\theta})$ with elements

$$cov(\hat{\mu}_{di}, \hat{\mu}_{dj}) = \sigma_{di}^2/n \quad \text{for } i=j$$

$$= \sigma_{dij}/n \quad \text{for } i \neq j$$

$$cov(\hat{\sigma}_{di}^2, \hat{\sigma}_{dj}^2) = 2(n - 1)\sigma_{di}^4/n^2 \quad \text{for } i=j$$

$$= 2(n - 1)\sigma_{dij}^2/n^2 \quad \text{for } i \neq j$$

and all other elements zero. The asymptotic covariance matrix of the maximum likelihood estimates evaluated at $\hat{\theta}$ is $V = H(\hat{\theta})'\Sigma(\hat{\theta})H(\hat{\theta})$.

Now define $h(\hat{\theta})$ as the kx1 vector of constraints evaluated at $\hat{\theta}$, the maximum likelihood estimates. Then Wald's statistic for testing H_0 is

$$W = h(\hat{\theta})'V^{-1}h(\hat{\theta})$$

which has a chi-square distribution with k-1 degrees of freedom when H_0 is true and the sample size is large (Nelson, 1982).

B. COMPUTING UNIVARIATE AND MULTIVARIATE RELIABILITIES USING RESIDUALS FROM A LINEAR MODEL

To demonstrate how probabilities are computed, define the following linear model:

$$D_{ijk} = \mu_{ik} + \epsilon_{ijk} ; \quad i=1,...,p; \; j=1,...,n; \; k=1,...,m$$

where D_{ijk} = the test-check difference for the ith trait, the jth environment and the kth entry with the m+1th entry being the check, μ_{ik} = mean quantity for the ith trait of the kth entry, ϵ_{ijk} = error where

$$E(\epsilon_{ijk}) = 0, \; V(\epsilon_{ijk}) = \sigma_{ik}^2 \; \text{and} \; cov(\epsilon_{ijk}; \; \epsilon_{ijk}) = \sigma_{i(m+1)}^2$$

and all other covariances zero. The covariance structure of ϵ_{ijk} is the result of each test entry being compared with the same check in each environment.

Residuals are estimated as $\hat{\epsilon}_{ijk} = D_{ijk} - \bar{D}_{i.k}$. Standardized residuals are estimated as $u_{il} = \hat{\epsilon}_{ijk} / s_{ik}$ where $l = n(j-1) + k$ and s_{ik} is the standard deviation of D_{ijk}. The u_{il} may then be arranged as a mnxp matrix of standardized residuals with a column for each trait. If variances are homogeneous across entries, $u_{il} = \hat{\epsilon}_{ijk}$. The lth pseudo-observation for the ith trait of the gth entry is

$$\hat{D}_{igl} = s_{ig}u_{il} + \bar{D}_{i.g}$$

when residuals are standardized and

$$\hat{D}_{igl} \cdot u_{il} \cdot \bar{D}_{ig} \text{ when } u_{il} \cdot \hat{\varepsilon}_{ijk}.$$

The multivariate probability of the gth entry performing at least as well as the check for all p traits is estimated as

$$\hat{P}_g \cdot (1/mn)\sum_{j1}^{n}\sum_{k-1}^{m} I_{jk}^{g}$$

where $I_{jk}^{g} \cdot \begin{cases} 1 & \text{if } 0 < D_{1gl}, ..., 0 < D_{pgl}; \\ 0 & \text{otherwise}. \end{cases}$ Univariate probabilities for

the ith trait and the gth entry are estimated by using only $\hat{D}_{igl}$ in I_{jk}^{g}.

Variances of $\hat{P}_g (V(\hat{P}_g), g \cdot 1, ..., m)$ and covariances for any pair of

entries $cov(\hat{P}_g; \hat{P}_h)$ are needed to compute approximate confidence intervals for the

reliabilities. Define a mnx1 vector of indicators for the gth entry as

$$I^g \cdot (I_{11}^{g}, ..., I_{1m}^{g}, I_{21}^{g}, ..., I_{mn}^{g})'.$$

Then $\hat{P}_g = (1/mn)\underline{1}'I^g$ and $cov(\hat{P}_g; \hat{P}_h) = (1/mn)^2(\underline{1}'\Sigma_{gh}\underline{1})$ where $\underline{1}$ is a mnx1

vector of 1s and Σ_{gh} is the covariance matrix of the vectors $\underline{I}^g$ and $\underline{I}^h$. Each I_{jk}^{g} is

a Bernoulli random variable with

$$E(I_{jk}^{g}) \cdot p_g \text{ and } V(I_{jk}^{g}) \cdot p_g(1 - p_g).$$

The elements of Σ_{gh} are

$$cov(I_{jk}^{g}; I_{jk'}^{h}) \cdot P(I_{jk}^{g} \cdot 1, I_{jk'}^{h} \cdot 1) - p_g p_h \text{ for all } j, j', k \text{ and } k'.$$

Since $\underline{I}^g$ is a function of residuals, assume that Σ_{gh} has a structure similar to the covariance matrix of $\underline{\varepsilon}_i$, where $\underline{\varepsilon}_i$ is the mnx1 error vector,

$\text{cov}(\underline{\epsilon}_i) = (\mathbf{H}_i + \sigma^2_{i(m+1)}\mathbf{J}_m)\otimes\mathbf{I}_n$ with $\mathbf{J}_m$ a mxm matrix of 1s and

$$\mathbf{H}_i = \begin{pmatrix} \sigma^2_{i1} & 0 & \cdots\cdots & 0 \\ 0 & \sigma^2_{i2} & \cdots\cdots & 0 \\ & & \cdots\cdots & \\ 0 & \cdots\cdots & \sigma^2_{im-1} & 0 \\ 0 & \cdots\cdots & 0 & \sigma^2_{im} \end{pmatrix}.$$

That is, let $\Sigma_{gh} = {}_{gh}\Phi_n \otimes \mathbf{I}$ where the kk'th element of Φ is $C^{gh}_{kk'} = \text{cov}(I^g_{jk}; I^h_{jk'})$ for all j, k and k'.

To estimate Φ_{gh}, let $1^{gh}_{jj} = \sum_{k}^{m} I^g_{jk}I^h_{jk}$ and $n_j^g = \sum_{k}^{m} I^g_{jk}$. Now if we define the subvectors of I^g as $I^g_j = (I^g_{j1}, \ldots, I^g_{jm})$, the total number of 1s in $I^g_j I^{h'}_j$ is $n_j^g n_j^h$, the sum of the diagonal elements of $I^g_j I^{h'}_j$ is 1^{gh}_{jj} and the sum of the off-diagonal elements of $I^g_j I^{h'}_j$ is $n_j^g n_j^h - 1^{gh}_{jj}$. $P(I^g_{jk} = 1, I^h_{jk} = 1)$ is then estimated as $(1/mn)\sum_j^n 1^{gh}_{jj}$, p_g is estimated with $(1/(mn))\sum_j^n n_j^g$ and $P(I^g_{jk} = 1, I^h_{jk'} = 1)$ is estimated as $(1/(mn(m-1)))\sum^n (n_j^g n_j^h - 1^{gh}_{jj})$, where $k \neq k'$. The elements in Φ_{gh} are then estimated as: $\hat{C}^{gh}_{kk} = (1/mn)\sum^n 1^{gh}_{jj} - [(1/mn)\sum_j^n n_j^g][(1/mn)\sum_j^n n_j^h]$ for all k, $\hat{C}^{gh}_{kk'} = (1/(mn(m-1)))\sum_j^n (n_j^g n_j^h - 1^{gh}_{jj}) - [(1/mn)\sum_j^n n_j^g][(1/mn)\sum_j^n n_j^h]$ for all k≠k'. Using these values in Φ_{gh} and computing $1'\Sigma_{gh}1$,

gives $\text{cov}(\hat{p}_g, \hat{p}_h) = (1/mn)^2[\sum^n n_j^g n_j^h - (1/n)\sum^n n_j^g \sum^n n_j^h]$, which may be expressed as $(1/n)^2[\sum^n \hat{p}_j^g \hat{p}_j^h - (1/n)\sum^n \hat{p}_j^g \sum^n \hat{p}_j^h]$ where $\hat{p}_j^g$ is the proportion of entry g pseudo-observations > 0 based on residuals from location j. The variance of $\hat{p}_g$ is obtained by setting $\hat{p}_j^h = \hat{p}_j^g$ in the covariance equation. The variance of $\hat{p}_g$ may be expressed as:

$$V(\hat{p}_g) = P(1-P)/mn + (m-1)C/mn ,$$

where P is the true reliability and the covariance C is the common off-diagonal elements of Φ_{gg}. When the pseudo-observations are uncorrelated and the off diagonals of Φ_{gg} are zero, then C=0 and $V(\hat{p}_g) = P(1 - P)/mn$.

Thus, for m entries in a trial, the approximate number of environments for approximate 95% confidence that the estimate will be within δ units of P is n = $4P(1-P)/(m\delta^2)$ where δ is the acceptable difference between P and the estimate and $t^2 = 4$ gives approximate 95% confidence. It should be noted that C can never be larger than P(1-P) and so $V(\hat{p}_g) \leq P(1 - P)/n$. But when there is little correlation among pseudo-observations within an environment, C is small and $V(\hat{p}_g)$ can be considerably smaller than the binomial variance P(1-P)/n.

C. RELATIONSHIP BETWEEN STABILITY PARAMETERS AND VARIANCE OF THE TEST-CHECK DIFFERENCES

Define the performances of the ith test entry and the check in environment j as Y_{ij} and Y_{cj} respectively.

Following Shukla (1972), for the hth entry

$$Y_{hj} = \mu + G_h + E_j + V_{hj} \quad \text{for h = i or c}$$

where μ is the grand mean, G_h is a fixed entry effect, E_j is an environmental effect and V_{hj} is the h, jth random deviation from the additive model. V_{hj} has expectation 0, variance σ^2_h and covariance $cov(V_{ij}, V_{cj}) = 0$. The difference between the ith entry and the check in environment j is

$$d_{ij} = Y_{ij} - Y_{cj} = G_i - G_c + V_{ij} - V_{cj}.$$

Using rules of expectation, the variance of d_{ij} over environments is

$$\sigma^2_{di} + \sigma^2_i + \sigma^2_c.$$

For Eberhart and Russell's model (1966), define

$$Y_{hj} = \mu_h + \beta_h I_j + \delta_{hj}$$

for h = i or c where μ_h is a fixed entry effect, β_h is the regression coefficient for the hth entry, I_j is a random environmental index and δ_{hj} is the deviation from regression of the hth entry in the jth environment. I_j has expectation 0 and variance σ^2_I while δ_{hj} has expectation 0, variance $\sigma^2_{\delta h}$ and covariance $cov(\delta_{ij}, \delta_{cj}) = cov(I_j, \delta_{hj}) = 0$. Given this model, the difference between the ith entry and the check is

$$d_{ij} = \mu_i - \mu_c + (\beta_i - \beta_c)I_j + \delta_{ij} - \delta_{cj}.$$

Using rules of expectation, the variance of d_{ij} over environments is

$$d_{ii}^2 \cdot (\beta_i - \beta_o)^2 \sigma_I^2 + \sigma_{\delta i}^2 + \sigma_{\&c}^2$$

The relationship between d_{ii}^2 and Finlay and Wilkinson's slope parameter (1963) is obtained using the Eberhart and Russell model with all δ's and their variances set to zero.

D. SAS PROGRAMS
1. Program 1

```
***********************************************************************
* THIS PROGRAM COMPUTES THE PROBABILITY OF A VARIETY OUTPERFORMING
* A CHECK. THESE RELIABILITIES ARE COMPUTED IN TWO WAYS : BASED
* ON THE ASSUMPTION OF NORMALLY DISTRIBUTED DIFFERENCES AND USING
* A BINOMIAL APPROACH.
***********************************************************************;
OPTIONS LS = 72 NODATE;
CMS FILEDEF IN1 DISK NULAND DATA *; * CHECK IS VAR=4 ;
DATA ALL1;  INFILE IN1;
INPUT VARIETY 1 LOC 3-4 YIELD 6-9 WM 11-14 SEEDS 16-19 BM 21-24
PLANTS 26-30; I=1; *YIELD = 1600/YIELD; ** SEED WEIGHT OZ PER 100;
***********************************************************************
THE FOLLOWING SEGMENT COMPUTES DIFFERENCES BETWEEN EACH VARIETY
AND THE CHECK AND SETS R = 1 IF DIFF > 0 OR R=0 IF DIFF <= 0.
***********************************************************************;
DATA ALL; SET ALL1;
PROC SORT DATA=ALL; BY LOC VARIETY;
DATA NEW(KEEP=VARIETY LOC V1-V4 );
ARRAY COMBINE(J) V1-V4 ; ** <<< CHANGE NUMBER OF VARIETIES;
DO OVER COMBINE; SET ALL; BY LOC VARIETY;
COMBINE=YIELD;
IF LAST.LOC THEN RETURN; END;
DATA NEXT;SET NEW;ARRAY D(II) D1-D4 ; ARRAY V(II) V1-V4 ;
ARRAY R(II) R1-R4 ;
DO II=1 TO 4 ; D=V-V4 ; *<<< CHANGE CHECK ENTRY;
IF D<=0 THEN R=0; IF D>0 THEN R=1;
IF II=4  THEN DO; R=.; D=.; END;  *<<< CHANGE CHECK ENTRY;
END;
DATA NEXT1; ARRAY ALLR(VARIETY) R1-R4 ; ARRAY ALLD(VARIETY) D1-D4;
SET NEXT;
DO VARIETY=1 TO 4; RESPONSE = ALLR; DIFF = ALLD;
OUTPUT; END;
***********************************************************************
```

```
THE FOLLOWING PART COMPUTES RELIABILITIES BASED ON RESIDUALS
******************************************************************;
DATA NEXT2; SET NEXT1; I=1; IF VARIETY < 4;  ** DROP CHECK;
PROC SORT DATA=NEXT2; BY VARIETY LOC;
PROC GLM; CLASSES VARIETY;
MODEL DIFF = VARIETY / NOUNI; OUTPUT OUT=XXR P=PDIFF R=RDIFF;
PROC MEANS NOPRINT; BY VARIETY ; VAR DIFF;
OUTPUT OUT=XX MEAN=MDIFF STD=SDIFF;
DATA XXX; ARRAY MD(VARIETY) MDIFF1 MDIFF2 MDIFF3;
ARRAY SD(VARIETY) SDIFF1 SDIFF2 SDIFF3;
DO VARIETY=1 TO 3; SET XX; I=1; MD=MDIFF; SD=SDIFF; END;
DATA CCC; MERGE XXR XXX; BY I;
DIFFP1=RDIFF+MDIFF1; DIFFP2=RDIFF+MDIFF2;  DIFFP3=RDIFF+MDIFF3;
IF DIFFP1 > 0 THEN R1=1; ELSE IF DIFFP1 <= 0 THEN R1=0;
IF DIFFP2 > 0 THEN R2=1; ELSE IF DIFFP2 <= 0 THEN R2=0;
IF DIFFP3 > 0 THEN R3=1; ELSE IF DIFFP3 <= 0 THEN R3=0;
PROC MEANS NOPRINT; VAR R1 R2 R3;OUTPUT OUT=CCCC MEAN=R1 R2 R3;
DATA CCCCC; SET CCCC; ARRAY R(VARIETY) R1-R3;
DO VARIETY=1 TO 3; PROBNP=R; OUTPUT; END; DROP R1-R3;
******************************************************************
THE FOLLOWING SEGMENT MERGES ALL DATA SET WITH DIFFERENCES AND R'S.
******************************************************************;
PROC SORT DATA=ALL;  BY VARIETY LOC;
PROC SORT DATA=NEXT1; BY ARIETY LOC;
DATA EE; MERGE ALL NEXT1; BY VARIETY LOC;
PROC MEANS NOPRINT;  BY  VARIETY; VAR YIELD RESPONSE DIFF;
OUTPUT OUT=XX MEAN = YIELD PROB DIFF N = LOCS  STD = SY SP SDIFF;
DATA B1; MERGE CCCCC XX; BY VARIETY;
PROBN = 1 - PROBNORM(-(DIFF/SDIFF));
*PROBN = 1 - PROBT(-(DIFF/SDIFF),LOCS-1);
PROC SORT; BY DESCENDING PROB;
PROC PRINT;TITLE1 ' BEAN TRIAL: FOR YIELD - CHECK IS VARIETY 4';
VAR VARIETY LOCS YIELD DIFF SDIFF  PROB PROBN  PROBNP;
TITLE2 'PROBN = RELIABILITY BASED ON NORMALITY ';
TITLE3 'PROB  = RELIABILITY BASED ON BINOMIAL';
TITLE4 'PROBNP= RELIABILITY BASED ON RESIDUALS';
TITLE5 'DIFF  = DIFFERENCE FROM CHECK';
TITLE6 'SDIFF = STANDARD DEVIATION OF DIFFERENCE';
**********************************************************
FOLLOWING PART PERFORMS  WALD TEST
FOR EQUALITY OF RELIABILITIES FOR ALL CULTIVARS .
BASED ON ASSUMPTION RESPONSES ARE NORMALLY DISTRIBUTED
**********************************************************;
```

```
DATA NEXT2; SET NEXT; KEEP D1-D3;    *<<<< CHANGE TO EXCLUDE CHECK;
PROC IML; USE NEXT2; READ ALL INTO D; *RESET PRINT ;
P =NCOL(D); N = NROW(D);
JJ=(J(N,1))`; M= ((JJ*D)/N)`; * M - UNRESTRICTED MEAN;
COV= (D`*D - N*M*M`)/N;    * COV - UNRESTRICTED COVARIANCE;
******* WALD TEST ********;
COVV = 2*(N-1) # (( COV#(1/N))##2 ) ; * COVARIANCES AMONG VARIANCES;
COVW1= ( (COV#(1/N)) || J(P,P,0) ) // ( J(P,P,0) || COVV );
BOT = SQRT(DIAG(COV))*J(P,1,1) ;  * VECTOR OF STDEVS ;
RATIO = -M / BOT;
DENS = (1/(SQRT(2*3.1416)))#EXP((-RATIO##2)/2); *VECTOR NORM DENSITIES ;
DENV = DENS/BOT;  DENV=DENV`;   * DENSITIES / STDEVS ;
H12 = DIAG(DENV(|,2:P|));
H11 = DENV(|1,1|) # J(1,P-1,1);
DENV2 = (-DENS#M)/(2*(BOT##3)) ; DENV2=DENV2`;*
PHI*MEAN/2(STDEV**3);
H22 = DIAG(DENV2(|,2:P|));
H21= DENV2(|1,1|) # J(1,P-1,1);
H = H11 // -H12 // -H21 // H22 ;  * MATRIX OF PARTIAL DERIVATIVES ;
COVW = H`*COVW1*H;  * ASYP. COV MATRIX OF RELIABILITIES;
FN =1-(PROBNORM(RATIO));
FN1=FN(|1,|)#J(P-1,1,1); FN2=FN(|2:P,|);
HH = FN1 - FN2;      * VECTOR OF CONSTRAINTS ON RELIABS;
WALD = HH`*INV(COVW)*HH; DFW=P-1;   * WALD STATISTIC ;
PROBWALD = 1 - PROBCHI(WALD,P-1);   * P-VALUE FOR WALD TEST;
PRINT ' BEAN TRIAL: FOR YIELD. CHECK IS VARIETY 4';
PRINT 'WALD STATISTIC TESTS ALL VARIETIES HAVE SAME RELIABILITY';
PRINT 'WALD  = WALD STATISTIC';
PRINT 'DFW  = DEGREES OF FREEDOM FOR WALD STATISTIC';
PRINT 'PROBWALD = P-VALUE FOR WALD STATISTIC';
PRINT LL  PROBLR DFLR; PRINT HH WALD PROBWALD DFW;
** WALD = WALD STATISTIC, PROBWALD = PVALUE, DFW = DEGREES OF FREEDOM;
```

2. Program 2

```
***************************************************************;
*THIS PROGRAM COMPUTES NONPARMETRIC RELIABILITIES BASED ON
*RESIDUAL'S FOR TRAITS - FLOUR PROT (FP), SDS SEDIM (SDS), MIXING
*TIME (MT), MIXING TOLERANCE (MTI), AND KERNEL HARDNESS (KH);
***************************************************************
OPTIONS LS=72;
CMS FILEDEF IN1 DISK PROB_WQ DATA A;
DATA ALL; INFILE IN1;
INPUT YR LOC REP EN WP FP SDS KWT MT MTI WGHT KH; I=1;
```

```
IF YR=88 THEN ENV=LOC; ELSE IF YR=89 THEN ENV=LOC + 7; IF ENV NE 6 THEN
YLD=WGHT*3.325; ELSE IF ENV=6 THEN YLD=WGHT*6.725;
IF EN=16 THEN DELETE;
PROC SORT; BY ENV EN REP;
MACRO QVAR   FP MT MTI SDS KH      %
MACRO SVAR  SFP SMT SMTI SSDS SKH       %
MACRO RVAR  RFP RMT RMTI RSDS RKH
MACRO PVAR  PFP PMT PMTI PSDS PKH       %
MACRO DVAR  DFP DMT DMTI DSDS DKH         %
PROC MEANS NOPRINT;  BY ENV EN;
VAR QVAR; OUTPUT OUT=X0  MEAN=QVAR STD=SVAR;
DATA BB(KEEP=ENV FPV1-FPV20 MTV1-MTV20 MTIV1-MTIV20 SDSV1-SDSV20
KHV1-KHV20 FPD1-FPD20 MTD1-MTD20 MTID1-MTID20 SDSD1-SDSD20
KHD1-KHD20 );
ARRAY FPV(EN) FPV1-FPV20;
ARRAY MTV(EN) MTV1-MTV20; ARRAY MTIV(EN) MTIV1-MTIV20;
ARRAY SDSV(EN) SDSV1-SDSV20; ARRAY KHV(EN) KHV1-KHV20;
ARRAY FPD(EN) FPD1-FPD20;
ARRAY MTD(EN) MTD1-MTD20; ARRAY MTID(EN) MTID1-MTID20;
ARRAY SDSD(EN) SDSD1-SDSD20; ARRAY KHD(EN) KHD1-KHD20;
DO EN=1 TO 20;  SET X0; BY ENV;
FPV=FP; MTV=MT; MTIV=MTI; SDSV=SDS; KHV=KH;
FPD=FPV-FPV1; MTD=MTV-MTV1; MTID=MTIV-MTIV1; SDSD=SDSV-SDSV1;
KHD=KHV-KHV1;
IF LAST.ENV THEN RETURN; END;
DATA CC;
ARRAY FPD(EN) FPD1-FPD20;
ARRAY MTD(EN) MTD1-MTD20; ARRAY MTID(EN) MTID1-MTID20;
ARRAY SDSD(EN) SDSD1-SDSD20; ARRAY KHD(EN) KHD1-KHD20;
SET BB;
DO EN=3 TO 15, 17 TO 20; I=1; DFP=FPD; DMT=MTD; DMTI=MTID;
DSDS=SDSD; DKH=KHD; OUTPUT; END; DROP FPD1--KHD20 FPV1--KHV20;
PROC SORT; BY EN ENV;
PROC GLM; CLASSES EN;
MODEL DVAR = EN /NOUNI;  OUTPUT OUT=XXR P=PVAR R=RVAR;
PROC MEANS NOPRINT; BY EN; VAR DVAR;ID I;
OUTPUT OUT=XX   MEAN=QVAR STD=SVAR;
DATA CCC; MERGE XXR  XX; BY EN I;
/* LEAVE IN FOLLOWING STATEMENTS IF HETEROGENEOUS VARIANCE
RFP=RFP/SFP; RMT=RMT/SMT; RMTI=RMTI/SMTI;
RKH=RKH/SKH;RSDS=RSDS/SSDS; */
DATA B0(KEEP=I ENV FPV1-FPV20 MTV1-MTV20 MTIV1-MTIV20 SDSV1-SDSV20
KHV1-KHV20 FPD1-FPD20 MTD1-MTD20 MTID1-MTID20 SDSD1-SDSD20
```

```
KHD1-KHD20 SFP1-SFP20 SMT1-SMT20 SMTI1-SMTI20 SSDS1-SSDS20
SKH1-SKH20 );
ARRAY FPV(EN) FPV1-FPV20;
ARRAY MTV(EN) MTV1-MTV20; ARRAY MTIV(EN) MTIV1-MTIV20;
ARRAY SDSV(EN) SDSV1-SDSV20; ARRAY KHV(EN) KHV1-KHV20;
ARRAY SSFP(EN) SFP1-SFP20;
ARRAY SSMT(EN) SMT1-SMT20; ARRAY SSMTI(EN) SMTI1-SMTI20;
ARRAY SSSDS(EN) SSDS1-SSDS20; ARRAY SSKH(EN) SKH1-SKH20;
DO EN=1 TO 20;  SET XX ;I=1;
FPV=FP; MTV=MT; MTIV=MTI; SDSV=SDS; KHV=KH;
SSFP=SFP; SSMT=SMT; SSMTI=SMTI ; SSSDS=SSDS; SSKH=SKH ; END;
DATA C; MERGE CCC B0; BY I;
ARRAY FPV(EN) FPV1-FPV20;
ARRAY MTV(EN) MTV1-MTV20; ARRAY MTIV(EN) MTIV1-MTIV20;
ARRAY SDSV(EN) SDSV1-SDSV20; ARRAY KHV(EN) KHV1-KHV20;
ARRAY SSFP(EN) SFP1-SFP20;
ARRAY SSMT(EN) SMT1-SMT20; ARRAY SSMTI(EN) SMTI1-SMTI20;
ARRAY SSSDS(EN) SSDS1-SSDS20; ARRAY SSKH(EN) SKH1-SKH20;
ARRAY FPD(EN) FPD1-FPD20;
ARRAY MTD(EN) MTD1-MTD20; ARRAY MTID(EN) MTID1-MTID20;
ARRAY SDSD(EN) SDSD1-SDSD20; ARRAY KHD(EN) KHD1-KHD20;
ARRAY FPR(EN) FPR1-FPR20;
ARRAY MTR(EN) MTR1-MTR20; ARRAY MTIR(EN) MTIR1-MTIR20;
ARRAY SDSR(EN) SDSR1-SDSR20; ARRAY KHR(EN) KHR1-KHR20;
ARRAY RRFP(EN) RRFP1-RRFP20;
ARRAY RRMT(EN) RRMT1-RRMT20; ARRAY RRMTI(EN) RRMTI1-RRMTI20;
ARRAY RRSDS(EN) RRSDS1-RRSDS20; ARRAY RRKH(EN) RRKH1-RRKH20;
ARRAY R(EN) R1-R20;
DO EN=1 TO 20;
FPD=RFP+FPV; MTD=RMT+MTV;
MTID=RMTI+MTIV; SDSD=RSDS+SDSV;
KHD=RKH+KHV;
/* USE FOLLOWING STATEMENTS INSTEAD OF ABOVE IF HETERO. VARIANCES
FPD=RFP*SSFP+FPV; MTD=RMT*SSMT+MTV;
MTID=RMTI*SSMTI+MTIV; SDSD=RSDS*SSSDS+SDSV;
KHD=RKH*SSKH+KHV;  */
IF FPD >0 AND MTD >0 AND MTID >0 AND SDSD >0
AND KHD >0 THEN R=1; ELSE R=0;     *** NOTE >, NOT => ;
IF 0< FPD  THEN  RRFP=1;  ELSE RRFP=0;
IF 0< MTD  THEN  RRMT=1;  ELSE RRMT=0;
IF 0< MTID THEN  RRMTI=1; ELSE RRMTI=0;
IF 0< SDSD THEN  RRSDS=1; ELSE RRSDS=0;
IF 0< KHD  THEN  RRKH=1;  ELSE RRKH=0;
```

```
END;
PROC MEANS NOPRINT; VAR R1-R20 RRFP1-RRFP20 RRMT1-RRMT20
RRMTI1-RRMTI20 RRSDS1-RRSDS20 RRKH1-RRKH20;
OUTPUT OUT=XXXX MEAN=R1-R20 RRFP1-RRFP20 RRMT1-RRMT20
RRMTI1-RRMTI20 RRSDS1-RRSDS20 RRKH1-RRKH20;
DATA D; SET XXXX;ARRAY R(EN) R1-R20;
ARRAY RRFP(EN) RRFP1-RRFP20;
ARRAY RRMT(EN) RRMT1-RRMT20; ARRAY RRMTI(EN) RRMTI1-RRMTI20;
ARRAY RRSDS(EN) RRSDS1-RRSDS20; ARRAY RRKH(EN) RRKH1-RRKH20;
DO EN=1 TO 20; PROB = R;
PFP=RRFP; PMT=RRMT; PMTI=RRMTI; PSDS=RRSDS; PKH=RRKH;
OUTPUT; END; KEEP EN PROB PFP PMT PMTI PSDS PKH;
PROC PRINT; VAR EN PFP PMT PMTI PSDS PKH PROB;
FORMAT PFP PMT PMTI PSDS PKH PROB 4.3;
```

3. Program 3

```
*****************************************************************;
*CREATES AND PLOTS RELIABILITY FUNCTIONS FOR CORN STRIP TEST DATA;
CMS FILEDEF IN DISK YIELD DATA;
*****************************************************************;
OPTIONS LS=72;
DATA A; INFILE IN;
INPUT PLOTKEY $ 1-15 YEAR 18-20 VARIETY $ 22-25
YIELD 35-39 MOISTURE 45-49 REPS 51;
YIELD = .06278*YIELD; ** CONVERTS TO METRIC TONS;
PROC SORT; BY PLOTKEY VARIETY;
PROC MEANS NOPRINT; BY PLOTKEY VARIETY;
OUTPUT OUT = XX MEAN = YIELD; VAR YIELD;
PROC SORT; BY VARIETY;
DATA C; SET XX;
IF VARIETY = 'VAR1' THEN V=1; IF VARIETY = 'VAR2' THEN V=2;
IF VARIETY = 'VAR3' THEN V=3; IF VARIETY = 'VAR4' THEN V=4;
PROC SORT; BY PLOTKEY;
DATA TURN (KEEP=V PLOTKEY YIELD V1-V4);
ARRAY COMBINE(V) V1-V4;
DO OVER COMBINE; SET C; BY PLOTKEY; COMBINE=YIELD;
IF LAST.PLOTKEY THEN RETURN; END;
DATA DA;
SET TURN;* HYBRID 4 IS CHECK VARIETY;
D1=V1-V4; B1=1*(D1<=0) + 0*(D1>0); * THE B VARIABLES ARE USED ;
D2=V2-V4; B2=1*(D2<=0) + 0*(D2>0); * TO COMPUTE RELIABILITIES ;
D3=V3-V4; B3=1*(D3<=0) + 0*(D3>0);
IF D1=. OR D2=. OR D3=. THEN DELETE;
```

306

```
IF D1=. THEN B1=.; IF D2=. THEN B2=.; IF D3=. THEN B3=.;
** NEXT PART COMPUTES RELIABILITY FUNCTIONS FOR EACH TEST HYBRID ;
DATA D; ARRAY D(V) D1-D4;   SET DA;
DO V=1 TO 4;   DIFF=D; OUTPUT; END; DROP D1-D3;
PROC SORT DATA = D; BY V; PROC FREQ; BY V;   TABLES DIFF/OUT=XXXX;
PROC SORT ; BY V DIFF;
DATA E; SET XXXX; BY V;  IF FIRST.V THEN CUM=0;
CUM+PERCENT; R=(100-CUM)/100;
LABEL V = 'TEST CULTIVAR'; LABEL R = 'PROBABILITY';
LABEL DIFF = 'TEST-CHECK DIFFERENCE (TONS PER HECTARE)';
PROC GPLOT;   PLOT R * DIFF=V;
SYMBOL1 C=RED  L=1 I= STEPL;
SYMBOL2 C=YELLOW  L=2 I=STEPL;
SYMBOL3 C=PINK L=22 I=STEPL;
```

REFERENCES

Anderson, T.W. 1984. *An Introduction To Multivariate Statistical Analysis* 2nd edn. Wiley, New York.

Barah, B.C., H.P. Binswanger, B.S. Rana, and N.G.P. Rao. 1981. The use of risk aversion in plant breeding: Concept and application. *Euphytica* 30:451-458.

Bradley, J.P., K.H. Knittle, and A.F. Troyer. 1988. Statistical methods in seed corn product selection. *Journal of Production Agriculture* 1:34-38.

Cochran, W.G. 1950. The comparison of percentages in matched samples. *Biometrika* 37:256-266.

Conover, W.J., M.E. Johnson, and M.M. Johnson. 1981. A comparative study of tests for homogeneity of variances, with applications to the outer continental shelf bidding data. *Technometrics* 23:351-361.

Eberhart, S.A., and W.A. Russell. 1966. Stability parameters for comparing varieties. *Crop Science* 6:36-40.

Eskridge, K.M. 1990a. Safety-first models useful for selecting stable cultivars. p. 151-168. *In* M.S. Kang (ed.) *Genotype-By-Environment Interaction In Plant Breeding.* Louisiana State Univ., Baton Rouge, LA.

Eskridge, K.M. 1990b. Selection of stable cultivars using a safety-first rule. *Crop Science* 30:369-374.

Eskridge, K.M. 1991. Screening cultivars for yield stability to limit the probability of disaster. *Maydica* 36:275-282.

Eskridge, K.M., P.F. Byrne, and J. Crossa. 1991. Selecting stable cultivars by minimizing the probability of disaster. *Field Crops Research* 27:169-181.

Eskridge, K.M., and B.E. Johnson. 1991. Expected utility maximization and selection of stable plant cultivars. *Theoretical and Applied Genetics* 81:825-832.

Eskridge, K.M., and R.F. Mumm. 1992. Choosing plant cultivars based on the probability of outperforming a check. *Theoretical and Applied Genetics* 84:494-500.

Eskridge, K.M., O.S. Smith, and P.F. Byrne. 1993. Comparing test cultivars using reliability functions of test-check differences from on-farm trials. *Theoretical and Applied Genetics* 87:60-64.

Eskridge, K.M., C.J. Peterson, and A.W. Grombacher. 1994. Probability of wheat quality traits falling within acceptable limits. *Crop Science* 34:866-869.

Finlay, K.W., and G.W. Wilkinson. 1963. The analysis of adaptation in plant-breeding programme. *Australian Journal of Agricultural Research* 14:742-754.

Hollander, M., and D.A. Wolfe. 1973. *Nonparametric Statistical Methods.* Wiley, New York.

Jones, T.A. 1988. A probability method for comparing varieties against checks. *Crop Science* 28:907-912.

Kang, M.S., and H.N. Pham. 1991. Simultaneous selection for high yielding and stable crop genotypes. *Agronomy Journal* 83:161-165.

Kotz, S., and N.L. Johnson. 1983. *Encyclopedia Of Statistical Sciences.* Wiley, New York.

Lin, C.S., M.R. Binns, and L.P. Lefkovitch. 1986. Stability analysis: Where do we stand? *Crop Science* 25:894-900.

Menz, K.M. 1980. A comparative analysis of wheat adaptation across international environments using stochastic dominance and pattern analysis. *Field Crops Research* 3:33-41.

Nelson, W. 1982. *Applied Life Data Analysis.* Wiley, New York.

Nuland, D.S., and K.M. Eskridge. 1992. Probability of outperforming a check. p. 17-20. *In* H.F. Schwartz (ed.) *Proceedings, 35th Bean Improvement Cooperative Meetings.* Colorado State Univ. , Fort Collins, CO.

Peterson, C.J., R.A. Graybosch, P.S. Baenziger, and A.W. Grombacher. 1992. Genotype and environment effects on quality characteristics of hard red winter wheat. *Crop* Science 32:98-103.

Schervish, M.J. 1984. Algorithm AS 195. Multivariate normal probabilities with error bound. *Applied Statistics* 34:81-94.

Schervish, M.J. 1985. Algorithm AS 195. Correction. *Applied Statistics* 34:103-104.

Shukla, G.K. 1972. Some statistical aspects of partitioning genotype-environmental components of variability. *Heredity* 29:237-245.

Winer, B.J. 1971. *Statistical Principles In Experimental Design.* 2nd edn. McGraw-Hill, New York.

Chapter 11

BREEDING FOR RELIABILITY OF PERFORMANCE ACROSS UNPREDICTABLE ENVIRONMENTS

P. J. Bramel-Cox [1]

I. INTRODUCTION

As world population increases and arable land for agricultural production either decreases or is shifted into more adverse environments, a need arises for more effective breeding of new cultivars of existing crop species for adaptation to these new conditions. Evans (1993) described two separate objectives for future breeding programs: the need by large-scale, globally oriented, public and commercial, research programs to develop new cultivars with very broad adaptation to a wide range of diverse environments and the need by local farmers to have new cultivars with reliable performance from year to year in a very specific locality. These two characteristics were termed adaptability and reliability, respectively. Evans (1993), Bramel-Cox et al. (1991), and Ceccarelli (1989) concluded that breeding for broad adaptation could limit the potential for genetic gain in specific environments and ultimately could result in greater gain in favorable environments at the expense of unfavorable environments.

The evolution of crop plants has been characterized by three distinct stages in the selection of modern cultivars from their wild progenitors (Evans, 1993). First came changes involved in domestication from their wild relatives; the second stage involved modifications in their adapted responses because of their movement out of their areas of domestication to different environments; and the third stage involved changes from an increase in their yield potential within each of these diverse environments. Bramel-Cox et al. (1991) concluded most of the selection during the first two stages involved improved adaptation to very specific environments. Selection in the third phase has been done for the benefit of larger target environments than individual farmer's fields. The shift to larger target areas and the success of selection have been assisted by the increased environmental potential and reduced heterogeneity of the individual environments resulting from the use of agronomic inputs to reduce production constraints.

The effectiveness of selection and the associated changes in various crop species brought about by domestication, broader adaptation and increased yield potential have been well documented (cf. Bramel-Cox et al., 1991; Evans, 1993). In most of these studies, the improvement in yield under both favorable and unfavorable environments was due to an increase in the yield potential of the

[1]Department of Agronomy, Throckmorton Hall, Kansas State University, Manhattan, KS 66506-5501.
Contribution No. 94-501-B, Kansas Agric. Exp. Stn.

improved cultivars. Yet, some evidence exists for differential adaptation as shown by crossovers in relative performance between favorable and unfavorable environments (cf. Evans, 1993). Evans (1993) concluded that this differential adaptation could be found across stresses such as moisture, temperature, irradiance, and soil fertility. Even though these crossovers were found most often when the range of environments was most extreme, continued selection for improved yield potential under the more favorable conditions, with little emphasis on adaptation to the specific stresses that characterize the more unfavorable environments, will result in preferential adaptation to the more optimal conditions (Bramel-Cox et al., 1991; Evans, 1993). Bramel-Cox et al. (1991) emphasized the need to develop cultivars with adaptation to the heterogeneous conditions resulting from lower agronomic inputs, whereas Evans (1993) described the need to balance high average yields, wide adaptability, and reliability under specific difficult environments. However, both concluded that future breeding programs will need to better characterize the target region, to develop a better strategy to allocate resources between test environments, to develop the optimal population type to buffer against a diversity of environmental constraints, to define the optimal selection criteria to enhance the mean performance and reduce environmental sensitivity, and to increase the use of genetic diversity within a crop species for specific adaptation to various stresses.

II. CHARACTERIZING TARGET ENVIRONMENTS

The impact of genotype x environment (GE) interaction on plant breeding is ultimately to reduce the observed gain from selection in relation to what is predicted (Dudley and Moll, 1969). This is due to the confounding effects of the three components of interaction between genotype and environment; genotype x location, genotype x year, and genotype x location x year. A number of studies (Hanson, 1964; Sprague and Federer, 1951; Horner and Frey, 1957) have used analyses of the components of the GE interaction to design testing procedures that allocate resources optimally into replications, locations, or years. The interaction of genotypes with test locations usually is classified as predictable if the locations are chosen to represent large-scale differences within the target region. In contrast, the interaction of genotypes and locations with years is seen as unpredictable and random because of seasonal fluctuations.

Horner and Frey (1957) demonstrated a procedure to partition the genotype x location interaction into subregions in which the within-region interaction was minimized based upon a prior understanding of the environmental heterogeneity. Abou-El-Fittouh et al. (1969) used cluster analysis to subdivide the cotton-growing regions in the U.S. into zones. The analysis was based upon the interaction of the testing locations with the genotypes, and similarity measures, which included a distance estimate, d_{ij} and the product moment correlation coefficient. In contrast, Campbell and Lafever (1977) used only the correlation coefficient, averaged over 3 years, to group test locations for the soft-wheat growing areas.

Pattern analysis that utilized both classification techniques, such as cluster analysis, and ordination techniques, such as principal coordinate analysis, were demonstrated by Byth et al. (1976). Shorter et al. (1977) compared the use of pattern analysis with the procedures suggested by Horner and Frey (1957), Abou-El-Fittouh et al. (1969), and Campbell and Lafever (1977). Although the previous techniques were useful for defining environments that differed for the genotype response and for minimizing the interaction component, pattern analysis could be used to delineate further those environments that identified genotype responses fitting the specific objectives of a breeding program by determining an environments' contribution to the classification and ordination of the lines' performance.

Brennan et al. (1981) found that expressing a cultivar's performance as a percentage of the environment mean gave equal weights to each environment and aided in the analysis for cultivar-response differences. Fox and Rosielle (1982a) further investigated the impact of various data transformations to remove large environmental effects that affect the analysis of interaction. The best option was to use a standardization procedure. Bull et al. (1992) suggested that the use of a procedure giving equal weights to all environments would bias the results, because all may not contribute equally to the interaction. They used a discrimination index, whereby the environments were weighted by a repeatability measure based upon the ratio of genetic variability to total phenotypic variability.

Goodchild and Boyd (1975) analyzed the variation between and within seasons caused by physical factor in the wheat growing areas of Western Australia. In an analysis of GE in this same region using the yield data standardized for a mean of 0 and standard deviation of 1, Fox et al. (1985) verified seasonal variation seriously hindered progress in breeding for this region, and subdivision of environments did not appear to be regionally based or repeatable between years. Thus, selection in any one year may have had limited advantage to the long-term gain, whereas selection under atypical conditions could have a negative impact on the optimum genetic advance. Fox and Rosielle (1982b) described a method to determine the relationship of any single environment to the long-term average using a set of representative cultivars. Bull et al. (1994) concluded the number of genotypes needed to describe a consistent environmental classification depended upon the strength of the pooled genetic correlation between environments.

Yau et al. (1991) described the target region for bread wheats represented by West Asia, North Africa, and Mediterranean Europe. They used the correlation coefficient as a distance measure in data from 2 years of the Regional Bread Wheat Yield Trials conducted by ICARDA/CIMMYT. This measure is equivalent to the standardized Euclidean distance measure of Fox and Rosielle (1982a). Alternatively, Ivory et al. (1991) used the GE deviation matrix to classify test environments for soybeans in Thailand in two seasons. The deviation matrix allowed the separation of environments based on their ability to discriminate among the genotypes. The relative genotype ranking index (RGRI) for each environmental group was used to discern the exact difference in environmental response for the specific regions represented by each group.

Westcott (1986) criticized the use of cluster analysis to classify environments or genotypes, because it puts a specific structure on a data set without a full understanding of its actual basis or even whether it exists. Furthermore, there are numerous possible dissimilarity measures and clustering strategies, and the use of any may result in a different grouping. Westcott (1986) suggested the use of principal coordinate analysis to quantify the relationship and to superimpose the results on a minimum spanning tree for identifying dissimilar genotypic or environmental groups. The use of this technique has been demonstrated by Westcott (1987) and Crossa et al. (1988). Another alternative to cluster analysis to group environments or genotypes is demonstrated by Cornelius et al. (1992) and Crossa et al. (1995), i.e., the shifted multiplicative model (SHMM), which is a step-down method based upon separability of genotype effects. This exploratory method differs from cluster analysis because it starts with the entire data set and then subdivides based upon the presence of crossover interactions in the resulting subdivision. The final grouping results in a set of environments with few or no crossover interactions; thus, the genotype effects are homogeneous within each group.

The first goal of any analysis of GE interaction for the target region of a breeding program is to delineate the regional grouping based on predictable attributes (Allard and Bradshaw,1964). If these attributes are large-scale conditions such as irrigation potential, growing-season duration, or photoperiod response, they can be classified as separate target regions or macro environments (Yau et al. 1991; Ceccarelli, 1989). This procedure would allow a more concentrated breeding effort in an area where broad adaptation might be a more realistic goal.

III. SELECTING OPTIMAL TESTING SITES

To choose the best test site, Frey (1964) suggested the use of the heritability estimate. Allen et al. (1978), however, concluded the heritability estimate at the selection environment and the genetic correlation between the mean genotypic value at the selection environment and the target environment would be better. Falconer (1989) described gain from selection in the target environment as a correlated response from indirect selection at the test environment. Thus, the value of a test site could be judged by the indirect response to selection for all the specific environments within a target region. The use of the relative efficiency of genetic gain from indirect selection in relation to direct selection at a specific environment was demonstrated by Zavala-Garcia et al. (1992b).

Allen et al. (1978), Falconer (1989), and Zavala-Garcia et al. (1992b) concluded that the best selection site would have a high heritability and optimize the gain from indirect selection for all the specific conditions within the target region. Frey (1964), Allen et al. (1978), and Shabana et al. (1980) found a positive relationship between the mean of the test site and its heritability, whereas Ceccarelli (1989) and Zavala-Garcia et al. (1992b) found no relationship. In both of the latter studies, heritability varied more as a function of genetic variability within breeding populations for adaptation to the range of target environments. Pederson and Rathjen (1981) also concluded that neither mean nor heritability reliably identified the best selection environment. They suggested the use of a coheritability (h^2_c), an estimate of the heritability of the correlated response to selection. Selection from environments with a large, positive h^2_c would be expected to result consistently in gain over all the target conditions.

Hamblin et al. (1980) suggested the use of the genotypic or phenotypic correlation, the discriminatory ability of the environment, its mean yield, and its probability of very low yield to identify the best test sites. All these characteristics need to be consistent over seasons or years. The identification of testing locations on the basis of these criteria was demonstrated by Brown et al. (1983) and Lin and Morrison (1992). Generally, all of these procedures require a large set of multiple location-year trials, with at least a small set of consistent genotypes for the target region. These analyses are retrospective but can be very useful in defining breeding strategies, as demonstrated in Brennan et al. (1981) for a multi-phase selection program that initially uses an early generation test at a limited number of sites which have been selected to represent the range of all possible genotype interactions in the target region. In the advanced stages of testing, more locations that better characterized specific adaptation would be used to determine the optimal range for the improved cultivars. This strategy could be very practical for a target range with very large predictable differences in genotype responses. This is not possible in target areas described by Fox et al. (1985), Yau et al. (1991), and Zavala-Garcia et al. (1992b), given the unpredictable nature of GE and our poor understanding of the physiological basis for differential genotypic responses (Blum, 1988). In these cases, the best strategy would involve better utilization of performance at variable selection environments to breed for stable or reliable responses and/or use alternative population types. Excellent reviews of criteria used to select the best test environments and to optimize gain in widely defined target regions can be found in Blum (1988), Bramel-Cox et al. (1991), and Evans (1993).

IV. DEVELOPING OPTIMAL POPULATION TYPES

If a target environment is characterized largely by seasonal fluctuation, it may be very difficult to identify a set of test locations that will predictably select a single genotypic response, or the range of adaptations may be beyond the limits of any single genotype. Allard and Bradshaw (1964) concluded that strategies to incorporate both specific adaptation to the predictable environmental differences and adaptation to the more unpredictable transient fluctuations of the environment will require a careful evaluation of the optimal genetic systems to develop a " well-buffered" variety. This buffering could be due to either developmental homeostasis of individual genotypes or genetic homeostasis between individuals in a variable population. These are referred to as individual buffering and population buffering by Allard and Bradshaw (1964). Varieties made up of pure lines or single-cross hybrids are genetically homogeneous populations that rely solely on individual buffering, whereas heterogeneous populations, such as mixtures of pure lines or double and three-way hybrids, rely on population buffering. In some variety types, such as open-pollinated synthetic varieties of cross-pollinated species like maize or pearl millet, both mechanisms are important.

The use of intrapopulation genetic diversity to buffer against seasonal fluctuations has been reviewed by Simmonds (1991) and Marshall and Brown (1973). Marshall and Brown (1973) determined that the yield of a mixture theoretically will always be less than its best pure line component, but its stability will always be greater as long as one of its components reacts differently in at least one environment, and the advantage widens as diversity of the component's responses increases. If intergenotypic interactions occur among components, this will reduce the stability of the mixtures, unless these interactions are less important than differences in mean performance of each component. Marshall and Brown (1973) predicted that mixtures would be of no advantage where the environments represented a fairly uniform target area, and the components were highly adapted varieties. This is the situation described in Walker and Fehr (1978) for soybeans in Iowa. In their study, mixtures of well adapted soybean varieties, which were similar in their stability or adaptation, did not differ from multiple plantings of the pure lines for stability.

Conversely, Marshall and Brown (1973) would have predicted that, in environments with a very wide range or extreme fluctuations, mixtures of pure lines with differences in their adaptation would be very stable but not necessarily the highest yielding under all the specific conditions. These were the results reported in barley by Soliman and Allard (1991) and in wheat by Fox et al. (1985). The former study found that the buffering ability of barley composites was better than that of pure lines, but pure lines always yielded better especially under favorable conditions. The stability of the composites decreased as their heterogeneity decreased in the more advanced generations. In contrast, Fox et al. (1985) found that unselected F_4 bulks were both more stable and higher yielding than pure line cultivars. In fact, 48 of 55 bulks were higher yielding than their highest parents. These bulks represented diverse parents with large differences in adaptation. This yield improvement could have been due to large positive intergenotypic fluctuations as predicted by Marshall and Brown (1973).

The stability and yield of both mixtures of homozygous pure lines or heterogeneous bulks are influenced by individual and population buffering. The benefit of population buffering and individual buffering can also be influenced by the degree of heterozygosity. The impact of heterozygosity has been documented for cross-pollinated species, such as maize or pearl millet, but is less important for self-pollinated crops, such as wheat or sorghum (cf. Becker and Léon, 1988). Becker and Léon (1988), in comparing GE interaction mean squares of homozygous versus heterozygous cultivars of various crop species, reported a general advantage for hybrids. On the other hand, Carver et al. (1987) in wheat and Blum et al (1992) in sorghum found that the advantage of hybrids in these two crops resulted more from higher yield potential than from wider adaptation. In wheat, Carver et al. (1987) concluded this resulted from comparing semidwarf hybrids with taller pure lines. In sorghum, Blum et al. (1992) determined that hybrids generally had a higher yield potential than varieties because of their earlier flowering, greater harvest index, higher leaf area index, and greater seed number but not because of better adaptation to favorable or unfavorable environments. These studies lack any genetic relationship between the pure lines and the hybrids tested, making it impossible to separate the effects of genetic potential and adaptation from that of heterozygosity. In sorghum, Marinesco (1992) and Chisi (1993) found that homozygous parent lines showed a greater interaction with the environment than their test crosses. Specific homozygous parents were more stable than their hybrids. No relationship existed between the stability of the parents and their F_1 hybrids. Unlike the sorghum and wheat hybrids in the studies of Carver et al. (1987) and Blum et al. (1992), these sorghum hybrids had a greater advantage under stress conditions, where heterosis was larger.

Schnell and Becker (1986) described a design to separate the relative importance of heterozygosity and heterogeneity to yield and stability. They compared a study in maize with a similar trial in sorghum. In both crops, only heterozygosity was important for yield, while heterozygosity and heterogeneity were of nearly equal importance for stability. In general, the impact of both types of buffering was of greater importance in maize than in sorghum; thus, the relative importance of individual and population buffering could be dependent upon the mating system of the crop and its previous evolution. Defining the optimal population type for a target area will depend upon the specific crop, its mating system, and its previous breeding. An example of this is demonstrated by Mahalakshmi et al. (1992) for pearl millet, a highly cross-pollinated crop, where topcross hybrids between variable landraces and open-pollinated inbred male steriles allowed optimal yield and stability from both heterozygosity and heterogeneity.

V. DETERMINING OPTIMAL SELECTION CRITERIA

A. IMPACT ON MEAN PERFORMANCE

Defining an optimal breeding strategy for highly variable target environments involves the careful characterization of the target area to identify both the selection environments and the selection criteria to best allocate resources, with regard to both number of sites and stages of selection. Gain from selection, realized by improved overall performance and adaptation, is a result of the efficiency of indirect selection at the specific selection environments for all environments that define the target area versus the gain that would be realized from direct selection for all the variable conditions that make up the target area (Falconer, 1989). Wright (1976) concluded that selection for broad adaptation could be equal to selection for specific adaptation but never greater. If the interaction of genotypes and environments was very large, a large disadvantage of selection for broad adaptation would exist, even with a large number of selection environments.

The impact of the level of productivity of the selection environment on potential gain in the same and different environments has been evaluated in a number of studies with contradictory results (see review in Atlin and Frey (1990)). Falconer (1990) described two types of selection that occur with regard to contrasting environments. The first is "synergistic" selection for high performance in high-productivity environments and for low performance in low-productivity environments. The second type is "antagonistic" selection for high performance in low-productivity environments and for low performance in high-productivity environments. For yield across environments, we are interested only in synergistic selection in high-productivity environments or antagonistic selection in low-productivity environments. The difference in performance under high- and low-productivity conditions was defined by Falconer (1990) as environmental sensitivity. The effect of antagonistic and synergistic selection on changes in the mean and sensitivity is described as the "Jinks and Connally rule" by Falconer (1990). Theoretically, this rule predicts that the overall expected response of antagonistic selection or synergistic selection will be asymmetrical, i.e., $R_l + CR_h \neq R_h + CR_l$ where R_l is the direct response from selection at low-productivity conditions, CR_h is the correlated response from selection at low-productivity environments for high-productivity environments, R_h is the direct response from selection at high-productivity environments and CR_h is the correlated response from selection at high-productivity environments for low-productivity environments. If more than one environment is involved, then it would also include the correlated response at all other test sites from selection at one site. This asymmetrical response would be expected for both mean and sensitivity; thus, only one type of selection would be the most effective over all conditions. In most of the cases re-evaluated by Falconer (1990), antagonistic selection was more effective for improving the overall mean than synergistic selection.

Realized gain from direct and indirect selection in low-, medium-, and high-yield environments for drosophila (Paleolog and Maciejowski, 1991), barley (Ceccarelli and Grando, 1991a and b), oats (Atlin and Frey, 1990), maize (Johnson and Geadelmann, 1989), and sorghum (Marinesco, 1992; Chisi, 1993) is given in Table 1.

The average responses in the mean to antagonistic and synergistic selection are given for each selection trial as well. The average response of the mean to selection is the sum of the direct response to selection in a specific environment and the gain from indirect selection at all other types of environments, divided by the total number of environments (Falconer, 1990). As predicted in that report, the responses from antagonistic selection and synergistic selection were not equal. In the S population of drosophila ebony flies studied by Paleolog and Maciejowski (1991) selected for fecundity under optimal versus alcohol-stressed conditions, the superiority of synergistic selection was due to a larger gain from direct selection

Table 1
Realized gains for direct and indirect selection at high-yield or low-yield environments for high-, low- or medium-yield environments for drosophila (Paleolog and Maciejowski, 1991); barley (Ceccarelli and Grando, 1991a and b); oats (Atlin and Frey, 1990); maize (Johnson and Geadelmann, 1989); and sorghum (Marinesco, 1992; Chisi, 1993) and the overall response of the mean to antagonistic selection at low environments and synergistic selection at high environments.

| | Selecting environment | | | | | | Response of mean | |
| | High | | | Low | | | | |
Organism	R_H	CR_{HM}	CR_{HL}	R_L	CR_{LM}	CR_{LH}	Low[a]	High[b]
Drosophila								
SH	+21.0		-5.3	+14.7		+7.3	11.0	7.8
S	+19.3		-3.7	+10.3		+5.3	6.8	7.8
Barley								
YS	6.6%		-25.0%	14.7%		-16.0%	-0.65	-9.2
Y	9.8%		-21.5%	23.9%		-20.5%	1.7	-5.8
Group 1	+13.4%		-7.0%	+11.4%		-5.7%	52.8	3.2
Group 2	+25.7%		-35.9%	+53.4%		-5.3%	24.1	-5.1
Oats	11.4	10.3	6.1	10.3	9.3	4.6	8.1	9.3
Maize								
MS	14.1	9.8	8.9	7.1	1.0	1.2	3.1	10.9
FS	11.2	10.1	5.1	7.9	5.3	6.2	6.5	8.8
Sorghum								
Parents	-0.1%	0.66%	0.85%	-0.4%	+1.63%	-1.8%	-0.19	0.17
Hybrids	-0.1%	0.56%	0.16%	-0.55%	+1.03%	+0.7%	-0.29	-0.17

[a]Antagonistic.
[b]Synergistic.

under optimal conditions, even though the correlated response under alcohol-stressed conditions was poor. In the other population (SH), a slightly higher gain from direct selection under alcohol stress and from indirect selection for optimal conditions resulted in a greater response of the mean from antagonistic selection.

In contrast, the superiority of synergistic selection in oats (Atlin and Frey, 1990) and maize (Johnson and Geadelmann, 1989) was a product of a generally higher response from both direct and indirect selection at high-yield versus low-yield conditions. · In barley, antagonistic selection was better than synergistic selection because of the generally greater gain over all environments from selection under low-yielding conditions. In sorghum, both types of selection made very small gains in both direct and indirect responses (Marinesco, 1992; Chisi,1993). But, synergistic selection was slightly more effective for parent lines because of more consistent indirect effects.

Ceccarelli and Grando (1991a) found that the magnitude of realized gain from direct response to selection in stress environments, but not in nonstress environments, was dependent upon the similarity between performance in the selection and test environments. This could be explained by differences in the magnitude of genetic correlations between more variable stress environments and the more homogeneous optimal conditions used as selection and testing environments. Ceccarelli and Grando (1991a) found that when results were summarized over a larger number of test location trials, the average effects of selection were more evident. Atlin and Frey (1989) subdivided environments into contrasts defined by specific constraints such as heat, nitrogen, or phosphorus stress. They concluded that the relative efficiency of indirect selection differed according to the stresses involved. Thus, determining the average relationship between low- and high-productivity environments, as did Atlin and Frey (1990), could mask certain environmental constraints, such as heat or phosphorus stress, with which synergistic selection would be a poor alternative.

The sorghum production area of Kansas and Nebraska in the U.S. Great Plains is characterized by very large GE interactions because of large seasonal fluctuations. Using the procedure described by Ceccarelli and Grando (1991a) to categorize environments as low-, medium- or high-yielding, I computed the average relative efficiency of predicted gain from indirect selection in relation to that from direct selection for KP9B C_0, and after one cycle of S_1 family selection under good, KP9B C_1 (M), and poor, KP9B C_1 (GC), conditions for grain sorghum (Table 2). Results are averages of four-location/2-year trials reported in Zavala-Garcia et al. (1992b) and three-location/2-year trials reported in Marinesco (1992). These stress environments, like those characterized by Ceccarelli and Grando (1991a), Fox et al. (1985), Yau et al. (1991), and Scott et al. (1994) are very heterogeneous in contrast to those reported for oats by Atlin and Frey (1990) and for maize by Johnson and Geadelmann (1989).

According to Zavala-Garcia et al. (1992b) and Marinesco (1992), no single environment was efficient for selecting indirectly for any other single environment, but any single environment was better than the mean over all environments to improve the average mean. Scott et al. (1994) found selection based on single environments was a poor measure of over all mean performance, irrespective of the mean yield or genetic variance of that environment for soybean in Kansas. In the sorghum populations, the relative efficiency of indirect selection was compared within low, medium or high productivity test environments (Table 2). In the base population, KP9B C_0, the indirect predicted response to selection was low in comparison to the direct predicted response, even within grouping of similar environments. Yet in both the low- and high-yielding environments, the average relative efficiency was higher within than between all of these groups. This was not as evident in the medium group, where the relative efficiency was more similar across all groups.

Table 2

Relative efficiency of predicted gain from indirect selection in relation to the predicted gain from direct selection in KP9B C_0, KP9B C_1 (M), and KP9B C_1 (GC) grain sorghum averaged for low- (L), medium- (M), and high- (H) productivity environments and overall environments (Zavala-Garcia et al., 1992a; Marinesco, 1992).

Selection Environment	Response Environment			Predicted Overall
	L	M	H	
			%	
C_0				
L	48.6	33.0	26.3	40.4
M	35.8	51.8	57.6	62.4
H	26.8	26.8	54.7	59.4
C_1 (M)				
L	100.0	35.6	---	67.8
M	32.3	19.5	---	25.9
C_1 (GC)				
L	100	53.0	---	76.5
M	86.5	24.9	---	55.7

Scott et al. (1994) compared the ranks of soybean lines in two maturity groups in 14 separate environments versus the rank from performance over all. They concluded a very large proportion of the lines at the individual environments had to be selected to insure the top 10 % of the lines with the best performance over all the environments were selected; the range of selection intensities needed at the individual test sites was 43-100 % within each group. If the test environments were subdivided to low- and high- yielding, the range of selection intensities was reduced to 14-34 % within each group. Thus, when the test environments were more homogeneous, the yield level of the individual environments used to select was less important.

Atlin and Frey (1990) concluded that one way to increase the relative efficiency of correlated response was to increase the number of replications. They defined replication to include both the number of trials evaluated at that yield level as well as the number of replications within a trial. In their experiments with oats, the correlated response from indirect selection at medium- or high-yielding environments for low-yielding environments was greater than that of direct selection. Their contention that using increased replication, and the resources needed to conduct them, would be an advantage for increased gain from direct selection in low-productivity environments would not be valid if the indirect response to selection at high-yielding environments is already greater than direct selection at the low-yielding site. The best allocation of resources in this situation is to select only in high-productivity environments. This was not the situation for sorghum experiments grown in Kansas, where the average correlated response was very low in relation to direct response in the lower yielding sites so the increased number of trials or replications within trials would have resulted in an even lower indirect gain (Zavala-Garcia et al., 1992a). The relative efficiency within environments at a similar productivity level did increase as the level of stress decreased (Table 2) which fit the results of Ceccarelli and Grando (1991a) and can be explained by the very diverse nature of environmental constraints within this target area.

In general, antagonistic selection in intermediate-stress environments was superior to synergistic selection (Zavala-Garcia et al., 1992a). When selection of S_1 families was conducted separately in the two types of environments, the effect of the previous selection environment could be seen in the next cycle (Marinesco, 1992). When the two selected populations were compared with the base population in low- and medium-productivity environments, antagonistic selection (high yield in stress conditions) resulted in a population, KP9B C_1 (GC), with a higher relative efficiency for further gain from selection under stress conditions. Synergistic selection in KP9B C_1 (M) resulted in loss of adaptation to stress conditions, as seen in its reduced relative efficiency under stress conditions.

This impact of selection in one type of environment on the breadth of adaptation to different types of conditions also can be seen in a comparison of the number of S_1 families that were high yielding across all test sites (Marinesco, 1992). In the base population, KP9B C_0, 18.3% of the top 20 families were highest in all six environments, whereas in the two selected cycles, KP9B C_1 (M) and KP9B C_1 (GC), 22.5% and 25%, respectively, the top families were highest overall. The slightly greater increase in gain per cycle from antagonistic selection demonstrates the increased frequency of more broadly adapted lines obtained from this type of selection. Scott et al. (1994) found 25-50% of the top 10 % of the best soybean lines over all were identified from high-yielding test sites while 50-75% were identified from low-yielding tests.

A similar phenomenon was reported by Ceccarelli and Grando (1991a) in barley and by Paleolog and Lorkiewicz (1991) in drosophila. The latter researchers evaluated two generations of *ebony* flies selected for improved fecundity in optimal or alcohol-stress conditions under heat stress. With this different stress, selection under alcohol stress resulted in a population that also showed better adaptation to heat-stress conditions. Based upon heterosis between the two divergently selected populations, they found evidence that these two populations were different after 28 generations of selection under optimal and stress conditions, due to different genetic architectures for the same quantitative trait. Antagonistic selection under stress conditions also resulted in wider adaptation of the population. Paleolog and Maciejowski (1991) and Paleolog and Lorkiewicz (1991) concluded that realized gain from selection was a product of the genetic properties of the base population, the kind of selection environment, and the cumulative history of previous selection, either artificial or natural.

Ceccarelli and Grando (1991a) came to similar conclusions when comparing various germplasm sources for broad adaptation using antagonistic and synergistic selection. They used improved cultivars, landraces, and their intercrosses. Improved germplasms comprised entries selected artificially in other breeding programs for improved performance, whereas landraces were adapted locally by both natural selection and artificial selection by farmers. When they selected only in high-yielding environments among improved germplasm, they obtained three entries that outyielded the best checks in low-yielding environments. In contrast, if they selected among landraces or crosses of landraces and improved cultivars for yield at low-yielding sites, they found that 33 entries outyielded the best check in high-yielding environments. The actual gain from selection in low-yielding environments did not differ between the best lines of these three groups. However, indirect selection in high-yielding environments using improved germplasm for low-yielding environments would require an initial test of about 50,000 entries to identify the same number of entries that did well in both conditions after selection among 1742 landraces or crosses of landraces and improved germplasm. Alternatively, for high-productivity conditions, direct selection in high-yielding environments with improved germplasm was about three times more efficient than indirect selection in low-yielding environments with landraces and their crosses.

B. IMPACT ON ENVIRONMENTAL SENSITIVITY

Falconer (1990) concluded that the Jinks and Connally rule predicted that either antagonistic or synergistic selection could be effective at improving the mean, but that total response to selection was a result of changes in both mean and sensitivity. He described this total response as $R_T = R_m \pm \frac{1}{2}DR_s$, where R_m is the response of the mean, D is the difference in performance between environments in the original population, and R_s is the response of the sensitivity. The impact of changes in sensitivity can be either positive or negative and the magnitude is dependent on the relative differences between environments (D). If differences are very wide, then D will be large and R_s will be more important than if D is small. Thus, the impact of changes in sensitivity increases as the diversity of the target region increases. The response of sensitivity is defined as the difference between correlated response and direct response relative to overall differences in the environments, i.e., $(CR_H - R_L)/D$ for antagonistic selection and $(R_H - CR_L)/D$ for synergistic selection. Falconer (1990) concluded antagonistic selection will decrease sensitivity and synergistic selection will increase it if the sum of the direct response, $R_H + R_L$, is greater than the sum of the correlated responses, $CR_H + CR_L$. Estimates of R_T, R_m, and R_s based on the examples given in Table 1 are presented in Table 3. In all the examples, antagonistic selection resulted in a reduced sensitivity, irrespective of the impact on response of the mean. Using the formula of Falconer (1990), the total response (R_T) to antagonistic selection is equal to the correlated response in high-productivity environments from selection in low-productivity environments (CR_H). The R_T for synergistic selection is equal to the direct response to selection at the high-productivity environment (R_H); thus, the difference between total responses from synergistic selection and antagonistic selection is equal to the difference between the correlated response and direct response in high-productivity environments. For the total response to be greater from antagonistic selection, the response for sensitivity also would have to be positive. Neither the examples in Table 3 nor those in Falconer (1990) fit this situation; thus, selection under stress conditions should result in genotypes with reduced sensitivity but not necessarily improved performance under optimal conditions.

Table 3

The response of the mean (R_M), response of sensitivity (R_S) and total response (R_T) to selection for experiments given in Table 1.

Organism	Antagonistic			Synergistic		
	R_M	R_S	R_T	R_M	R_S	R_T
Drosophila						
SH	11.0	-0.69	7.3	7.8	2.47	20.96
S	6.8	-0.66	3.3	7.8	2.15	19.26
Oats	8.1	-0.16	4.6	9.3	0.19	11.4
Maize						
MS	3.1	-2.86	1.2	10.9	2.52	13.49
FS	6.5	-0.83	6.2	8.8	2.96	11.85
Barley						
1991a	101.24	-0.66	79.88	93.46	0.49	109.3
1991b	137.9	-3.86	94.55	98.9	1.86	119.8

Falconer (1990) described the correlation between the mean and sensitivity, r_{ms}, and its impact on the total response to antagonistic or synergistic selection. He concluded that antagonistic selection theoretically results in reduced sensitivity, but its impact on the mean was not predictable (see also Tables 1 and 3). Falconer (1990) illustrated four possible configurations of four genotypes with different sensitivities and means. If the genetic correlation between environments is positive, then either type of selection will improve the mean in both environments, because the same genotype is best in both, and sensitivity will remain the same irrespective of the magnitude or direction of r_{ms} or the magnitude of genetic variance in each environment. If the genetic correlation between environments is negative and the correlation between r_{ms} is positive, then antagonistic selection will reduce or not change performance in the high-productivity environment, whereas sensitivity will be reduced. Alternatively, if r_{ms} is negative, and the genetic variance under the low-productivity environment is greater than under the high-productivity environment, antagonistic selection would lead to $R_L \cong CR_H$ and reduced sensitivity. The most common configurations for a typical range of a crop will be either those in which no crossovers occur in a narrow range, as in the first example, or where crossovers occur in the lower range. Thus, antagonistic selection would either not change or reduce performance in the high-productivity environments.

Ceccarelli and Grando (1991a and b) reported under high-yielding environments lines selected under stress (antagonistic selection) were always lower yielding than those directly selected in a high-yielding environment but nevertheless were still better than the checks. Ceccarelli and Grando (1991b) used three stability parameters- the linear regression coefficient (b), the mean squares of deviations from regression (S^2_d), and the determination coefficient (R^2)- to measure environmental sensitivity. The lines selected at the poor sites (antagonistic) had lower values for all parameters, which indicated their reduced responsiveness or environmental sensitivity. Use of three "safety-first" indexes as described by Eskridge (1990) resulted in a similar response to selection, so Ceccarelli and Grando (1991b) concluded that antagonistic selection would give barley cultivars with high yield under poor conditions and acceptable, but not maximum, yield under good conditions. These cultivars would better fit the needs of farmers for reliable performance than would more broadly adapted cultivars developed from synergistic selection, which were responsive to optimal conditions but poor yielders under stress conditions.

St-Pierre et al. (1967) evaluated the F_2 to F_5 generations of barley using one stress and one nonstress environment in 16 different selection pathways. These ranged from four generations of selection at the nonstress site to four generations at the stress site. Environmental sensitivity was evaluated using average mean yield, linear regression coefficient, and percent adaptability (Finlay and Wilkinson, 1963). Using these measures, St.-Pierre et al. (1967) found that the best selection pathway for improved average mean yield, stability, and adaptability was stress-nonstress-stress-nonstress sequence or, alternatively, four generations of stress. The poorest pathway included four generations of nonstress. Thus, antagonistic selection resulted in reduced sensitivity and the best response of the mean. Simmonds (1991) simulated the effect of multiple cycles of selection on mean and

324

sensitivity, using consistent selection at high- or low-environments versus alternating high-low. The use of consistent selection in low-productivity environments reduced sensitivity and in high-productivity environments increased it in relation to the original population.

Zavala-Garcia et al. (1991) evaluated the environmental sensitivity of sorghum lines selected under stress (GC) and nonstress (M) conditions. Using two estimates of sensitivity, the linear regression coefficient (b) and the principal component score from the additive multiplicative model (AMMI) (Zobel et al., 1988). Lines selected under stress had an b-value of .94 compared to 1.19 for selection at a nonstress site. The ranges of the b-values were actually the same for both selection types, but a higher frequency of lines (15/20 vs 9/20) were below or equal to 1.0 from antagonistic selection. The AMMI analysis was used to describe the complexity of the interaction of selected lines with the 14 test site-year combinations. The original population had three significant principal components in the interaction matrix, as did the lines selected at the nonstress site, whereas those selected under stress had only one. Selection under stress had a more simplified pattern of interaction between the lines and environments. In maize, Johnson and Geadelmann (1989) reported that synergistic selection resulted in increased environmental sensitivity, as measured by the linear regression coefficient (b), whereas the effects of selection under stress either reduced or did not change sensitivity in relation to the original population. Their results concurred with those of Marek and Gardner (1979) and Moll et al. (1978). Thus, in all the studies reported here and in Table 3, antagonistic selection resulted in a reduced sensitivity as compared with synergistic selection. The impacts of the two types of selection on responses of the mean were more variable and less predictable. If the main interest of a breeding program is to maximize selection response under optimal conditions and to reduce sensitivity, these goals may be very difficult to achieve (Falconer, 1990). The key to progress toward both goals will be to determine the relationship between the mean and sensitivity and select for both simultaneously, or use a scheme that selects indirectly for one or the other criterion.

VI. DEFINING OPTIMAL SELECTION STRATEGY

A. SELECTION INDICES

Rosielle and Hamblin (1981) evaluated two alternative procedures to select for performance in stress and nonstress environments simultaneously. Their focus was on two selection criteria: mean productivity and tolerance. Selection for mean productivity always resulted in an increase in nonstress yield, but not necessarily an increase in stress yield, unless the genetic correlation between stress and nonstress yield, r_g, is negative and the genetic variance, σ_g^2, is greater under nonstress than in stress environments. Selection for tolerance will increase yield under stress in all conditions except when σ_g^2 is larger in nonstress and r_g is positive. Under no circumstances would one expect selection for tolerance to increase nonstress yield. In the situation most commonly found for a fairly narrow target region, σ_g^2 is greater under nonstress and r_g is low or positive. In these cases, selection for tolerance or sensitivity would result in a reduction in nonstress yield and little change in the stress yield.

Selection for sensitivity or tolerance as defined by Rosielle and Hamblin (1981) has very limited potential to improve performance in optimal environments. Zavala-Garcia et al. (1992a) studied the potential to select for tolerance in the sorghum population, KP9B, using the squared Euclidean distances among the eight test environments for each genotype. When this estimate of tolerance (DT) was compared with overall mean productivity or single- environment mean, the relative efficiency of indirect selection at all levels was very poor in relation to direct selection (Table 4).

Zavala-Garcia et al. (1992a) defined mean productivity using only the two extreme environments (MBW) or two extreme and one intermediate environment (MBIW) and developed a Smith (1936)-Hazel (1943) index, in which environments were weighed to optimize gain across the range of the target area. Four different distance matrices were analyzed to group the eight environments using a complete hierarchical agglomerative cluster procedure. After clustering, one environment with the highest heritability was selected, and an index developed that gave equal economic weights to each of these. The relative efficiency of indirect selection in relation to direct selection in sorghum is shown in Table 4, the selection criteria reported in Zavala-Garcia et al. (1992a) and Chisi (1993). The effectiveness of indirect selection was separated for various subgroups of test sites. Selection using only one type of environment (nonstress, intermediate, or stress) was best in the test environments most closely related, as is also evident in Table 2. All the relative efficiencies indicated that these indirect selection criteria were better than direct selection.

Van Sanford et al. (1993) reported similar evaluations for both wheat and soybean in Kentucky. They estimated mean productivity using the average of the two primary selection sites which was better than either individual environment in any of the 4 years, and always was predicted to be slightly better than use of the mean overall. A selection index requires the estimation of genetic variance and covariance within and between environments. Van Sanford et al. (1993) used cultivar trials that contained highly selected lines to estimate these genetic parameters but concluded that this was valid given that these are usually the only performance data for a target area and should represent the breeding populations to be tested there. Cooper et al. (1993a and b) and Calhoun et al. (1994) used highly selected wheat cultivars to estimate genetic parameters. In all of these papers, violation of a key assumption in the estimation of the genetic (PC1)

TABLE 4

Relative efficiency from indirect selection in high- (H), medium- (M), and low- (L) environments, using five measures of mean productivity, five measures of tolerance when tested at high-, medium-, and low-yielding environments (Zavala-Garcia et al., 1992; Chisi, 1993).

Selection criteria	Testing environments		
	High	Medium	Low
		%	
Zavala-Garcia et al. (1992)			
Single environment			
H	84.7	61.5	77
M	44.7	24.0	100
L	62.3	76.0	29
Mean productivity[a]			
MBW	78.7	50.5	104
MBIW	85.3	71.0	88
SI1	101.3	72.8	94
SI2	90.7	89.5	63
SI3	104.7	92.5	64
Tolerance[b]			
DT	33.0	37.3	18
RT	109.0	102.5	65
RBIW	85.0	70.3	90
B	28.3	11.8	-27
PC1	19.3	41.5	-53
Chisi (1993)			
MBIW	134.5	121.5	148
RSBIW	130.0	92.0	142
SI	156.5	110.5	171

[a] Mean productivity defined by the mean of high and low (MBW), mean of high, medium and low (MBIW), selection index 1 with two high and one medium environment, SI2 with 2 medium and 1 low environments, and SI3 with one high and one medium environments.
[b] Tolerance defined as the squared Euclidean distance between all eight environments (DT), the sum of the rank at all eight environments (RT), or at one high, medium and low environments (RSBIW), the linear regression coefficient (B) and the first principal component score.

parameters, which relates to random sampling of a defined reference population, does not seem justified. If breeders are interested in developing selection criteria to optimize genetic gain, the appropriate trials need to be conducted using random lines from a well-defined reference population.

An alternative to the selection index is to weight environments using a repeatability estimate, as illustrated by Bull et al. (1992). This index, DI, is calculated using a random model and is defined as the ratio of genetic variance to total variance for each environment. This was assumed to represent the power of an environment to discriminate among genotypes. Other procedures have been suggested to weight the selection environments. St-Pierre et al. (1967) and Cooper et al. (1993b) described the potential use of probe genotypes to describe biologically or bioassay the environmental constraints. This information could be used to determine the emphasis to be given the selection site. Fox and Rosielle (1982b) described the use of a reference set of genotypes to determine the weight to be given an environment based upon its distance from the long-term average. In all cases, the definition of the standard or reference genotypes is critical to enhance selection.

B. STABILITY

After reviewing relationships between various estimates of stability, Becker and Léon (1988), Lin et al. (1986), and Léon and Becker (1988) concluded that the best estimate of stability or responsiveness was the linear regression coefficient. This was true despite its many faults, described by these authors as well as Westcott (1986). To overcome these deficiencies, a number of other alternatives have been suggested, including the use of nonparametric measures with ranks (Nassar and Hühn, 1987) and pattern analysis (Byth et al.,1976), which involved both a classification and ordination techniques. Pattern analysis was useful to reduce the interaction matrix and describe the patterns of the interaction. Zobel et al. (1988) described the application of an additive and multiplicative interaction model to the analysis of the main effects and GE interaction of an analysis of variance. They compared this procedure with linear regression and principal component analysis. They found the model to be superior because it was based on analysis of the GE matrix and described the interaction separately from main effects. Crossa et al. (1988) compared stability using a modified linear regression analysis (Verma et al., 1978) and a spatial method using principal coordinate analysis (Westcott, 1987). They found that the two procedures agreed only partially, because responses of genotypes to environments were not only linear. The spatial method was superior because (1) it could be used over a wide range of environments; (2) it was model-free and described multidimensional interactions; and (3) stable genotypes could be identified by a sequence of graphical displays. When Crossa et al. (1988) applied this technique to an additional set of trials, they were able to characterize a set of genotypes for stability within and between high- and low-yielding environments.

One other procedure has been suggested to describe GE interactions and identify those genotypes that were less sensitive to environmental fluctuations. This is the analysis of crossover GE interactions described by Gregorius and Namkoong (1986), Baker (1988), and Seyedsadr and Cornelius (1992). Baker (1988) applied the Azzalini and Cox (1984) test for crossover interactions to identify those genotypes that show few rank changes over the entire test area or a subset of test area. He suggested that this test could have application in breeding for broad versus specific adaptation. Gregorius and Namkoong (1986) described the need to test for the "separability" of cultivars from environmental effects and identify groups of environments or genotypes that show no rank changes.

Cornelius et al. (1993) demonstrated the application of the shifted multiplicative model (SHMM) described in Seyedsadr and Cornelius (1992). The technique was used to cluster cultivars of wheat into groups without rank changes, because crossover points occurred outside the range of test sites for genotypes within each group. SHMM is used to define a distance measure from the residual sums of squares, where no crossover interactions occur within the data. This distance measure is used to cluster genotypes into groups that differ in their response and do not have any internal crossovers. This partition of interaction into groups of cultivars that are homogeneous with regard to interaction with environments and to most interaction between groups is also an objective of pattern analysis as described by Byth et al. (1976). The difference is that the technique demonstrated by Cornelius et al. (1993) was made more effective by using an F_{GH1} and F_1 test to test for the significance of rank changes within a cluster.

In all the multivariate procedures described, comparison of the genotypes is not as directly quantifiable as the stability parameters described by Becker and Léon (1988) or Lin et al. (1986), nor is the optimal response pattern as simply described. The use of these techniques requires a much better understanding of the basis for differences in the environments and the genotypes sensitivity to these differences. This is what Blum (1988) described as the differences between dealing with the GE interaction using a biometrical approach versus a stress physiological approach. Our application of statistical tools to constraints to breeding for broader adaptation clearly has a limit and progress ultimately will require a better understanding of the physiological basis of the interaction. The use of this knowledge to define the optimal response will be a key to progress in the future. The application of these alternative multivariate procedures, such as AMMI (Zobel et al., 1988), SHMM (Cornelius et al., 1993), or principal coordinate analysis (Crossa et al., 1988), will be very important to delineate those genotypes with the desirable responses for selection and cultivar recommendations. Chisi (1993) used AMMI to cluster 100 full-sib families of KP9B into five groups when the principal component scores were used. These five groups were characterized for yield, flowering date, and plant height at the six environments. One group was found to be fairly consistent in its performance across the environments and genotypes selected from this group would be expected to be better adapted across the target region.

Falconer (1990) and Simmonds (1991) discussed the use of stability parameters to select for sensitivity. Léon and Becker (1988), Pham and Kang (1988), and Scott et al. (1994) found that stability parameters had a poor repeatability across years and environments at different levels of productivity. This was despite the fact that a high rank correlation occurred between yield in different years and types of test environments. This lack of repeatability, estimated by a rank correlation, could be interpreted as an indicator of poor heritability for the sensitivity measure, limiting its use for selection.

Zavala-Garcia et al. (1992a and b) compared the use of various estimates of sensitivity for S_1-family selection in KP9B. One was a rank summation, a nonparametric measure of stability that used all eight environments (RS), the two extreme environments only (RSBW), or the two extremes and an intermediate environment (RSBIW). The relative efficiency of these for indirect-selection strategies in high-, medium- and low-yielding environments in relation to direct selection is shown in Table 4. Using all eight environments to estimate ranks (RT) was much more efficient than the distance measure (DT) in all three types of test sites. However, the greatest efficiency was in high and medium environments. This is probably due to the fact that only one out of eight environments was classified as low yielding; so it was underrepresented. If the criteria were to include one environment of each type, the relative efficiency of these stability parameters was better balanced and very similar to mean productivity measures of both Zavala-Garcia et al. (1992a and b) and Chisi (1993).

Zavala-Garcia et al. (1992b) compared the use of the rank summation with the linear regression coefficient (b), the mean squares for deviations from regression (S), and the first principal components score (PC) from the AMMI analysis. Heritabilities were moderate to low for these stability parameters but slightly higher than heritability for yield itself. When the indirect effect of selection using the better of the two criteria is compared with rank summation (Table 4), both b and PC are very poor in relation to direct selection, especially in low-yielding environments. Chisi (1993) also evaluated the use of AMMI to enhance selection for stability. In their trials, Zavala-Garcia et al. (1992b) and Chisi (1993) both concluded that the only stability parameter that had potential use for selection was rank summation.

C. MEAN PERFORMANCE AND STABILITY

Mean and sensitivity could be selected for simultaneously, especially if yield potential and adaptation are separate traits determined by different sets of genes. Brumpton et al. (1977) attempted to determine whether mean performance and sensitivity were independent by selecting for both in a population of *Nicotiana rustica*. They found the direct response to selection for mean performance was greater than correlated response to selection for sensitivity; thus, the two criteria showed some interdependence. Mean performance and sensitivity were calculated as described in Rosielle and Hamblin (1981) for mean productivity and tolerance.

Hughes et al. (1987) demonstrated the use of an early generation breeding scheme to select for both mean performance and sensitivity. The selection criteria, used in the F_4 and F_5 generations, were mean performance measured as the total yield over two planting dates at one site and another drier site and sensitivity as the sum of differences between yields at the two planting dates at one site and the differences between the two moisture regimes from one planting date and the drier site. Other traits such as disease resistance and agronomic characteristics also were included. They concluded that this procedure was successful in identifying genotypes that were higher yielding and more stable than the controls.

Barah et al. (1981) reported the application of risk aversion to breeding for genotypes that perform well overall and have a reliable yield level for farmers at specific locations. This is what Evans (1993) described as the balance of mean yield, adaptability, and reliability. Witcombe (1988) compared the use of various criteria to select for mean yield and adaptability. He suggested using the mean standard deviation analysis with a choice -- theoretic framework to select the top yielding entries and then testing them further to determine stability. Eskridge (1990) also suggested the use of the safety-first decision-making procedure to develop an index that weights stability and mean yield. These indices then would be used to select the optimal genotype. A slightly different procedure using a yield-stability statistic (Ys_i) was described by Kang (1993). Ceccarelli and Grando (1991b) demonstrated the use of safety-first models on the impact of selection environment on the stability and adaptability of selected genotypes. The indices derived could be used to select genotypes that had a low probability of failure and an acceptable economic threshold in the poorest conditions. In none of these examples would the genotype selected for both mean yield and stability have the best yield under the best conditions.

Zavala-Garcia et al. (1992b) calculated a Smith-Hazel index using yield and the regression coefficient (b) or yield and the first principal component score (PC). Selection for both yield and stability also was done using independent culling at a 25% intensity. Both selection indices were more efficient than indirect selection for the stability parameters alone, except in the poorest environment. When the number of families in common with those selected directly was compared in the eight environments, the indices were superior to those from independent culling and equal to the two measures of mean productivity and rank summation, except in the poorest environment. Zavala-Garcia et al. (1992b) concluded that although indirect selection for a wide range of unpredictable environments was improved most with the use of an index containing both yield and a stability parameter, the gain over mean productivity or a rank summation using only a subset of environments did not justify the extra resources needed to estimate stability.

Chisi (1993) studied realized gains from the selection among S_1 families using single environments, two measures of mean productivity, rank summation, and selection at different productivity levels in alternating generations as described by St-Pierre et al. (1967), Ceccarelli (1989), Simmonds (1991), Campbell and Lafever (1977), and Brennan et al. (1981). St-Pierre et al. (1967) and Ceccarelli (1989) demonstrated that this breeding scheme had potential for improving both mean performance and stability, whereas Simmonds (1991) concluded that the use of alternating selection environments for each cycle of selection was ineffective. Chisi (1993) observed that the most consistent gain was realized by using independent culling only in stress, only in nonstress, or in stress followed by nonstress environments.

The use of stability parameters as direct estimates of sensitivity requires extensive testing in early generations, when the number of entries is very high and seed supplies are low. The best allocation of resources in the critical earlier generations would be to identify a limited number of test sites that consistently represented the range of target conditions for the entire region. These multiple locations could be used to select for mean performance or sensitivity, using a criterion such as rank summation. Then in the more advanced generations, these selections would be tested more widely to determine their adaptability for later recommendation or release. These are the procedures outlined by Fox et al. (1985), Yau et al. (1991), Brennan et al. (1981), and others to breed for highly variable environments characterized by unpredictable fluctuations.

Determining the optimal set of selection environments and their application can be affected by the germplasm or breeding populations used. This was demonstrated by Ceccarelli and Grando (1991a) for barley and in the comparison of two sorghum populations by Zavala-Garcia et al. (1992a and b). Delineating the breadth of adaptation and the optimal procedure to improve it can be influenced by the level of adaptation of the germplasm to target environments. Thus, gain across diverse environments can be increased by a germplasm enhancement program to introduce better adaptation to stress from exotic sources. In hybrid crops, the impact of selection criteria on combining ability is critical in maximizing response of selection in hybrids. Marinesco (1992) and Chisi (1993) found that selection based upon S_1 family performance was a very poor indicator of combining ability. No selection criterion in either study resulted in a consistent improvement in adaptation or performance of resulting hybrids. If the objective of a breeding program is to produce hybrids with wider adaptation and better yielding ability, selection should be based on test crosses in an early generation using a procedure similar to that described previously with multiple locations.

D. ALTERNATIVE TRAITS

A number of studies in diverse species have looked at the potential for using alternative traits to breed for both stress tolerance and performance in favorable environments. An excellent review of this topic can be found in Blum (1988). None of these attempts has resulted in any clearly identified alternative to yield under contrasting conditions. Both Blum (1988) and Ceccarelli et al. (1991) presented summaries of problems associated with attempts to identify alternative traits that select indirectly for yield under stress conditions. The basic principles for use of alternative traits or alternative environments are exactly the same. The effectiveness of indirect selection from correlated response depends upon the genetic correlation between the trait of interest and yield under all conditions that define the target region. Many studies that have been done to evaluate alternative traits have not been able to demonstrate a good relationship with yield under more favorable conditions or did not estimate any of these criteria, as described by Blum (1988) and Ceccarelli et al. (1991).

Nageswara Rao et al. (1989) found that the relationship between drought resistance and yield potential under optimal conditions depended upon the drought pattern. In some types of drought, such as a midseason stress, breeding for both drought tolerance and yield potential was feasible, but for most other drought patterns, yield potential was related negatively to resistance. Thus, even through use of alternative morphological or physiological traits, improved resistance would result in a lower yield potential under optimal conditions. These results could be interpreted as evidence that the breadth of adaptation required to combine resistance to some types of very severe stresses and performance in optimal conditions is not genetically feasible. Thus, the target region needs to be subdivided into narrower ranges that are more practical, as discussed by Yau et al. (1991). Within this range, the use of alternative traits to characterize and select for a level or type of stress tolerance that results in the best yield potential in the better conditions may offer an improvement over selection for yield under stress conditions. This is particularly true when the trait has a higher heritability under variable stress conditions and can be selected in earlier generations. Ceccarelli et al. (1991) stated that it is "the variation in adverse conditions that is the real challenge and not the adverse conditions *per se*." Thus, although a single trait or genotype may be superior at one year or stress condition, those that are consistently superior across all the stress conditions are desirable. This is what Nageswara Rao et al. (1989) was able to demonstrate with 12 known patterns of drought and 60 genotypes.

Wallace et al. (1993), Blum (1988), and Evans (1993) described yield as the result of three physiological components: 1) net accumulated biomass, 2) harvest index, and 3) time needed to develop to harvest maturity. Each of these components is the result of numerous biochemical and physiological processes. Wallace et al. (1993) diagrammed how these components interact to define yield and how the four components of yield, environmental potential, and duration of the growing season ultimately determine the pathway to yield by modifying or terminating the other processes. Wallace et al. (1993) concluded that, in environments in which duration of season varies because of environmental constraints, the GE interaction is due to the impact of these constraints on the timing of the physiological processes and their total duration. Under these conditions, the optimal pathway would vary from season to season, and constraints to maximum production from cultivars for each maturity would be different. This is what Blum (1988) and Ceccarelli et al. (1991) described as the impact of location or year on the correlation between multiple traits and yield. This interaction makes it very difficult to identify single traits or even multiple complexes of traits that can be used as alternatives to select for yield under stress conditions or optimal conditions. Ceccarelli et al. (1991) gave an excellent example of the impact of trait associations and seasonal differences in defining the optimal genotype under variable conditions. The complex nature of trait associations and the influence of environment and genetic background led Ceccarelli et al. (1991) and Blum (1988) to conclude that most studies conducted to date on indirect selection criteria have little application in breeding programs.

Ceccarelli et al. (1991) were particularly critical of the use of isogenic lines to determine the impact of single genes or traits on stress tolerance or stability and the application of these to designing breeding approaches. Romagosa et al. (1993) looked at the impact of three mutant genes of barley on yield stability and the morpho-physiological basis of stability. Unlike genes in the previous studies using isogenic lines, these genes had effects on a complex of traits. Thus, among the three mutant lines, their three recombinant types, and the parent lines, there were various combinations of five phenotypic characteristics: spike density, plant height, earliness, leaf area, and grain filling period. They used 15 environments to determine yield stability and four environments to measure the morpho-physiological traits. The authors concluded that the technique of using isogenic lines to characterize the relationship between particular environmental responses and complexes of morpho-physiological traits would allow identification of optimal types or genes for a given set of conditions. Even though this study included a large set of diverse environments, and the three genes had impacts on a diversity of traits, application was still limited by the impact of the narrow genetic base and lack of a clear general recommendation.

Another method used by physiologists to investigate the impact of specific characteristics of stability or stress resistance involves a limited set of genotypes that can be classified as "resistant" or "susceptible", or differ in their stability. These diverse reactions are sometimes just "known" or poorly described prior to their use in the study (Blum, 1988). This limited diversity for overall response or for the trait of interest will bias the correlation between traits and between the trait and yield (Blum, 1988; Ceccarelli et al., 1991). The predictions made from these trials will have very limited use by breeders for indirect selection.

An alternative procedure to identify traits for indirect selection is to use a retrospective historical analysis of the changes that have occurred through natural selection and artificial selection in a crop species. Blum (1988) and Evans (1993) both described the physiological changes that have resulted from selection for yield. Evans (1993) included an extensive review of this topic for the main physiological processes and their components. He found that the predominant improvements in yield potential have been from changes in regulatory processes that control patterns of partitioning and the timing of development. These components are related to harvest index and harvest maturity (Wallace et al., 1993). No changes have been made in the efficiency of the major metabolic and assimilatory processes that result from changes in photosynthesis, respiration, translocation, or growth rate. Continued selection for yield potential under optimal conditions will only continue to alter these regulatory processes, according to the hypothesis of both Evans (1993) and Wallace et al. (1993). Evans (1993) concluded that continued selection for modifications in regulatory processes without increased efficiency in metabolic or assimilatory processes will result in improvement only if innovations in agronomy allow greater agronomic support of the crop to ameliorate environmental stresses. Changes in crop plants since domestication have been results of empirical selection for yield, but the future may depend more on selection by design (Evans, 1993). Unfortunately, most of the research reviewed by Blum (1988), Evans (1993), Ceccarelli et al. (1991), and Wallace et al. (1993) has little application in designing the optimal genotype for stress and nonstress conditions. Evans (1993) stated "Empirical selection is an extremely powerful agent of change but selection by design may yet prove to be even more powerful when our understanding of the physiology of crop yield is more comprehensive than it is at present." Not only does our understanding of the physiology of yield need to be broadened, but it needs to be applied in a manner that will increase our efficiency of selection for adaptation to a wide range of environmental conditions.

334

One of the keys is to describe better the relationship between complexes of traits and between environmental responses. Romagosa et al. (1993) and van Oosterom and Ceccarelli (1993) used principal component analysis to describe a complex of traits to be used for selection. Both studies also used AMMI to describe the environmental response for yield of the genotypes. The application of multivariate techniques to describe better the direct and indirect characteristics of the GE interaction will be critical in the future, especially when applied to diverse genotypes. In van Oosterom and Ceccarelli (1993), the direct effect of winter plant ideotype, which was calculated from a complex of three traits, was determined for heading data and yield at low- and high-yielding environments for barley. Although different ideotypes were required for optimal yield in the two conditions, selection for winter plant ideotype under favorable conditions in early generations would improve adaptation; selection then could be concentrated on yield at variable environments for stress tolerance.

Wallace et al. (1993) observed that selection for yield only will select indirectly for maturity and harvest index, but long-term gain in yield potential is dependent on gains in the rate of accumulation of biomass, which is not affected by indirect selection for yield. Thus, these authors concluded that the best alternative is to select using a yield-system analysis of trials. This analysis required the evaluation of four traits: days to flowering, days to maturity, aerial biomass at harvest, and yield. From these, an additional five traits that measure accumulation rates and partitioning were calculated. Selection then could be conducted for yield and an optimal ideotype for the other eight traits. Wallace et al. (1993) suggested the use of AMMI to quantify and describe the environmental responses of genotypes for yield and other traits. They did not describe clearly how this simultaneous selection would be conducted, but this method would allow for a better characterization of the optimal genotypic response.

Fischer et al. (1989), in a study of maize, compared the use of an index for drought resistance, which incorporated a number of attributes known to be related to drought resistance with selection for yield under various water stresses. This index was computed as the Euclidean distance of a genotype from a target genotype. The selection target was defined in terms of numbers of standard error limits from the mean and by allocating arbitrary weights to the attributes. Attributes used were grain yield at all three moisture levels, relative leaf elongation, days from anthesis to silking, leaf death score, and, in one trial, canopy temperature. The selected progeny had both above-average yield potential and an index value in the resistant range. Simultaneous selection of integrative, drought-adaptive plant characteristics, along with yield potential, was the only procedure that effectively improved yield under severe stress and maintained yield under mild stress conditions. This index contained traits that were known to be related to stress tolerance, had a higher heritability than that of yield under stress, and were less affected by environmental variability. This use of an index to define the optimal genotype under stress conditions allowed incorporation of a number of attributes to define a complex of responses and, thus, increased the efficiency of selection for both stress tolerance and yield potential. Byrne et al. (1995) extended the study of Fischer et al. (1989) to include 8 cycles of the population selected using the index. These were compared with 6 cycles selected using multiple location testing. The index selected population was better until the eighth cycle when its gain was reduced. They concluded the best procedure was to initially select for the index of drought resistance estimated from evaluations in controlled environments; then in later cycles yield test at multiple locations where the environmental conditions were more unpredictable and varied.

A final alternative to improve the efficiency of selection for tolerance to the various types of stresses that characterize a target region, as well as yield potential under the more favorable environments, is to use molecular markers to select indirectly for stress tolerance, yield potential, or both. Paterson et al. (1991) discussed the use of "quantitative trait loci" or QTL mapping to select indirectly for specific attributes. They predicted that the use of marker-assisted selection with QTL's would be of the greatest benefit for traits with a low heritability. These QTL's would be more effective as indicators of an individual breeding value and, thus, would result in greater correlated response to selection. These could be used to select indirectly for yield under variable environmental conditions if they were more consistent, i.e., less subject to genotype x environment interaction than yield itself. Paterson et al. (1991) and Hayes et al. (1993) demonstrated that QTL's could show the same range of sensitivities to environments as yield itself. In most cases, the impact of the QTL x environment was through significant effects in only a portion of the environments or differences in magnitude of effect at each environment. Two classes of genetic markers were identified. The first included QTL's that were constant, having no environmental interactions; the second were environment-specific QTL's. Paterson et al. (1991) concluded that the best strategy would be to select for genotypes with both classes of QTL's to identify individuals with both broad adaptation and site-specific adaptation. Use of the genetic markers would give enhanced opportunities to combine these attributes without the need to test in all specific conditions. The key to progress for this strategy is in the careful identification of the environment-specific QTL's and their impact on both adaptation and yield potential. Identification requires testing under the specific conditions to link markers consistently with a trait related to specific adaptation or performance. Although this approach to indirect selection has potential for future gain in broad adaptation, it may suffer from the same limitation as the previously described alternatives. Final yield is a result of a complex interaction of biochemical, morphological, and physiological traits with the environment within the entire genetic background. Attempts to identify simple procedures that do not account for these interactions may be of little actual benefit except in rare situations.

REFERENCES

Abou-El-Fittouh, H.A., J.O. Rawlings, and P.A. Miller. 1969. Classification of environments to control genotype x environment interactions with an application to cotton. *Crop Science* 9:135-140.

336

Allard, R.W., and A.D. Bradshaw. 1964. Implications of genotype-environment interactions in applied plant breeding. *Crop Science* 4:503-508.

Allen, F.L., R.E. Comstock, and D.C. Rasmusson. 1978. Optimal environments for yield testing. *Crop Science* 18:747-751.

Atlin, G.N., and K.J. Frey. 1989. Predicting the relative effectiveness of direct versus indirect selection for oat yield in three types of stress environments. *Euphytica* 44:137-142.

Atlin, G.N., and K.J. Frey. 1990. Selecting oat lines for yield in low-productivity environments. *Crop Science* 30:556-561.

Azzalini, A., and D.R. Cox. 1984. Two new tests associated with analysis of variance. *Journal of the Royal Statistical Society Bulletin* 46:335-343.

Baker, R.J. 1988. Tests for crossover genotype-environmental interactions. *Canadian Journal of Plant Science* 68:405-410.

Barah, B.C., H.P. Binswanger, B.S. Rana, and N.G.P. Rao. 1981. The use of risk aversion in plant breeding: Concept and application. *Euphytica* 30:451-458.

Becker, H.C., and J. Léon. 1988. Stability analysis in plant breeding. *Plant Breeding* 101:1-23.

Blum, A. 1988. Plant breeding for stress environments. p. 15-42. CRC Press, Inc. Boca Raton, FL.

Blum, A., G. Golan, J. Mayer, B. Sinmena, and T. Obilana. 1992. Comparative productivity and drought response of semi-tropical hybrids and open-pollinated varieties of sorghum. *Journal of Agricultural Science (Cambridge)* 118:29-36.

Bramel-Cox, P.J., T. Barker, F. Zavala-Garcia, and J.D. Eastin. 1991. Selection and testing environments for improved performance under reduced-input conditions. p. 29-56. *In* D. Sleper, P.J. Bramel-Cox, T. Barker (ed.) Plant breeding and sustainable agriculture: Considerations for objectives and methods. CSSA Special Publ. 18. ASA, CSSA, SSSA, Madison, WI.

Brennan, P.S., D.E. Byth, D.W. Drake, I.H. DeLacy, and D.G. Butler. 1981. Determination of the location and number of test environments for a wheat cultivar evaluation program. *Australian Journal of Agricultural Research* 32:189-201.

Brown, K.D., M.E. Sorrells, and W.R. Coffman. 1983. A method for classification and evaluation of testing environments. *Crop Science* 23:889-893.

Brumpton, R.J., H. Boughey, and J.L. Jinks. 1977. Joint selection for both extremes of mean performance and of sensitivity to a macro-environmental variable. I. Family selection. *Heredity* 38:219-226.

Bull, J.K., K.E. Basford, I.H. DeLacy, and M. Cooper. 1992. Classifying genotypic data from plant breeding trials: A preliminary investigation using repeated checks. *Theoretical and Applied Genetics* 85:461-469.

Bull, J.K., M. Cooper, and K.E. Basford. 1994. A procedure for investigating the number of genotypes required to provide a stable classification of environments. *Field Crops Research* 38:47-56.

Byrne, P.F., J. Bolaños, G.O. Edmeades, and D.L. Eaton. 1995. Gains from selection under drought versus multilocational testing in related Tropical maize populations. *Crop Science* 35:63-69.

Byth, D.E., R.L. Eisemann, and I.H. DeLacy. 1976. Two-way pattern analysis of a large data set to evaluate genotypic adaptation. *Heredity* 37:215-230.

Calhoun, D.S., G. Gebeyehu, A. Miranda, S. Rajaram, and M. van Ginkel. 1994. Choosing evaluation environments to increase wheat grain yield under drought conditions. *Crop Science* 34:673-678.

Campbell, L.G., and H.N. Lafever. 1977. Cultivar x environment interactions in soft red winter wheat yield tests. *Crop Science* 17:604-608.

Carver, B.F., E.L. Smith, and H.O. England, Jr. 1987. Regression and cluster analysis of environmental responses of hybrid and pureline winter wheat cultivars. *Crop Science* 27:659-664.

Ceccarelli, S. 1989. Wide adaptation: How wide? *Euphytica* 40: 197-205.

Ceccarelli, S., E. Acevedo, and S. Grando. 1991. Breeding for yield stability in unpredictable environments: Single traits, interaction between traits, and architecture of genotypes. *Euphytica* 56:169-185.

Ceccarelli, S., and S. Grando. 1991a. Environment of selection and type of germplasm in barley breeding for low-yielding conditions. *Euphytica* 57:207-219.

Ceccarelli, S., and S. Grando. 1991b. Selection environment and environmental sensitivity in barley. *Euphytica* 57:157-167.

Chisi, M. 1993. Evaluation of selection procedures in multiple environments to identify parent lines for grain sorghum hybrids (*Sorghum bicolor* (L.) Moench). Ph.D. diss. Kansas State Univ.Manhattan (Dissertation Abstract -   )

Cooper, M., D.E. Byth, I.H. DeLacy, and D.R. Woodruff. 1993a. Predicting grain yield in Australian environments using data from CIMMYT International Wheat Performance Trials. 1. Potential for exploiting correlated response to selection. *Field Crops Research* 32:305-322.

Cooper, M., I.H. DeLacy, D.E. Byth, and D.R. Woodruff. 1993b. Predicting grain yield in Australian environments using data from CIMMYT International Wheat Performance Trials. 2. The application of classification to identify environmental relationships which exploit correlated response to selection. *Field Crops Research* 32:323-342.

Cornelius, P.L., M. Seyedsadr, and J. Crossa. 1992. Using the shifted multiplicative model to search for "separability" in crop cultivar trials. *Theoretical and Applied Genetics* 84:161-172.

Cornelius, P.L., D.A. Van Sanford, and M.S. Seyedsadr. 1993. Clustering cultivars into groups without rank-change interactions. *Crop Science* 33:1193-1200.

Crossa, J., P.L. Cornelius, K. Sayre, and J.I. Ortiz-Monasterio R. 1995. A shifted multiplicative model fusion method for grouping environments without cultivar rank changes. *Crop Science* 35:54-62.

Crossa, J., B. Westcott, and C. Gonzalez. 1988. Analyzing yield stability of maize genotypes using a spatial model. *Theoretical and Applied Genetics* 75:863-868.

Dudley, J.W., and R.H. Moll. 1969. Interpretation and use of estimates of heritability and genetic variances in plant breeding. *Crop Science* 9: 257-262.

Eskridge, K.M. 1990. Selection of stable cultivars using a safety-first rule. *Crop Science* 30:369-374.

Evans, L.T. 1993. Crop evolution, adaptation, and yield. Cambridge University Press. Cambridge, Great Britain.

Falconer, D.S. 1989. Introduction to quantitative genetics. p. 322-325. Third edition. Longman Scientific and Technical. Essex, England.

Falconer, D.S. 1990. Selection in different environments: effects on environmental sensitivity (reaction norm) and on mean performance. *Genetical Research Cambridge* 56:57-70.

Finlay, K.W., and G.N. Wilkinson. 1963. The analysis of adaptation in a plant-breeding programme. *Australian Journal of Agricultural Research* 14:742-754.

Fischer, K.S., G.O. Edmeades, and E.C. Johnson. 1989. Selection for the improvement of maize yield under moisture-deficits. *Field Crops Research* 22:227-243.

Fox, P.N., and A.A. Rosielle. 1982a. Reducing the influence of environmental main-effects on pattern analysis of plant breeding environments. *Euphytica* 31:645-656.

Fox, P.N., and A.A. Rosielle. 1982b. Reference sets of genotypes and selection for yield in unpredictable environments. *Crop Science* 22:1171-1175.

Fox, P.N., A.A. Rosielle, and W.J.R. Boyd. 1985. The nature of genotype x environment interactions for wheat yield in Western Australia. *Field Crops Research* 11:387-398.

Frey, K.J. 1964. Adaptation reaction of oat strains selected under stress and nonstress environmental conditions. *Crop Science* 4:55-58.

Goodchild, N.A., and W.R.J. Boyd. 1975. Regional and temporal variations in wheat yields in Western Australia and their implications in plant breeding. *Australian Journal of Agricultural Research* 26: 209-217.

Gregorius, H.R., and G. Namkoong. 1986. Joint analysis of genotype and environment effects. *Theoretical and Applied Genetics* 72:413-422.

Hamblin, J., H.M. Fisher, and H.I. Ridings. 1980. The choice of locality for plant breeding when selecting for high yield and general adaptation. *Euphytica* 29:161-168.

Hanson, W.D. 1964. Genotype-environment interaction concepts for field experimentation. *Biometrics* 20:540-552.

Hayes, P.M., B.H. Liu, S.J. Knapp, F. Chen, B. Jones, T. Blake, J. Franckowiak, D. Rasmusson, M. Sorrells, S.E. Ullrich, D. Wesenberg, and A. Kleinhofs. 1993. Quantitative trait locus effects and

338

environmental interaction in a sample of North American barley germplasm. *Theoretical and Applied Genetics* 87:392-401.

Hazel, L.N. 1943. The genetic basis for constructing selection indices. *Genetics* 28:476-490.

Horner, I.W., and K.J. Frey. 1957. Methods of determining natural areas for oat varietal recommendations. *Agronomy Journal* 49: 313-315.

Hughes, W.G., B. Westcott, and P.L. Sharp. 1987. Joint selection for high yield and low sensitivity in winter wheat (Triticum aestivum L.). *Plant Breeding* 99:107-117.

Ivory, D.A., S. Kaewmeechai, I.H. DeLacy, and K.E. Basford. 1991. Analysis of the environmental component of genotype x environment interaction in crop adaptation evaluation. *Field Crops Research* 28:71-84.

Johnson, S.S., and J.L. Geadelmann. 1989. Influence of water stress on grain yield response to recurrent selection in maize. *Crop Science* 29: 558-564.

Kang, M.S. 1993. Simultaneous selection for yield and stability in crop performance trials: Consequences for growers. *Agronomy Journal* 83:754-757.

Léon, J., and H.C. Becker. 1988. Repeatability of some statistical measures of phenotypic stability-correlations between single year results and multi years results. *Plant Breeding* 100: 137-142.

Lin, C.S., and M.J. Morrison. 1992. Selection of test locations for regional trials of barley. *Theoretical and Applied Genetics* 83: 968-972.

Lin, C.S., M.R. Binns, and L.P. Lefkovitch. 1986. Stability analysis: Where do we stand? *Crop Science* 26:894-900.

Mahalakshmi, V., F.R. Bidinger, K.P. Rao, and D.S. Raju. 1992. Performance and stability of pearl millet topcross hybrids and their variety pollinators. *Crop Science* 32:928-932.

Marek, J.H., and C.O. Gardner. 1979. Responses to mass selection in maize and stability of resulting populations. *Crop Science* 19:779-783.

Marinesco, O.A. 1992. Effectiveness of selection under stress or low-stress environments for grain yield in a sorghum population (*Sorghum bicolor* (L.) Moench). Ph.D. Diss. Kansas State Univ.,Manhattan. (Dissertation Abstract B. 53-1118).

Marshall, D.R., and A.H.D. Brown. 1973. Stability of performance of mixtures and multilines. *Euphytica* 22: 405-412.

Moll, R.H., C.C. Cockerham, C.W. Stuber, and W.P. Williams. 1978. Selection responses, genetic-environmental interactions, and heterosis with recurrent selection for yield in maize. *Crop Science* 18:641-645.

Nageswara Rao, R.C., J.H. Williams, and M. Singh. 1989. Genotypic sensitivity to drought and yield potential of peanuts. *Agronomy Journal* 81: 887-893.

Nassar, R., and M. Hühn. 1987. Studies on estimation of phenotypic stability: Test of significance for nonparametric measures of phenotypic stability. *Biometrics* 43:45-53.

Paleolog, J., and M. Lorkiewicz. 1991. Selection for female fecundity in *Drosophila* test crosses and lines selected in different environments. II. Response to heat stress and top-crossing. *Journal of Animal Breeding and Genetics* 108: 363-368.

Paleolog, J., and J. Maciejowski. 1991. Selection for female fecundity in *Drosophila* test crosses and lines selected in different environments. I. Response to selection. *Journal of Animal Breeding and Genetics* 108: 355-362.

Paterson, A.H., S. Damon, J.D. Hewitt, D. Zamir, H.D. Rabinowitch, S.E. Lincoln, E.S. Lander, and S.D. Tanksley. 1991. Mendelian factors underlying quantitative traits in tomato: Comparison across species, generations, and environments. *Genetics* 127: 181-197.

Pederson, D.G. and A.J. Rathjen. 1981. Choosing trial sites to maximize selection response for grain yield in spring wheat. *Australian Journal of Agricultural Research* 32: 411-424.

Pham, H.N., and M.S. Kang. 1988. Interrelationships among and repeatability of several stability statistics estimated from international maize trials. *Crop Science* 28: 925-928.

Romagosa, I., P.N. Fox, L.F. García del Moral, J.M. Ramos, B. García del Moral, F. Roca de Togores, and J.L. Molina-Cano. 1993. Integration of statistical and physiological analysis of adaptation of near-isogenic barley lines. *Theoretical and Applied Genetics* 86: 822-826.

Rosielle, A.A., and J. Hamblin. 1981. Theoretical aspects of selection for yield in stress and non-stress environments. *Crop Science* 21: 943-946.

Schnell, F.W., and H.C. Becker. 1986. Yield and yield stability in a balanced system of widely differing population structures in *Zea mays* L. *Plant Breeding* 97: 30-38.

Scott, R.A., M. Champoux, and W.T. Schapaugh, Jr. 1994. Influence of environmental productivity levels and yield stability on selection strategies in soybean. *Euphytica* 78:115-122.

Seyedsadr, M., and P.L. Cornelius. 1992. Shifted multiplicative models for nonadditive two-way tables. *Communications of the Statistical Bulletin of Simulation and Computing* 21:807-832.

Shabana, R.T., T. Bailey, and K.J. Frey. 1980. Production traits of oats selected under low, medium, and high productivity. *Crop Science* 20:739-744.

Shorter, R., D.E. Byth, and V.E. Mungomery. 1977. Genotype x environment interactions and environmental adaptation. II. Assessment of environmental contributions. *Australian Journal of Agricultural Science* 28: 223-235.

Simmonds, N.W. 1991. Selection for local adaptation in a plant breeding programme. *Theoretical and Applied Genetics* 82: 363-367.

Smith, H.F. 1936. A discriminant function for plant selection. *Annual of Eugenics* 7:240-250.

Soliman, K.M., and R.W. Allard. 1991. Grain yield of composite cross populations of barley: Effects of natural selection. *Crop Science* 31: 705-708.

Sprague, G.F., and W.I. Federer. 1951. A comparison of variance components in corn yield trials. II. Error, year x variety, location x variety, and variety components. *Agronomy Journal* 43: 535-541.

St-Pierre, C.A., H.R. Klinck, and F.M. Gauthier. 1967. Early generation selection under different environments as it influences adaptation of barley. *Canadian Journal of Plant Science* 47:507-517.

van Oosterom, E.J., and S. Ceccarelli. 1993. Indirect selection for grain yield of barley in harsh Mediterranean environments. *Crop Science* 33: 1127-1131.

Van Sanford, D.A., T.W. Pfeiffer, and P.L. Cornelius. 1993. Selection index based on genetic correlations among environments. *Crop Science* 33: 1244-1248.

Verma, M.M., G.S. Chahal, and B.R. Murty. 1978. Limitations of conventional regression analysis: A proposed modification. *Theoretical and Applied Genetics* 53:89-91.

Walker, A.K., and W.R. Fehr. 1978. Yield stability of soybean mixtures and multiple pure stands. *Crop Science* 18: 719-723.

Wallace, D.H., J.P. Baudion, J. Beaver, D.P. Coyne, D.E. Halseth, P.N. Masaya, H.M., J.R. Myers, M. Silbernagel, K.S. Yourstone, and R.W. Zobel. 1993. Improving efficiency of breeding for higher crop yield. *Theoretical and Applied Genetics* 86: 27-40.

Westcott, B. 1986. Some methods of analyzing genotype-environment interaction. *Heredity* 56: 243-253.

Westcott, B. 1987. A method of assessing the yield stability of crop genotypes. *Journal of Agricultural Science* 108:267-274.

Witcombe, J.R. 1988. Estimates of stability for comparing varieties. *Euphytica* 39: 11-18.

Wright, A.J. 1976. The significance for breeding of linear regression analysis of genotype-environment interactions. *Heredity* 37: 83-93.

Yau, S.K., G. Ortiz-Ferrara, and J.P. Srivastava. 1991. Classification of diverse bread wheat-growing environments based on differential yield responses. *Crop Science* 31: 571-576.

Zavala-García, F., P.J. Bramel-Cox, and J.D. Eastin. 1992a. Potential gain from selection for yield stability in two grain sorghum populations. *Theoretical and Applied Genetics* 85: 112-119.

Zavala-García, F., P.J. Bramel-Cox, J.D. Eastin, M.D. Witt, and D.J. Andrews. 1992b. Increasing the efficiency of crop selection for unpredictable environments. *Crop Science* 32: 51-57.

Zavala-García, F., O.A. Marinesco, P.J. Bramel-Cox, J.D. Eastin, and M.D. Witt. 1991. Choosing selection environments in sorghum for high variable target areas. p. 122 *In* Agron Abstracts. ASA, Madison, WI.

Zobel, R.W., M.J. Wright, and H.G. Gauch, Jr. 1988. Statistical analysis of a yield trial. *Agronomy Journal* 80: 388-393.

Chapter 12

CHARACTERIZING AND PREDICTING GENOTYPE BY ENVIRONMENT INTERACTION: VARIABLE ACTIVE GENIC *VS.* PARENTAL GENIC PROPORTIONS MODEL

L. P. Subedi [1]

I. THEORETICAL MODEL

A computer simulation model referred to as Variable Active Genic (VAG) model for specifying the phenotype of a trait for populations of varied genetic structures was developed by Subedi and Carpena (1988), the modified version of which is given below.

A. A VARIABLE ACTIVE GENIC (VAG) MODEL
1. The Model

The phenotypic value (Y_{ijt}) of ith (i= 1,2,...,n) entry at j^{th} (j= 1,2,....,p) environment in t^{th} (t= 1,2,.....,r) replication within environment, can be represented as:

$$Y_{ijt} = \sum_{w=1}^{K_{ijt}} u_{wjt} + \sum_{w=1}^{K_{ijt}} \sum_{u=1}^{2} \sum_{v=1}^{2} P_{iwuv}\{a_{wjt}\,\Theta_{wi} + d_{wjt}(1-\Theta^2_{wi})\} \quad (1)$$

where:

u_{wjt} = the average of the phenotypic values of homozygous dominant and homozygous recessive genotypes of wth locus at j^{th} environment and t^{th} replication within j^{th} environment. u_{wjt} is zero for all inactive loci.

K_{ijt} = the number of active loci of ith entry at j^{th} environment and t^{th} replication within j^{th} environment. The number of active loci vary among entries or genotypes because of hypostatic relation (interaction) among some genes.

P_{iwuv} = the frequency of A_uA_v genotype at wth locus in the ith entry. Following relation holds good for different populations:

Purelines: P_{iwu} = 0 or 1; P_{iwuv} = 0 if u≠v
Mixtures : $0 \le Piwu \ge 1$; P_{iwuv} = 0 if u≠v
Hybrids : P_{iwu}=0 or 0.5 or 1; P_{iwuv}= 0 or 1 if u≠v
Synthetics: P^2_{iwu}= P_{iwuu}; $2P_{iwu}\,P_{iwv}$ = P_{iwuv} and
$\qquad P^2_{iwv}$ = P_{iwvv}
in all cases P_{iwu} + P_{iwv} = 1.0 $\qquad\qquad$ (2)

---

[1]IAAS, Tribhuvan University, Rampur, Chitwan, Nepal.

342

a_{wjt} =   the mean of the difference between the phenotypic values of homozygous dominant and homozygous recessive genotypes of wth locus at j^{th} environment and t^{th} replication within j^{th} environment. a_{wjt} is zero for all inactive loci. A hypostatic gene (allele) is considered inactive.

d_{wjt} =   the difference between the phenotypic values of a heterozygous genotype of wth locus and its corresponding u_{wjt} at the jth environment and t^{th} replication within the jth environment.

Θ_{wi} =   1, 0, -1 for homozygous dominant, heterozygote and homozygous recessive genotype, respectively, of wth locus in the ith entry.

2. Assumptions
1. Some basic loci and some other variable loci control the character of interest.

2. The species under consideration is diploid and it has only 2 alleles per locus.

B. METHOD OF GENERATING POPULATIONS

1. Purelines: To generate the genotype of a particular pureline, the following steps were followed:
a) Generate a random number from a uniform distribution ranging from 0 to 1.

b) Generate one random number each for K loci from the same distribution.

c)   Compare the random number generated for a locus to the one generated in (a) and label the locus as homozygous dominant if the locus random number is larger. Declare the locus homozygous recessive, otherwise.

d)   Repeat the process for all loci in each genotype.

e)   Repeat the whole process until 10 entries are generated.

2. Mixture of Purelines: While generating the mixtures of purelines, steps (a) to (d) of generating the purelines were duplicated. The remaining steps consist of generating 10 genotypes per entry with the locus random number already generated. Ten such entries per mixtures of purelines were generated.

3. Hybrids: To generate a particular hybrid, the following steps were followed:

a) Generate a random number from a uniform distribution ranging from 0 to 1.

b) Generate two random numbers each for K loci from the same distribution.

c) Compare the random numbers generated for each locus to the one generated in

(a) and label the locus as homozygous dominant if both are smaller than the locus random number. Label the locus as homozygous recessive if both are greater than the locus random number. Declare the locus as heterozygous, otherwise.

d) Repeat the process for all loci in each genotype.

e) Repeat the whole process until all entries are generated.

4. Synthetics: Steps (a) and (b), used in generating hybrid genotypes, are also applicable to synthetics. The remaining steps consist of generating 10 genotypes per entry using the same locus random number. Ten such entries per synthetic were generated.

For the purpose of generating phenotypic values as in section E below, the value is averaged over 10 entries in case of mixture of purelines and synthetics to make it comparable with that of purelines and hybrids.

C. PROCEDURE FOR SPECIFYING PHENOTYPIC VALUES
To specify the phenotypic value of an entry, the following assumptions are made:

1. all a_{wjt}'s for a given environment are equal and greater than zero.

2. all u_{wjt}'s for a given environment are equal and greater than zero.

Under these assumptions, the phenotypic value, Y_{ijt}, can be expressed as:

where
$$Y_{ijt} = K_{ijt}\, u_{jt} + A_{ijt}\, a_{jt} + D_{ijt}\, d_{jt}$$

$$u_{1jt} = u_{2jt} = .. = u_{wjt} = u_{jt}.$$

Similarly for a_{jt} and d_{jt},

$$A_{ijt} = \sum_{w=1}^{K_{ijt}} \sum_{u=1}^{2} \sum_{v=1}^{2} P_{iwuv}\, \Theta_{wi} \tag{3}$$

$$D_{ijt} = \sum_{w=1}^{K_{ijt}} \sum_{u=1}^{2} \sum_{v=1}^{2} P_{iwuv}(1-\Theta^2_{wi}) \tag{4}$$

D. METHOD OF GENERATING GEI
Three levels of variability in GEI are generated by varying the active loci of each entry, separately, over entries and environments through 3 different distributions, namely, normal, uniform, and inverted normal. The variances of

these distributions were 1, 3 and 4.5, respectively. These distributions are expected to produce low, medium and high levels of GEI, respectively.

E. GENERATING PHENOTYPIC VALUES AND COMPUTING ANALYSIS OF VARIANCE

For this purpose following steps were used:

1. Generate n entries with K number of loci each based on the genetic structures of their populations. See section (B) above.

2. Specify the number of environments or locations, p

3. Generate $\bar{a}_j$'s with a mean of $\bar{a}_.$ and standard deviation of σ_a

4. If there are hybrid entries, generate $\bar{d}_j$'s with a mean of $\bar{d}..$ and standard deviation of σ_d

5. Generate $\bar{u}_j$'s with a mean of $\bar{u}..$, and standard deviation of σ_u

6. Recall number of entries, n

7. Use normal or uniform or inverted normal distribution with a mean (say 10) and a range (say 8-13) to generate number of active loci, ($\bar{K}_{ij.}$, $\bar{A}_{ij.}$, $\bar{D}_{ij.}$). This will generate 7 basic loci and the rest variable loci.

8. Specify the number of replications, r

9. Generate error with a mean of $\bar{a}_j$'s, $\bar{u}_j$'s and $\bar{d}_.$'s and their respective standard deviations. This will generate a_{jt}'s, u_{jt}'s and d_{jt}'s.

10. Compute phenotypic values: $Y_{ijt} = \bar{K}_{ij}.u_{jt} + \bar{A}_{ij}.a_{jt} + \bar{D}_{ij}. d_{jt}$

11. Repeat the process to satisfy no. of replications, no of entries, and no of environments.

12. Compute the analysis of variance

13. Print (whatever you want from items specified above)

14. STOP and END

F. SUMMARY OF RESULTS

1. Any multi-environmental analysis of variations of any population can be perfectly simulated with this model/method.

2. The genotypic variance was found to be mainly affected by the means of distributions generating $\bar{a}_j$'s and $\bar{d}_j$'s and diversity of entries under test.

3. The environmental variance was found to be mainly affected by the variance of the distribution generating $\bar{u}_j$'s. A minor change in environmental variance was obtained by varying the levels of distributions generating the $\bar{a}_j$'s and $\bar{d}_j$'s.

4. The G x E interaction variance was mainly affected by the variability of active loci. It was also found to be affected by the extent of diversity of entries, diversity of environmental values per locus, and the mean and variance of distributions generating the $\bar{a}_j$'s and $\bar{d}_j$'s.

II. FITTING ACTUAL DATA INTO THE THEORETICAL MODEL

In practice, computer simulation model($\approx$VAG) has limited value except for simulating analysis of variance as well as understanding factors affecting the components of the phenotypic variance. Since it was found that various breeding lines developed by national and international crop research centers had varied parental lines incorporated, it was thought that a practically useful model could be developed equating varied loci in VAG model with varied parental lines. Ultimately good parental lines identified could be intercrossed to develop an optimum breeding line. Therefore, eighteen rice breeding lines that were tested at 21 locations in International Rice Yield Nursery-Medium (IRYN-M) of the test year 1984 (IRRI,1985) were selected for this study. The following Variable Active Parental Genic Proportions (VAPGeP) model (Subedi, 1992) was developed.

A. THE VAPGeP MODEL

$$Y^{(1)}_{ij} = I_j + a_j \bar{A}_{\cdot j} + B_g (a_i - \bar{a}.) + B_{ge,j}(W_{ij} - \bar{W}_{\cdot j})$$
$$\text{Model (I)} \tag{5}$$

where, $Y^{(1)}_{ij}$ = Yield in tons/ha of ith breeding line in jth (j=1,2 ..,p) location, and
I_j = Intercept of the prediction equation.

The prediction equation : $Y^{(2)}ij = I_j + W_{ij}$
$$= I_j + A_{ij} a_j$$
$$\text{Model (II)} \tag{6}$$

$$W_{ij} = \sum_{k=1}^{Q_j} B_{kj} X'_{ki} \qquad \text{where } X'_{ki} = (100 X_{ki} + 1/2)^{1/2}$$

B_{kj} = multiple (partial) regression coefficient of jth location yields of breeding lines (Y_{ij}) on genic proportions (X'_{ki}) of kth parental line.

X_{ki} =   genic proportion of kth (k = 1,2, ..., q) parental line included in ith breeding line

$$A_{ij} = \sum_{w=1}^{Q_j} X'_{ki}$$ = Sum of the genic proportion (transformed) of Q active parental lines at jth location. An active kth parental line at jth location is that has significant B_{kj} at that location.

$$a_j = \sum_{i=1}^{n} W_{ij} / \sum_{i=1}^{n} A_{ij} \qquad a_i = \sum_{j=1}^{p} W_{ij} / \sum_{j=1}^{p} A_{ij} \qquad (7)$$

B_g = Regression coefficient of the genotypic effects (G_i's) on a_i's.
$B_{ge,j}$ = Regression coefficient of genotype by environment interaction effects of breeding lines at jth location on W_{ij} at the same location.

$\bar{I}. + \bar{a}. \bar{A}..$ = Grand mean

$\bar{Y}_{ij} = Y_{ij} - d_{ij}$ = predicted yield of ith breeding line at jth location. d_{ij} refers to deviation of actual yields from predicted.

To effectively understand this model, the following elaboration is given:

1. Definition of Genic Proportion
        The genic proportion is the value calculated for genetic contribution of each of the parental lines included in the breeding line (Delannax et al; 1983). This does not account for the fact that breeding lines result from intense selection and that more genes may come from one parent of a cross than another. However, St. Martin (1982) showed that shifts are not likely to be large, and with a biparental mating 88 percent of offspring would receive between 40 to 60 percent of their alleles from each parent.

2.Method of Computing Genic Proportion
        An example of computing genic proportions of different parental lines in a breeding line is given below:
        Breeding line: BR 51-282-8
        Parents   : IR 20/IR5-114-3-1
        Proportion : 1/2   1/2
        But IR20 = Peta *3/TN1//TKM6
This means Peta was used as a recurrent parent for 2 backcrosses with TN1 (Taichung Native 1) as a non-recurrent parent and the resulting line was mated with TKM6. This resulted in:
        Parental line : Peta   TN1   TKM6

Proportion : 7/32   1/32   1/4

Since   TN1   = DGWG/   Tsan-Yuan-Chan
Proportion   = 1/64   1/64

Again   IR 5-114-3-1   = Peta / T. rotan
Proportion   = 1/4   1/4

Now summing under different parental lines, we have;

Parental line	Peta	DGWG	TYC	TKM6	T.rotan
Proportion(X_{ki})	0.468	0.015	0.015	0.250	0.250
$(100\, X_{ki}+1/2)^{1/2}$	6.877	1.414	1.414	5.050	5.050

3. Computation of Genotypic Worth (W_i)

The genotypic worth of a breeding line (Subedi, 1989) is computed as sum of the products between the mean of standardized partial regression coefficient (SPRC's) of each of the parental lines of ith breeding line and the respective genic proportions of the parental lines in that breeding line. In other words,

$$W_i = \sum_{k=1}^{q} X_{ki}\,\bar{B}_{k^\wedge.}$$

where $B_{k^\wedge.} = 1/p^\wedge \sum_{j=1}^{p^\wedge} B_{kj}\, S_{xk} / S_{yj}$

=   mean of the significant SPRC's of kth parental line from $p^\wedge$ active sites. S_{xk} refers to standard deviation of genic proportions of kth parental line and, S_{yj} refers to the standard deviation of yields of breeding lines at jth location.

Subedi (1989) obtained a high and significant relationship between genotypic worth and mean yields of rice breeding lines.

B. PROCEDURE OF SELECTING POSITIVE OR SUPER POSITIVE PARENTAL LINES

The kth parental line that has positive mean of the standardized partial regression coefficients between its genic proportions and yield of breeding lines over environment(s) is considered a positive parental line. There may be several such parental lines. A parental line having consistently positive SPRC values over environments is defined as super positive parental line. A breeding line constructed by incorporating super positive parental lines is expected to be more stable and higher yielding than that constructed by positive parental lines. When there are more than one positive or super positive parental lines, an optimum breeding line is thought to be developed by intercrossing those parental lines in proportions to their mean SPRC values.

348

C. VALIDATING THE VAPGeP MODEL

To validate the model, two methods are adopted: a) analysis of variance method and b) yield prediction method.

a). Analysis of Variance Method.

In this method, actual GEI effects obtained by testing 18 breeding lines at 21 locations were separately regressed on W_{ij} values at 21 locations each. Predictability of GEI at each location is given in Table 1.

The predictability ranged from over 5 percent to less than 80 percent. The overall predictability was 38 percent and was significant at 1% level of significance. Moreover, the model predicted 73 percent variation in genotype and 99 percent variation in environment. The overall predictability of components of phenotypic variance was 90 percent (Subedi, 1992).

Furthermore, the variations in the mean genic proportions of active parental lines ($\overline{A}._j$) and that in number of active parental lines (Q) jointly explained 47% variation in predictability of GEI. Variations in number of active parental lines (Q_j) explained more variations (r^2=40%) in predictability than that by mean genic proportions (r^2=28%) of active parental lines.

Table 1

Predictability of G x E interaction as explained by mean proportions of active parental lines ($\overline{A}._j$) and no. of active parental lines (Q_j)

Location	$\overline{A}._j$	Q_j	Predictability of GEI (%)
Mandalay	5.07	3	57.8
Bumbong	0.78	1	17.8
Tuaran	3.17	2	37.1
Rangsit	1.83	2	5.5
S. Buri	3.15	3	61.2
L. Xuyen	3.05	3	27.3
Comilla	5.68	2	21.0
Habiganj	3.01	3	37.0
Patna	1.33	1	10.4
Mandya	0.89	1	24.0
A. Samudram	4.72	2	36.5
Navsari	5.54	3	49.6
Sabour	3.20	2	58.1
Aduthurai	5.20	1	38.9
Raipur	1.29	1	13.2
Pusa	1.76	2	29.7
Birganj	3.56	4	73.0
Dokri	0.95	1	28.2
Badeggi	4.79	2	79.0
Ibadan	6.04	2	33.3
Yagoua	1.72	2	11.3
Mean			38.0

2). Yield Prediction Method.

Using the model I and II, the yields of X.3-D.T. breeding line which was tested in IRYN-M, 1984 but was not included in the computation of parameters of the model, were predicted at 21 test locations, the results of which are presented in Table 2. The models have predictability of 82% and 91%, respectively.

Table 2

The actual and predicted yields of X.3-D.T. breeding line over 21 test locations included in the study.

Location	Environmental value				Actual yield
	I_j	$\overline{A}_{.j}a_j$	$(Y^{(1)}_{ij})$ (Model I)	$(Y^{(2)}_{ij})$ (Model II)	$\overline{Y}_{ij}.$
Mandalay	6.79	0.772	5.20	7.25	8.00
Bumbong	6.11	-1.214	4.51	5.02	5.20
Tuaran	4.90	-0.198	4.25	4.81	4.50
Rangsit	3.00	-0.325	2.27	2.75	2.60
Suphan Buri	4.87	0.308	5.01	5.20	4.90
Long Xuyen	3.58	-0.705	2.26	3.26	3.40
Comilla	6.34	-0.615	4.52	5.82	5.70
Habiganj	8.11	-1.828	6.14	6.88	6.30
Patna	3.00	1.119	3.97	4.56	3.90
Mandya	5.08	0.555	5.01	5.52	5.50
A.samudram	1.49	1.809	1.54	2.02	1.70
Navsari	5.68	-2.898	2.21	2.86	2.10
Sabour	3.19	-0.269	2.40	3.07	2.70
Aduthurai	7.96	-3.798	2.34	2.46	3.20
Raipur	3.14	0.318	2.85	3.16	2.40
Pusa	1.62	0.419	1.50	1.96	2.30
Birganj	6.88	-2.149	4.21	5.18	5.30
Dokri	1.14	0.669	1.14	1.63	0.90
Badeggi	3.20	1.499	4.20	4.66	5.10
Ibadan	2.76	0.875	1.26	4.00	3.10
Yagoua	3.50	1.216	4.05	4.50	5.10
Mean	4.397	-0.214	3.40	4.10	4.00

An example is given below for a particular location:

Breeding line: X.3 - D.T. location : Mandalay

Parental line:	Peta	DGWG	Taducan	TYC
Proportions:	0.5625	0.1562	0.2500	0.0312

Active Parental lines at Mandalay: TYC   Sigadis   NMS19
Multiple partial regression :   -1.32   0.916   0.184
coefficients(B_{kj})

$$W_{ij} = \sum_{k=1}^{Q_j} B_{kj} X'_{ki} = -1.32 (3.12+0.5)^{1/2} + 0.916(0.5)^{1/2} + 0.184(0.5)^{1/2}$$
$$= -1.734$$

$a_i = -0.2206;\ \bar{a}. = -0.071,\ I_j = 6.79,\ a_j = 0.14;\ \bar{A}._j = 5.157$
$B_g = 3.422,\ B_{ge,j} = 0.732$

$$Y^{(1)}_{ij} = I_j + \bar{A}._j\, a_j + B_g\,(a_i - \bar{a}.) + B_{ge,j}(W_{ij} - \bar{W}._j)$$
$$= 6.79 + 0.722 - 0.512 - 1.798 = 5.20$$

$$Y^{(2)}_{ij} = I_j + A_{ij}\, a_j$$
$$= 6.79 + \{(3.12 + .5)^{1/2} + 0.5^{1/2} + 0.5^{1/2}\}0.14 \qquad = 7.25$$

D. IMPLICATIONS OF THE VAPGeP MODEL

1. GEI is a function of variable numbers and proportions of active parental lines
2. Multilocation testing of some breeding lines can generate information to predict the yields of large number of untested but parentally related breeding lines (Subedi, 1990; Subedi, 1992). Also see Table 2.
3. There exists a significant and positive relationship between genotypic worth of breeding lines and their multilocation mean yields (Subedi, 1989).
4. Optimum breeding line, which is more stable and higher yielding than the existing ones can be constructed with few selected parental lines with their predetermined genic proportions to be recombined (Subedi, 1989).
5. The method developed to assess and enhance the genetic architecture of crop species will be more useful if mitotic reduction and genetic engineering is simultaneously used. An example with such a scheme for integration of 8 parental lines into a breeding line is given below. This is to develop a multiparental and transgenic rice genotype.

   Eight parental rice lines: TKM6, Ptb21, Ptb18, Sigadis, Tadukan, Tetep, NMS19 and BG1 were identified for stability and higher yield (Subedi 1989, Subedi 1990). The conventional breeding method takes 10-15 years before a variety can be developed from these parental lines. Anther culture technique can reduce the breeding period by early fixation of genome. However, the period is lengthened if the number of parental lines incorporated in a breeding line is increased. Fusion of protoplasts from haploid parental lines can reduce the breeding cycle given that, a). hybrid protoplast can be selected in cell culture itself, b). somatic reduction can be induced in tissue/cells obtained from hybrid seedling regenerated from protoplast fusion. Sodium nucleate and p-fluorophenylalanine (Day and Jones,1971; Huskin,1948) are some of the chemicals that are known to induce somatic reduction in fungi and plants. Incorporation of selection markers may be necessary for efficient selection of hybrid protoplasts.

ACKNOWLEDGMENTS

The author expresses his deep sense of gratitude to Dr. A. L. Carpena, Professor, University of the Philippines at Los Banos, and to Dr. D. V. Seshu, International Rice Research Institute (IRRI), Philippines, in connection with this paper.

REFERENCES

Day, A.W., and J.K. Jones. 1971. p-Fluorophenylalanine-induced mitotic haploidization in Ustilago voilacea. *Genet. Res.* 18:299-309.

Delannax, X., D.M. Rotgers, and R.G. Palmer. 1983. Relative genetic contributions among ancestors lines to N. American soybean cultivars. *Crop Science* 23: 944-949.

Huskin, G.L. 1948. Segregation and reduction in somatic tissues. 1. Initial observations on Allium cepa. *Journal of Heredity* 39:311-325.

IRRI. 1984. *Final report of the eleventh international rice yield nursery, medium 1983.* IRRI, Los Banos, Philippines.

IRRI. 1985. *Final report of the twelveth international rice yield nursery, medium 1984.* IRRI, Los Banos, Philippines.

Subedi, L.P. 1989. A method of parental selection along with their genic proportions for the construction of a breeding line. p. 665-668. *In* S. Iyama and G. Takeda (eds.) Proceedings of 6th International Congress of SABRAO, Tsukuba, Japan. 21 25 Aug. 1989. Organizing Committee, The 6th International Congress of SABRAO.

Subedi, L.P. 1990. A new strategy for multilocation testing and breeding optimum variety of medium growth duration in rice (Q. sativa L.) for irrigated lowland global environment. A paper presented at International Symposium on Rice Research - New Frontiers, Hyderabad, India. 15-18 Nov. 1990.

Subedi, L.P. 1992. A variable active parental genic proportions (VAPGeP) model, I. experimental verification. p. 141-146. *In* S. C. Huang et al. (eds.) Proceedings of SABRAO International Symposium on 'Impact of Biological Research on Agricultural Productivity', Taichung, Taiwan, 10-13 March, 1992. Taichung District Agricultural Improvement Station.

Subedi, L.P., and A.L. Carpena. 1988. Characterization and prediction of genotype by environmental (G x E) interaction for populations of varying genetic structures. p. 371-380. *In* Proceedings of National Conference on Science & Technology, 24-29 April, 1988, Kathmandu, Nepal. Royal Nepal Academy, Kathmandu, Nepal.

St. Martin, S. K. 1982. Effective population size for the soybean improvement program in maturity group OO to IV. *Crop Science* 22: 151-152.

Chapter 13

SELECTION OF GENOTYPES AND PREDICTION OF PERFORMANCE BY ANALYSING GENOTYPE-BY-ENVIRONMENT INTERACTIONS

W. E. Weber,[1] G. Wricke,[2] and T. Westermann [1]

I. INTRODUCTION

Genotype-by-environment interaction is a wide-spread phenomenon. It indicates that genotypes react in different ways if environmental conditions change and is of significance for breeders, official test stations and growers.

The breeder is interested to know in which way the interaction can be incorporated into a selection programme. First, parts of the interaction can serve to describe genotypes. For example, a genotype with high ecovalence (low W_i value) may be considered as stable. The slope of the regression line can be used to characterize the response of the genotype to good and to less favourable conditions in comparison with other genotypes. If the interaction is very large, the development of local varieties for specific conditions may be indicated. More often the breeder is looking for stable varieties which behave well under all conditions. Several concepts exist for stability. Genotypes may be considered as stable if they yield the same under all conditions, including unfavourable ones. Another concept is to regard those genotypes which respond like the average genotypes as stable, that is, they yield more under better conditions and vice versa. This concept seems to be more realistic.

In official trials, locations must be chosen in such a way that they represent the region for which the trial is conducted. No genotype should ever be preferred or handicaped in a systematic manner. Interactions also occur with years, but years cannot be selected. There is always a risk that a series of years is not representative (Schmalz, 1992). Large interactions with years restrict the possibility of making good selection decisions, in which case trials over many years are necessary.

Finally, the grower wants to know which data are relevant for his specific location. Usually he cannot conduct variety trials on a large scale in his own fields, but he can use the information from official trials. He is then interested to know whether prediction for his sites can be improved by using information from other sites.

Several methods have been proposed in the literature to analyse genotype-by-environment interactions and some of them have been severely

[1] Institut für Pflanzenzüchtung und Pflanzenschutz, Martin-Luther Universität Halle-Wittenberg, Berliner Str. 2, D-06188 Hohenthurm, Germany.
[2] Institut für Angewandte Genetik, Herrenhäuser Str. 2, D-30419 Hannover, Germany.

criticized by other authors. For some methods it is difficult to see how they could be applied. Often only biometrical problems are treated, especially with tests in a series of experiments. Some measures are completely correlated and differ only with respect to the expected values. Examples are the ecovalence (Wricke, 1962) and the stability variance (Shukla, 1972) or the regression coefficients of Finlay and Wilkinson (1963) or Tai (1971). Other methods are highly correlated, such as the ecovalence and the sum of squares of deviations from the regression by Eberhart and Russell (1966), as was found in many experiments (Kang et al., 1987, Pham and Kang, 1988, Weber and Wricke, 1990, Piepho and Lotito, 1992). In this paper, only those approaches which could help to improve the selection of good genotypes and the prediction of their value are discussed.

II. SELECTION OF GENOTYPES

In a series of experiments the genotype-by-environment interaction serves as the error (see for example Wricke and Weber, 1986, page 45). The environments included in the series are a sample of all possible environments. The conclusions are only valid if the environmental conditions in future trials are comparable. Therefore differences between genotypes must be measured by the interaction.

Several replications are usually conducted in each environment. Replications serve to estimate the error variance within environments, reduce bias, and improve accuracy. The error variance is expected to be smaller than the interaction variance and cannot be used for the test of genotypic differences in a series. In the following, means from replications are used as basic values. The linear model is:

$$y_{ij}=\mu+g_i+p_j+e_{ij}$$

(with y_{ij}=observed value
    μ =general mean
    g_i=effect for genotype i, $i=1, n_g$
    p_j=effect for environment j, $j=1, n_p$
    e_{ij}=interaction).

(1)

The interaction is partitioned to obtain specific information on the genotypes in addition to the main effects g_i, which may be useful for selection.

A. VARIANCE OF GENOTYPES

By fusing the main effects of environments and the interaction of equation (1), we get the reduced model

$$y_{ij}=\mu+g_i+f_{ij}$$

(with $f_{ij}=p_j+e_{ij}$).

(2)

A separate error variance can now be calculated as measure of stability for each genotype:

$$s_i^2 = \frac{\sum\limits_{j}(y_{ij}-y_i)^2}{n_p-1} - \frac{\sum\limits_{j}f_{ij}^2}{n_p-1} \qquad (3)$$

The variance contains interactions and the main effects of environments and was proposed by Römer as early as 1917. A stable genotype has a small variance. Ideally a genotype yields the same in all environments. This measure alone does not allow any conclusion with regard to the main effect g_i.

B. ECOVALENCE

The term "ecovalence" was introduced by Wricke (1964). It includes only interaction effects. This way of partitioning the interaction sum of squares was proposed independently by Wricke (1960, 1962) and Calinsky (1961). Wricke proposed the ecovalence as a measure of the phenotypic stability of genotypes. It is calculated as follows:

$$W_i = \sum_j (y_{ij}-y_i - y_j + y_{..})^2 = \sum_j e_{ij}^2 \qquad (4)$$

Small values indicate a high ecovalence. The analysis can be compared with an analysis of residuals in a complete block design. Genotypes with, on average, small residuals are to be preferred since they show less unpredictable variability.

As explained at the beginning of this section, the interaction serves as an error term for the F-test in the analysis of variance and it is assumed that the interaction variance is homogeneous for all genotypes. Differences in W_i indicate that this is not true. The problems resulting from this inhomogeneity with regard to the usual F-test are discussed in Cochran and Cox (1957).

Shukla (1972) proposed a measure which he called "stability variance". The stability variance differs from the ecovalence only in a linear transformation and thus leads to exactly the same genotype order. The same holds for the measures proposed by Plaisted and Peterson (1959) and Plaisted (1960). A comparison was made by Lin et al. (1986), whose paper also includes appropriate formulae. In the literature, these relationships have not always been realized, and ecovalence and stability variance have been presented side by side although the second measure contains no new information. The variance s_i^2 and the ecovalence W_i in equations (3) and (4) can lead to different ranking orders of the genotypes. For the whole sample there is no covariance between the environmental and the interaction effects but this does not hold for single genotypes (Comstock and Moll, 1963). This specific covariance explains why s_i^2

356

and W, may lead to different ranking orders.

The ecovalence strongly depends on the environments included in the test and the breeder can manipulate the ecovalence by choosing specific locations. Lin and Binns (1988) therefore proposed estimating stability from a series of locations and levels of an unpredictable environmental factor such as years. The stability measure for a specific genotype then results from the variation of years within locations and cannot be manipulated by the breeder.

C. REGRESSION COEFFICIENTS

In analysing genotype-by-environment interactions the regression approach is very often used. In this approach the yield of a specific genotype in a given environment is regressed on another measure of the environment. It is assumed that the regression coefficients for the genotypes will differ and give specific characters for each genotype. To calculate the regression coefficients, a parameter for the environment is needed as an environmental index which is independent of the specific experiment. Normally such an independent measure is not available and, therefore, the average of all genotypes is used. With this method the mean of the regression coefficients is exactly one. The model is then:

$$y_{ij} = \mu + g_i + b_i * p_j + e_{ij}$$

(5)

where the regression coefficient b_i is estimated by

$$b_i = \frac{\sum_j (y_{ij} \cdot (y_{.j} y_{..}))}{\sum_j (y_{.j} - y_{..})^2}$$

(6)

The interaction e_{ij} of equation (1) is split into two parts:

$$e_{ij} = (b_i - 1) * p_j + \varepsilon_{ij}$$

(7)

Some authors have used physical characteristics of the environments to avoid an environmental index which is dependent on the trait measured (Gorman et al., 1989; Tai and Young, 1989). The results so far presented do not show much superiority of these independent measures over the environmental mean. This all indicates that characterisation of the environmental production capacity by using special environmental parameters such as temperature, rainfall, altitude or latitude is not sufficient in most cases.

The regression approach was proposed by Yates and Cochran (1938) but Westcott (1986) and Becker and Leon (1988) mention some earlier applications of this approach. It was rediscovered for plant breeding by Finlay and Wilkinson

(1963). Eberhart and Russell (1966) extended the approach and included the deviations from regression to characterize the genotypes. The sum of squares of deviations $\sum \sigma_{ij}^2$ are:

$$\sum_j \sigma_{ij}^2 = \sum_j (y_{ij} - y_i)^2 - b_i^2 \cdot \sum_j (y_j - y_.)^2 \qquad (8)$$

For good genotypes this measure should be small. The following relationship exists between ecovalence, regression, and deviation:

$$W_i = (b_i - 1)^2 \cdot \frac{s.s.environments}{n_g} + \sum \sigma_{ij}^2 \qquad (9)$$

In the literature some other definitions for the regression coefficient and the deviations are proposed which are merely linear transformations. For example, the regression coefficients are diminished by one to get a mean of zero (Perkins and Jinks, 1968; Shukla, 1972) but this has no effect upon the interpretation.

The regression coefficient can be considered as a trait under selection and the question as to which of the regression coefficients is ideal is a matter of controversy in the literature. If $b_i=1$ is ideal, the genotype will respond to environments as does the average. Another approach, with $b_i=0$, considers a genotype as ideal when it shows no correlation with the average of genotypes. However in this case it is possible that there is a large sum of squares of deviations around the regression line so that the yield of this genotype is not constant over all environments. Hanson (1970) considered the genotype with the smallest regression coefficient as being ideal.

Several authors, for example Hanson (1970), Wright (1971), Utz (1972), Freeman (1973), Hill (1975), Becker (1981), Becker and Leon (1988) and Ceccarelli (1989), discuss regression coefficients with respect to agronomic consequences but this point will not be discussed here.

D. GROUPING BY CLUSTERS

Another approach discussed in the literature is cluster analysis. With this method the genotypic population is divided into groups of similar genotypes. Cluster analysis also provides for the analysis of more than one character. Many measures of similarity and distances exist (see the review by Lin et al., 1986 and textbooks on cluster analysis). The distance from an ideal type has already been discussed by Hanson (1970).

If distinct classes of genotypes can be distinguished, such classification of genotypes by specific types of reaction to environmental conditions would be useful in plant breeding. This would perhaps allow one to classify new genotypes because it would be possible to compare them with known varieties for which enough experimental data are available and which represent different "prototypes".

E. MEASURES BASED ON RANK INFORMATION

The interaction is very important for the selection of genotypes if it is connected with a rank order change. Therefore Hühn (1979, 1990a and 1990b) solely developed criteria which are proposed to use rank order changes as a criterion for selecting the best genotype. He was able to show that they measure something different from the ecovalence or the regression coefficients. One of these measures is:

$$S_i^{(1)} = \frac{2 \sum_j \sum_{j > j'} |r_{ij} - r_{ij'}|}{(n_g * (n_g - 1))}$$

(10)

with r_{ij} as rank of genotype i in environment j. If this measure is zero, a genotype has the same rank, for example the third position, in each environment. The value of such measures also depends on the repeatability over a series of experiments.

F. TESTS

It is possible to test the parameters estimated from partitioning the interaction. In the literature, such tests are given frequently and they show that significant differences between genotypes are common, for example in the ecovalence, the regression coefficients or the sum of squares of deviations. Such tests have only limited impact from a breeding point of view since nothing can be said about whether the results will be similar in future experiments.

Several authors have investigated genotypes with different population structures, for example inbred lines versus hybrids, single cross versus three-way versus double cross hybrids and pure stands versus mixtures (for a review see Becker and Leon, 1988). There is some indication that the genotype-environment mean square is larger for inbreds than for the corresponding single crosses and that mixtures are more stable than pure stands.

G. EXPERIMENTAL RESULTS

Two points will be discussed in greater detail in this paper: the relation between different parameters and their repeatability.

1. Relation Between Different Stability Parameters and the Mean

Since only one series of experiments is necessary, the relation between different parameters has been reported in several papers and Table 1 provides a summary. With one exception, Heine and Weber (1982), the table shows the Spearman rank correlation coefficient. In some cases the authors analysed several experiments; in this case only averages are given. The number of genotypes and environments varies greatly from experiment to experiment.

As mentioned in the introduction, several parameters are completely correlated. Therefore only the mean, the ecovalence and the regression coefficient were examined. Piepho and Lotito (1992) investigated five crops (faba beans, fodder beet, sugar beet, oats and winter rape) and found, in all cases, that the stability parameters cluster into two groups. Within each group there is a high correlation. One group consists of the regression coefficient b and the variance, the other contains the ecovalence, the sum of squares of deviations from the regression and other parameters including the rank parameter $S_i^{(1)}$.

The average correlation between the mean and the regression coefficient is positive, which means that a higher mean is more often connected with a steeper regression. This is because, in most cases, genotypes with little reaction to environment cannot make use of the better conditions associated with more favourable environments.

There is no consistent correlation between the two groups of stability parameters represented by the ecovalence and the regression in Table 1.

2. Repeatability

The breeder is interested in parameters which can be reproduced in further experiments. Even the estimates from each single year must be reliable and should not depend too much on the population of the genotypes tested. There are only a few reports about investigations covering this topic. Experiments have been conducted over many years but, since the composition of the genotypes changes rapidly, only a few are common to a number of years, as for example the standard varieties in official trials.

Becker (1987) estimated the heritability of stability parameters for yield which was generally much lower than for the mean. These findings accord with other authors' experimental results. Table 2 contains some results from the literature. Since the ecovalence W_i is a sum of squares, the square root transformation has been used for this measure to improve the distribution for the analysis of variance. Since the degrees of freedom vary from series to series, the probability of the F-value is given. A value <0.05 indicates significant differences between genotypes.

Data from official German trials for winter wheat and maize were evaluated by Heine and Weber (1982). For winter wheat, 24 series at 6 to 16 sites were evaluated between 1969 and 1978. Three standard varieties were tested in all series so that for each of these varieties 24 estimates for the mean, the ecovalence and the regression coefficient are available. For maize two groups of genotypes were analysed over 12 and 14 series at 4 to 10 sites with two standard varieties in each. The results of an analysis of variance for the mean, ecovalence and regression are given in Table 2. A square root transformation was made prior to the analysis of ecovalence.

Table 1
Rank correlation coefficients (%) between the mean (y), ecovalence (W), and regression (b) for different crops

Crop	Rank correlation (%)			Reference
	y,W	y,b	W,b	
winter wheat	-	25	-	Campbell and Lafener, 1977
	-17	23	-6	Heine and Weber, 1982
	-	30	-	Becker, 1983
	-13	16	-7	Leon, 1985
maize	-	13	-	Becker 1981, 1983
	-14	11	9	Heine and Weber, 1982
	-6	22	9	Leon, 1985
oats	-13	-	-	Wricke,1965
	-	84	-	Qualset and Granger, 1970
	-25	63	-16	Langer et al., 1979
	-	28	-	Becker 1981, 1983
sorghum	1	62	16	Jowett, 1972
	-	54	-	Francis et al., 1984
potatoes	41	-	-	Wricke,1965
	-	4	-	Tai, 1979
	-	-	54	Wricke and Weber, 1980
rape	1	14	2	Hühn and Leon, 1985
	14	38	-9	Leon,1985
sugar beet	-	-	5	Piepho and Lotito, 1992
fescue	-38	27	8	Ngyen et al., 1980
orchard grass	4	49	-9	Gray, 1982
red canary grass	-	56	-	Casler and Hovin, 1984
soybeans	-	47	-	Hanson, 1970
faba beans	76	76	60	Dantuma et al., 1983
sugar cane	-7	-	-	Kang and Miller, 1984
peanut	-	22	-	Schilling et al., 1983

The table indicates that only the differences between variety means are significant and not those between stability measures. The results from other authors do not establish significant differences for stability parameters either (except the ecovalence for potato yield in the study of Wricke, 1965), while several cases with significant differences between varieties have been found for the mean. The series had not been set up for such studies, and only some varieties were tested in all the experiments. Therefore there is a variation between experiments with regard to the regression coefficient. This would not be so if all varieties were tested in all experiments.

Table 2
Analysis of variance for a series of trials for mean (y), ecovalence (square root transformed values, $\sqrt{W}$), and regression (b), measured by the probability of F=ms varieties/ms error

Crop	d.f. varie- ties	d.f. error	Probability y	$\sqrt{W}$	b	Reference
winter wheat	2	46	0.000	0.177	0.558	Heine and Weber, 1982
	1	21	0.000	-	0.854	Becker, 1983
maize (early)	1	22	0.067	0.843	0.051	Heine and Weber, 1982
maize (late)	1	22	0.000	0.061	0.840	Heine and Weber, 1982
rape	11	17	0.001	0.470	0.764	Hühn and Leon, 1985
oats	10	10	0.003	0.500	0.042	Langer et al., 1979
orchard grass	8	8	0.212	0.687	0.820	Gray, 1982
potatoes	13	65	-	0.000	-	Wricke, 1965

The results in Table 2 show that the repeatability of stability parameters is low compared with the mean in wheat and maize. The other results are only first indications, there not being enough data available to allow final conclusions. Taking into account the fact that even results from a single year should be reliable for use in selection, the results are not very encouraging.

III. PREDICTION FOR A SET OF ENVIRONMENTS

Breeders know from experience that for specific selection purposes differences exist between locations. In early generations, locations are preferred which allow one to select for disease resistance or tolerance to abiotic stress conditions. Later on, results must be comparable with those of the official trials in which case a location is good if the following requirements are fulfilled:

a) the rank order of genotypes in this environment corresponds to the average rank order and
b) the environment can differentiate the genotypes well.

The regression approach would therefore appear to be useful. The results of a specific environment are regressed on the average over all environments. The best environment has a large gradient with small deviations. The model is:

$$y_{ij} = \mu + p_j + c_j * g_i + e_{ij}$$

(11)

where for each location a regression coefficient c_j is estimated:

$$c_i = \frac{\sum\limits_{i} (y_{ij} \cdot (y_{i.} - y_{..}))}{\sum\limits_{i} (y_{i.} - y_{..})^2}$$

(12)

Similar to regression approach for genotypes the interaction of equation (1) is split:

$$e_{ij} = (c_j - 1) * g_i + \varepsilon_{ij}$$

The coefficients c_j can be used as weights in a selection index for genotypes. The components of the index are the yields at the different locations. Two cases must be considered. First, the regression coefficient can be considered as a parameter of a specific location, truly independent of years. The breeder can then use this information a priori and can select locations with a high regression coefficient for his breeding programme. An argument against this use is the dependence of the regression coefficient upon the set of genotypes. However, this argument is not too strong if the sample of genotypes is large enough.

Second, the regression coefficients can be used a posteriori. Then the locations are weighted with the regression coefficients obtained in the specific series, taking into account that the coefficients may change from year to year. Such a procedure has been proposed by Rundfeldt (1984).

A. PRINCIPAL COMPONENTS

The regression approach uses a specific partitioning of the interaction but a more general one is partitioning by principal components. This approach goes back to Williams (1952) and has been described extensively by Mandel (1969). The model has become popular under the name AMMI by Gauch (1988).

The estimation procedure is as follows. The interaction components e_{ij} can be arranged in a matrix X with n_g rows and n_p columns. Eigenvalues λ and standardized eigenvectors u and v are calculated from XX' (λ and u) and $X'X$ (λ and v), the K nonzero eigenvalues $\lambda > 0$ being the same for both matrices. Then the e_{ij} can be partitioned as follows:

$$e_{ij} = \sum_{k=1}^{K} (\sqrt{\lambda_k} * u_{ik} * v_{jk})$$

(13)

with $K \leq \min(n_g - 1, n_p - 1)$. Not all K principal components are chosen, but only $K^* < K$. Then Gauch (1988) calls

$$e_{ij}^* = \sum_{k=1}^{K^*} (\sqrt{\lambda_k} u_{ik} v_{jk}) \tag{14}$$

"pattern" and

$$e_{ij} - e_{ij}^* = \sum_{k-K^*+1}^{K} (\sqrt{\lambda_k} u_{ik} v_{jk}) \tag{15}$$

"noise".
If this is true,

$$y_{ij}^* = \mu + g_i + p_j + e_{ij}^* \tag{16}$$

would be a better estimator than y_{ij} itself. Therefore an index for genotype i based on K^* principal components would be:

$$y_{i.}^* = \frac{\sum_j y_{ij}^*}{n_p} \tag{17}$$

giving location weights depending on the K^* principal components. An extreme case is $K^*=1$. Then $[(\sqrt{\lambda_1})*v_{j1}]*u_{i1}$ replaces $(c_j-1)*g_i$ in the regression approach. From this comparison one can see that the first principal component can give a better fit than the regression approach, since u_{i1} and $[(\sqrt{\lambda_1})*v_{j1}]$ are estimated in addition to g_i and p_j, while in the regression approach only c_j is estimated additionally. Often a better fit is claimed as being a better approach.

The principal components are often used for graphical presentation. The eigenvectors u_{i1} and v_{j1}, belonging to the first principal component, are given on the x-axis. On the y-axis either the means for genotypes and locations or the eigenvectors for the second principal component can be given. Such graphical presentations are used to find clusters of similar genotypes or locations. Connections between mean performance and interactions can also be seen.

B. CLUSTER ANALYSIS

The aim of the cluster analysis is to find groups with similar behaviour so that within a group almost no interaction is found. If such a classification is stable over the years, the breeder can reduce the number of locations in the future, since he needs only one location from each group. Grouping is done with statistical methods, but it may be possible that underlying physical or chemical factors can be identified. In suger beet, preliminary studies have been conducted by Müller et al. (1994) and, even in one year, the same locations formed different clusters when other genotypes were used. Similar problems have been reported by Hamblin et al. (1980).

C. EXPERIMENTAL RESULTS

There are only a few investigations of these problems. The data from official German trials, which has already been described, was analysed with regard to the evaluation of locations by Weber and Vanselow (1985). The criterion used was the regression coefficient. A large coefficient in combination with small deviations from regression indicates a special capacity of the place to differentiate between genotypes. A similar investigation had been made by Utz (1973) for winter wheat. The results of these papers are summarized in Table 3. The table also includes data given by Brown et al. (1983), who investigated numerous places in Asia, Africa and South America to collect data on heading date in rice. Numerous experiments in Germany have been analysed by Giebel (1988).

The results show that, in several cases, differences in the regression coefficients for places could be partially reproduced. But there are only a few places with constant high or low regression coefficients over all years. Weber and Vanselow (1985) also found that there is a weak positive correlation between the regression coefficient and the sum of squares of deviation.

IV. PREDICTION FOR SPECIFIC LOCATIONS

So far, we have been considering prediction for regions and the predictive value of a single location has been discussed in this sense. Now prediction for the specific location j is to be discussed. Results from other locations may be included in such an index to improve the prediction but results from independent experimental trials are necessary to check its value. Several types of linear indices I_{ij} are considered as predictors for genotype i and environment j. As a control the observation y_{ij} itself is used as the index.

A. MAIN EFFECTS ONLY

The index for genotype i and location j does not include any interactions:

$$I_{ij} = \mu + g_i + p$$

(18)

Predictors for different locations differ only by a constant and are therefore completely correlated.

Table 3
Analysis of variance for regression coefficients for places as measured by the probability of F=ms places/ms error

Crop	d.f. places	d.f. error	Prob.	Reference
winter wheat	14	114	0.000	Utz, 1973
	18	176	0.000	Weber and Vanselow, 1985
	11	114	0.020	Giebel, 1988
	21	71	0.874	Giebel, 1988
	14	87	0.027	Giebel, 1988
winter barley	17	133	0.065	Giebel, 1988
spring barley	15	103	0.534	Giebel, 1988
oats	15	120	0.078	Giebel, 1988
rape	12	61	0.496	Giebel, 1988
rice	16	19	0.040	Brown et al., 1983
maize (early)	15	84	0.000	Weber and Vanselow, 1985
	5	25	0.001	Giebel, 1988
	11	63	0.043	Giebel, 1988
	6	27	0.000	Giebel, 1988
	11	57	0.002	Giebel, 1988
maize (late)	13	95	0.815	Weber and Vanselow, 1985
	5	27	0.788	Giebel, 1988
	10	53	0.193	Giebel, 1988
	6	28	0.000	Giebel, 1988
	11	62	0.000	Giebel, 1988

B. REGRESSION

The index uses only the part due to regression:

$$I_{ij}=\mu+p_j+c_j{*}g_i$$

(19)

with c_j as regression coefficient, as explained in equation (12). In this case it is assumed that the deviation from regression is "noise".

C. PRINCIPAL COMPONENTS

The first K^* principal components are included in the index, K^* being between 1 and $K=min(n_g-1, n_p-1)$. The index is:

$$I_{ij}=\mu+p_j+g_j\sum_{k=1}^{K^*}(\sqrt{\lambda_k}{*}u_{ik}{*}v_{jk})$$

(20)

With this approach the remaining $K-K^*$ components are considered as "noise".

D. PARTIAL REGRESSION COEFFICIENTS

The locations are more or less correlated with each other. While in the regression approach (equation 19) the regression of location j on the mean over all locations is used; now partial regression coefficients of location j on location j' are computed. The index is:

$$I_{ij} = \frac{\sum\limits_{j'=1}^{n_p} (b_{jj'} \cdot y_{ij'})}{\sum\limits_{j'=1}^{n_p} b_{jj'}} \qquad (21)$$

with weights $b_{jj'}$ as regression coefficients of location j on location j'. The regression coefficient is:

$$b_{jj'} = \frac{\sum (y_{ij} - y_j) \cdot (y_{ij'} - y_{j'})}{\sum (y_{ij'} - y_{j'})^2} \qquad (22)$$

More closely related locations get more weight. Instead of all locations only those with the highest weights could be used (Weber and Westermann, 1994).

E. EXPERIMENTAL RESULTS

The data used are yield data for winter wheat from official trials of the Bundessortenamt in Germany: most genotypes were experimental lines in the testing phase. Each trial also contained standard varieties. The data used were means from randomized block designs with 4 replications. From each pair of years only the genotypes and locations common to both years could be used since one set served for estimation and the other for validation.

In all, 5 pairs of years could be used; the number of genotypes and locations being given in Table 4. The experimental error variance was taken from all genotypes included in the official trial in the corresponding years.

For comparison this variance has been divided by the number of replications. In all cases the variance component for the second order interaction was largest (see Table 4). This component cannot be estimated if single years are evaluated inflating the variance component between genotypes and locations (see Wricke and Weber, 1986).

Table 4
Variance components for yield (dt/ha) of winter wheat trials. s^2_g, s^2_{gl}, s^2_{gy}, s^2_{gly} and s^2 = variance components for genotypes, genotypes by locations, genotypes by years, second order interaction and experimental error, divided by the number of replications

Year	n_g	n_l	s^2_g	s^2_{gl}	s^2_{gy}	s^2_{gly}	s^2
1979	21	11	7.9	16.1	-	-	3.3
1980	21	11	10.0	17.0	-	-	3.4
combined	21	11	5.5	-1.1	3.5	17.6	-
1980	27	11	9.9	14.2	-	-	2.5
1981	27	11	12.6	21.4	-	-	3.2
combined	27	11	10.5	1.7	0.8	16.1	-
1981	32	7	19.8	17.5	-	-	3.7
1982	32	7	8.5	23.4	-	-	3.7
combined	32	7	12.6	3.1	1.6	17.3	-
1982	29	7	6.3	11.7	-	-	2.7
1983	29	7	27.8	15.7	-	-	4.3
combined	29	7	7.5	3.6	9.5	10.0	-
1983	31	10	25.3	25.3	-	-	5.4
1984	31	10	40.8	27.9	-	-	2.9
combined	31	10	27.6	11.3	5.4	15.3	-

Predictions based on the different types of indices have been correlated with the observed yields in the other year of the corresponding pair. The results are summarized in Table 5. For principal components only the first was included, since further components did not improve the prediction. This is in agreement with results of other studies by Crossa et al. (1990, 1991) and Nachit et al. (1992), regardless of the crop (maize, wheat, durum wheat).

These results show that the control is not the best predictor and that the best predictor makes no use of any information from interactions because of the large second order interaction. The results may be different for trials in which the locations used cover a more heterogeneous region than found in Germany.

V. CONCLUSIONS

The experimental results show that genotype-by-environment interaction is important. Therefore large series of experiments are necessary to get good estimates for genotypic means. As has been stated by many research workers, stability parameters can be classified into two main groups, represented in this paper by the regression coefficient and the ecovalence. Stability parameters as an additional source of information to characterize genotypes suffer from the low

368

repeatibility. This holds for both classes. Two reasons can explain this fact. First, the measures are second order statistics with larger statistic errors. Second, interactions with years disturb estimates of a series over one year only. As indicated in Table 4, the second order interaction is very important in Germany so that single year results always must be regarded with caution.

Table 5
Mean correlation coefficients between prediction and the independent year for different types of indices compared with the use of the observed value at the single location as control

Year	Control	Main effects only	Re-gression	First principal component	Partial re-gression
1979	0.219	0.346	0.323	0.360	0.336
1980	0.219	0.365	0.351	0.320	0.338
1980	0.427	0.581	0.549	0.572	0.567
1981	0.427	0.572	0.562	0.539	0.564
1981	0.453	0.509	0.469	0.475	0.500
1982	0.453	0.632	0.545	0.561	0.540
1982	0.386	0.451	0.446	0.416	0.418
1983	0.386	0.343	0.339	0.330	0.362
1983	0.641	0.677	0.674	0.669	0.704
1984	0.641	0.628	0.616	0.595	0.629
Mean	0.425	0.510	0.487	0.484	0.496

Breeders know which locations allow selection for specific purposes. In this paper only yield is considered. Especially for wheat and maize differences between regression coefficients were found, which were significant over years. Large regression coefficients are of specific interest, since they differentiate genotypes better than the average. Nevertheless, no improvement could be reached by using the regression coefficients as weights for prediction.

The farmer wants to know how different genotypes react at his location. As has been shown, prediction can be improved by using results from other locations as additional information, but the best predictor was the unweighted average over all locations. This again indicates that it is not possible to estimate specific weights for different locations useful for prediction, since estimation of such weights is disturbed by the second order interaction, which cannot be taken into account.

REFERENCES

Becker, H.C. 1981. Correlations among some statistical measures of phenotypic stability. *Euphytica* 30:835-840.

Becker, H.C. 1983. Züchterische Möglichkeiten zur Verbes-serung der Ertragssicherheit. *Vortr. Pflanzenzüchtg.* 3:203-225.

Becker, H.C. 1987. Zur Heritabilität statistischer Maßzahlen für die Ertragssicherheit. *Vortr. Pflanzenzüchtg.* 12:134-144.

Becker, H.C., and J. Leon. 1988. Stability analysis in plant breeding. *Plant Breeding* 101:1-23.

Brown, K.D., M.E. Sorrells, and W. R. Coffman. 1983. A method for classification and evaluation of testing environments. *Crop Science* 23:889-893.

Calinski, T. 1961. On the application of analysis of variance to the results of series of varietal experiments. Thesis Poznan Agric. College.

Campbell, L.G., and H.N. Lafever. 1977. Cultivar x environment interaction in soft red winter wheat yield tests. *Crop Science* 17:604-608.

Casler, M.D., and A.W. Hovin. 1984. Genotype x environment interaction for red canarygrass forage yield, *Crop Science* 24:633-636.

Ceccarelli, S. 1989. Wide adaption: How wide? *Euphytica* 40:197-205.

Cochran, W.G., and G. M. Cox. 1957. Experimental Designs, 2nd ed. John Wiley, New York.

Comstock, R.E., and R.H. Moll. 1963. Genotype-environment interaction. p.164-196. In W.D. Hanson and H. F. Robinson (eds.) Statistical Genetics and Plant Breeding. NAS NRC Pub. 982.

Crossa, J., H.G. Gauch, and R.W. Zobel. 1990. Additive main effects and multiplicative interaction of two international maize cultivar trials. *Crop Science* 30:493-500.

Crossa, J., P.N. Fox, W.H. Pfeiffer, S. Rajaram, and H.G. Gauch. 1991. AMMI adjustment for statistical analysis of an international wheat yield trial. *Theoretical and Applied Genetics* 81:27-37.

Dantuma, G., E. von Kittlitz, M. Frauen, and D.A. Bond. 1983. Yield, yield stability and measurements of morphological and phenological characters of faba beans (Vicia faba L.) varieties grown in a wide range of environments in Western Europe. *Z. Pflanzenzüchtg.* 90:85-105.

Eberhart, S.A., and W.A. Russell. 1966. Stability parameters for comparing varieties. *Crop Science* 6:36-40.

Finlay, K.W., and G.N. Wilkinson. 1963. The analysis of adaptation in a plant-breeding programme. *Australian Journal of Agricultural Research* 14:742-754.

Francis, C.A., M. Saeed, L.A. Nelson, and R. Moomaw. 1984. Yield stability of sorghum hybrids and random-mating populations in early and late planting dates. *Crop Science* 24:1109-1112.

Freeman, G.H. 1973. Statistical methods for the analysis of genotype-environment interactions. *Heredity* 31:339-354.

Gauch, H.G. 1988. Model selection and validation for yield trials with interaction. *Biometrics* 44:705-715.

Giebel, I. 1988. Untersuchungen zur Aussagefähigkeit von spezifischen Ortskenngrößen für die Sortendifferenzierung, Diplomarbeit. Univ. Hannover.

Gorman, D.P., M.S. Kang, and M.R. Milam. 1989. Contribution of weather variables to genotype x environment interaction in grain sorghum. *Plant Breeding* 103:299-303.

Gray, E. 1982. Genotype x environment interactions and stability analysis for forage yield of orchardgrass clones. *Crop Science* 22:19-23.

Hamblin, J., H.M. Fisher, and H.I. Ridings. 1980. The choice of locality for plant breeding when selecting for high yield and general adaptation. *Euphytica* 29:161-168.

Hanson, W.D. 1970. Genotypic stability. *Theoretical and Applied Genetics* 40:226-231.

Heine, H., and W.E. Weber. 1982. Die Aussagekraft statistischer Maßzahlen für die phänotypische Stabilität in amtlichen Sortenprüfungen bei Winterweizen und Körnermais. *Z. Pflanzenzüchtg.* 89:89-99.

Hill, J. 1975. Genotype-environment interaction - a challenge for plant breeding. *Journal of Agricultural Science Cambridge* 85:477-493.

Hühn, M. 1979. Beiträge zur Erfassung der phänotypischen Stabilität, I. Vorschlag einiger auf Ranginformationen beruhenden Stabilitätsparameter. *EDV in Mediz. u. Biol.* 10:112-117.

370

Hühn, M. 1990a. Nonparametric measures of phenotypic stability, Part 1: Theory. *Euphytica* 47:189-194.

Hühn, M. 1990b. Nonparametric measures of phenotypic stability, Part 2: Applications. *Euphytica* 47:195-201.

Hühn, M., and J. Leon. 1985. Genotype x environment inter-actions and phenotypic stability of *Brassica napus*. *Z. Pflanzenzüchtg.* 95:135-146.

Jowett, D. 1972. Yield stability parameters for sorghum in East Africa. *Crop Science* 12:314-317.

Kang, M.S., and J.D. Miller. 1984. Genotype x environment interactions for cane and sugar yield and their implications in sugarcane breeding. *Crop Science* 24:435-440.

Kang, M.S., B. Glaz, and J.D. Miller. 1987. Interrelationships among stabilities of important agronomic traits in sugarcane. *Theoretical and Applied Genetics* 74:310-316.

Langer, I., K.J. Frey, and T. Bailey. 1979. Associations among productivity, production response and stability indexes in oat varieties. *Euphytica* 28:17-24.

Leon, J. 1985. Beiträge zur Erfassung der phänotypischen Stabilität unter besonderer Berücksichtigung unterschied-licher Heterogenitäts- und Heterozygotiegrade sowie einer zusammenfassenden Beurteilung von Ertragshöhe und Ertragssicherheit. Diss. Kiel, Germany.

Lin, C.S., M.R. Binns, and L.P. Lefkovitch. 1986. Stability analysis: Where do we stand? *Crop Science* 26:894-900.

Lin, C.S., and M.R. Binns. 1988. A method of analyzing cultivar x location x year experiments: A new stability parameter. *Theoretical and Applied Genetics* 76:425-430.

Mandel, J. 1969. The partitioning of interaction in analysis of variance. *J. Res. Nat. Bureau of Standards B. Math. Sci.*, 73B:309-328.

Müller, P., W.E. Weber, G. Steinrücken, and G. Diener. 1994. Classification of locations in sugar beet trials. Poster Abstracts, EUCARPIA, 4.-6.7.1994, Wageningen, Netherlands.

Nachit, M.M., G. Nachit, H. Keteta, H.G. Gauch, and R.W. Zobel. 1992. Use of AMMI and linear regression models to analyse genotype-environment interaction in durum wheat. *Theoretical and Applied Genetics* 83:597-601.

Ngyen, H.T., D.A. Sleper, and K.L. Hunt. 1980. Genotype x environment interactions and stability analysis for herbage yield of tall fescue synthetics. *Crop Science* 20:221-224.

Perkins, J.M., and J.L. Jinks. 1968. Environmental and genotype-environmental components of variability. *Heredity* 23:339-356.

Pham, H.N., and M.S. Kang. 1988. Interrelationships among and repeatibility of several stability statistics estimated from international maize trials. *Crop Science* 28:925-928.

Piepho, H.-P., and S. Lotito. 1992. Rank correlation among parametric and nonparametric measures of phenotypic stability. *Euphytica* 64:221-225.

Plaisted, R. L. 1960. A shorter method for evaluation the ability of selections to yield consistently over locations. *American Potato Journal*. 37:166-172.

Plaisted, R.L., and L.C. Peterson. 1959. A technique for evaluating the ability of selections to yield consistently in different locations or seasons. *American Potato Journal* 36:381-385.

Qualset, C.O., and R.M. Granger. 1970. Frequency dependent stability of performance in oats. *Crop Science* 10:386-389.

Römer, Th. 1917. Sind die ertragreichen Sorten ertragssicher? Mitt. DLG 32:87-89.

Rundfeldt, H. 1984. Some remarks on the biometrical evaluation of unorthogonal series of field trials. *Vortr. Pflanzenzüchtg.* 7:259-265.

Schilling, T.T., R.W. Mozingo, J.C. Wynne, and T.G. Isleib. 1983. A comparison of peanut multilines and component lines across environments. *Crop Science* 23:101-105.

Schmalz, H. 1992. Beiträge zur Kenntnis der Variabilität der Witterungserscheinungen. Martin-Luther-Universität, Halle.

Shukla, G.K. 1972. Some statistical aspects of partitioning genotype-environmental components of variability. *Heredity* 29:237-245.

Tai, G.C.C. 1971. Genotypic stability analysis and its application to potato regional trials. *Crop Science* 11:184-190.

Tai, G.C.C. 1979. Analysis of genotype-environment interaction of potato yield. *Crop Science* 19:434-438.

Tai, G.C.C., and D.A. Young. 1989. Performance and prediction of potato genotypes tested in international trials. *Euphytica* 42:275-284.

Utz, H.F. 1972. Die Zerlegung der Genotyp x Umwelt-Inter-aktion. *EDV in Medizin und Biologie* 3:52-59.

Utz, H.F. 1973. Eignung von Orten für die Selektion. *Z. Pflanzenzüchtg.* 69:30-41.

Weber, W.E., and M. Vanselow. 1985. Die Eignung von Prüforten zur Selektion von Sorten auf Ertrag, ermittelt aus amtlichen Prüfungen bey Winterweizen und Mais. *Z. Pflanzenzüchtg.* 94:64-73.

Weber, W.E., and G. Wricke. 1990. Genotype x environment interaction and its implication in plant breeding. p. 1-19. In M.S. Kang (ed.) *Genotype-By-Environment Interaction and Plant Breeding*. Louisiana State Univ., Baton Rouge, Louisiana.

Weber, W.E., and T. Westermann. 1994. Prediction of yield for specific locations in German winter wheat trials. *Plant Breeding* 113:99-105.

Westcott, B. 1986. Some methods of analysing genotype-environment interaction. *Heredity* 56:243-253.

Williams, E.J. 1952. The interpretation of interactions in factorial experiments. *Biometrika* 39:65-81.

Wricke, G. 1960. Einige Betrachtungen zur ökologischen Streubreite und der Möglichkeit ihrer exakten Erfassung in Feldversuchen. Rundschreiben Arbeitsgem. *Biometrie in der DLG*: 1-5.

Wricke, G. 1962. Über eine Methode zur Erfassung der ökologischen Streubreite in Feldversuchen. *Z. Pflanzenzüchtg.* 47:92-96.

Wricke, G. 1964. Zur Berechnung der Ökovalenz bei Sommerweizen. *Z. Pflanzenzüchtg.* 52:127-138.

Wricke, G. 1965. Die Erfassung der Wechselwirkung zwischen Genotyp und Umwelt bey quantitativen Eigenschaften. *Z. Pflanzenzüchtg.* 53:266-343.

Wricke, G., and W.E. Weber. 1980. Erweiterte Analyse von Wechselwirkungen in Versuchsserien. p.87-95. In W. Köpcke und K. Überla (eds.) *Biometrie - heute und morgen, Med. Inform. und Statistik*, 17.

Wricke, G., and W.E. Weber. 1986. *Quantitative Genetics and Selection in Plant Breeding*, W. de Gruyter, Berlin-New York.

Wright, A.J. 1971. The analysis and prediction of some two factor interactions in grass breeding. *Journal of Agricultural Science* 76:301-306.

Yates, F., and W.G. Cochran. 1938. The analysis of groups of experiments. *Journal of Agricultural Science* 28:556-580.

Chapter 14

SPATIAL ANALYSIS OF FIELD EXPERIMENTS: FERTILIZER EXPERIMENTS WITH WHEAT (*Triticum aestivum*) AND TEA (*Camellia sinensis*)

B.E. Eisenberg,[1] H.G. Gauch Jr.,[2] R.W. Zobel,[3] and W. Kilian[4]

I. INTRODUCTION

The transmission and reception of information is commonly described as the transmission of a message or signal which may be distorted to a greater or lesser extent by noise. The data emanating from scientific experiments are considered to be measurements which are composed of a signal, or pattern, plus noise and, indeed, it was R.A. Fisher's emphasis on the need for good design so as to maximise the information from an experiment (and the corollary of minimising random variation or noise), which is considered to be one of his major contributions to applied statistics. Field trials, whether agricultural, botanical or ecological, are carried out over relatively large areas of land, and more often than not, must develop over at least one season (often a few seasons) before meaningful results become available. Measurements such as grain yield are therefore vulnerable to many influences which may increase the noise. A few examples of such influences which may play a role are as follows:

(i) Inherent soil variability over the land area which might not have been previously expressed and therefore not have been visible in the planning stage. Seasonal precipitation level and distribution is an important cause of such factors entering into a field trial.

(ii) Insect pests, such as aphids in wheat, which may be carried in from one direction by a prevailing wind.

(iii) Plant diseases, which develop outwardly from a focus or foci in the experiment.

Clearly, bearing such factors in mind during the design phase of the experiment, that is both in the treatment design as well as in the experiment design, is quite critical to successful experimentation. Often, a great deal of care is given to the planning, preparation of materials and execution of a field trial, but uncontrolled variables may make their appearance and affect the trial to the extent that estimates of treatment effects are made with a low degree of accuracy.

[1] A.R.C., Agrimetrics Institute, Private Bag X640, Pretoria 0001, South Africa.
[2] Dept. Soil, Crop and Atmospheric Sciences, Cornell University, Ithaca, New York 14853, USA.
[3] USDA-ARS, Dept. Soil, Crop and Atmospheric Sciences, Cornell Univeresity, Ithaca, NewYork 14853, USA.
[4] A.R.C., Small Grain Institute, Private Bag X29, Bethlehem 9700, O.F.S., South Africa.

Concern about the effect of soil variability on yield trials has prompted researchers from at least as early as 1910 to attempt to describe such spatial or positional variation, for instance by means of uniformity trials (Mercer and Hall, 1911), and then to plan experiments on the same area in the following season such that the observed trends are taken into account. Nevertheless, some 73 years later, Pearce (1983) notes that:

"One of the chief features of an experiment is the field in which it takes place, but its characteristics are often taken for granted. In fact, surprisingly little is known about the spatial patterns with which investigators have to contend."

This author's further comments on patterned and non-patterned field variation provide an excellent summary of those features which should be taken into account when planning field trials.

As pointed out above, a large number of factors may act independently or interactively to create uncontrolled spatial (or positional) variation in field trials. This variation, generally referred to as fertility gradients, may take the form of a true gradient or of an irregular patch spread over the area of the field trial. The study of this kind of variation has especially received attention in the context of trials designed to test a number of genotypes (e.g., cultivar trials) as these often include large numbers of treatments and are routinely carried out over a number of environments (Kempton, 1984; Wilkinson et al., 1983). The phenomenon of systematic spatial variation in field trials, which does not affect the treatments at random, is widespread and is a matter for considerable concern in countries as widely disparate as Greece, Japan and Australia (Wilkinson, 1984), Great Britain, Germany and Central Africa (Kempton, 1984, 1990; Pearce et al., 1988), parts of the United States of America (Ball et al., 1993; Brownie et al., 1993; Scharf and Alley, 1993; Stroup et al., 1994), France (Pichot, 1993) and certainly also in South Africa. Two important reasons for the heightened interest in the phenomenon, as well as in methods which may be available to adjust treatments for spatial effects, have come to the fore. Firstly, the greatly increased plant populations which modern plant breeding programs demand for selection purposes require larger areas of land, with the concomitant of a greater chance of uncontrolled variables entering into the experiment. Secondly, the greatly increased cost of agricultural research in general and of field trials in particular make it essential to examine individual trials very carefully indeed before discarding the data as irrelevant.

The diagnosis of whether a fertility gradient is present is usually made by examination of the residuals which have been written into the randomized field plan, whereupon corrective analyses may be decided upon (Atkinson, 1969; Pearce and Moore, 1976). A less well known diagnostic is the calculation of a reference variance for a trial based on blocks of size two; this variance is theoretically the smallest variance in a trial and if the actual variance is large by comparison, corrective analyses might be necessary (Warren and Mendez, 1981).

The classical device for controlling local variation over the experimental site is to divide the area into blocks such that the block represents the most homogeneous area for grouping treatment plots. However, it is the application of diagnostics to such experiments which has emphasized the ubiquity of fertility gradient problems as well as bringing to the fore the near universal recognition that simple blocking in the classical sense is inadequate to deal with it (Warren and Mendez, 1981, 1982; Pearce, 1978; Tamura et al., 1988; Scharf and Alley, 1993; Pichot, 1993). More sophisticated designs, such as incomplete block designs, are undoubtedly useful, but there is evidence that they do not necessarily allow the complete assessment of, and therefore proper adjustment for, spatial effects (Kempton, 1990).

The consequences of ineffective blocking, which is quite often due to circumstances which arise during the course of the season, are an inflated experimental error mean square as well as the distinct possibility that treatment effects may be distorted (Warren and Mendez, 1981; Ball et al., 1993; Brownie et al., 1993; Stroup et al., 1994). An example of how this may happen as a result of an insect infestation is given in the thorough case study by Ellis et al. (1987) on carrots.

A. THE RATIONALE OF THE APPLICATION OF SPATIAL ANALYSIS TO PLANT NUTRITION EXPERIMENTS

Field experiments with different fertilizers or with differing levels of plant nutrients are a *sine qua non* of agricultural research aimed at improving crop production. Specifically the objectives of such experiments are to obtain reliable information about (a) which nutrients may be limiting in a particular area; (b) crop response to nutrients for calibration of soil testing and thus in support of recommendations based on soil tests (Nelson, 1987); and (c) to provide useful information for soil fertility data bases (Sumner, 1987).

The recommendation of levels of fertilizers for economic crop production based on such trials is no simple matter. Rates of application have to be chosen with due allowance for the costs involved, the value of the crops, expected responses, interactions among nutrients, risks and more. It is common knowledge that recommendations which are too low or too high may result in severe damage to the crop and consequent financial loss to producers.

In the pursuit of these goals, a greater or lesser number of nutritional and/or fertilizer experiments are, of necessity, carried out in a particular area representing the recommendation domain. This area is largely determined by soil characteristics, weather patterns and especially management practices. Factorial combinations of different levels of nitrogen, phosphate and potassium fertilizers are commonly employed and many such trials are laid out as randomized complete block (RCB) designs, as is also the case with many other comparative field experiments such as cultivar trials. These trials generate a great deal of data, especially when they are planned and executed so as to be of multidisciplinary interest.

Whilst there is a body of literature which indicates that much reliable information is to be obtained from such experiments, there is also evidence that the results can be so unreliable as to negate any recommendations emanating from them (Sumner, 1987). Thus the considerable cost involved in such experimentation as well as the critical consequences which derive from the results make it imperative that the statistical analysis of results meets the following goals:

(i) to obtain accurate crop response data from the individual sites of the experiments;

(ii) to determine those levels which will benefit the domain of recommendation;

(iii) to identify those areas or sites which merit special attention.

Statistical analysis of these experiments tends to follow a well established route. Individual experiments are analyzed as RCB's with the breakdown of the degrees of freedom (df) into main effects and interactions. On occasion the trend components are also calculated whilst interaction is sometimes examined with the analysis of simple effects (Keppel, 1982). To meet the goals set out above, an analysis of the combined set of experiments must also be undertaken. Methods of carrying out such analyses are readily available in standard texts such as those of Cochran and Cox (1957), Snedecor and Cochran (1980) and Gomez and Gomez (1984). Colwell (1978) provides a detailed description of the analysis of single trials as well as that of combined trials and shows how response surface analysis might be employed. Commonly, the results from these analyses are reported in tables or graphs of means over the replicates or localities; these might be accompanied by standard errors of the means or confidence limits of regression analyses.

As stated earlier all measurements of response variables in field experiments (and certainly also in many experiments under more controlled conditions) are made with error, or, synonymously, that data are noisy. This is most certainly also the case with yield data from fertilizer experiments where the much quoted aphorism "data = pattern + noise" (Freeman, 1973) is entirely applicable. The noise in these experiments is generated by the physico-chemical characteristics of the soil and the atmospheric environments as well as various interactions. Soil analyses of each plot of the experiment are often carried out before the nutrients are applied with the aim of enhancing the planning of the experiment as well as of using the data later as part of a multiple regression analysis of the effects of such variables on yield. However, these variables may prove to be of little value for such analysis because the applications of the treatments themselves might simply suppress the action of such variables and so prevent some of them playing any measurable role in plant nutrition, or, conversely, result in one or a few playing a dominant role. Spatial variation may thus become an important yield determining factor and this may result in relatively poor prediction models. It is surmised that these effects are far more prevalent than is generally thought to be the case and are contributory to vague fertilizer recommendations.

The analysis of nutritional trials must therefore take account of the possibility that there may be a fertility gradient which could markedly affect the observed yield values and result in an inaccurate estimate of treatment response.

An important objective of the Small Grain Institute (SGI) is the determination of nutritional requirements of wheat in the Orange Free State. The trials which have been carried out have produced some general guidelines, but fertilizer recommendations are nevertheless plagued by ambiguity, a problem which extends to the recommendations for other important crops.

This study was carried out in co-operation with the Institute, with the following goals:

(i) to determine whether the choice of a multivariate model, based on predictive criteria, might successfully assess the pattern of spatial variation which may exist in these trials;

(ii) to take such pattern into account in the determination of the final model and compare it with other methods of assessing spatial variation which have been proposed for field experiments, but have not been applied to crop nutrition experiments.

II. MATERIALS AND METHODS

A. THE EXPERIMENTAL DATA
1. Wheat Nutrition Experiments of the S.G.I.

The data from 42 wheat nutrition experiments were included in this study (see Table 1). They formed part of a project carried out at the Small Grains Institute by one of the authors (W.K.) which was designed to determine the nutritional requirements of wheat under rainfed conditions in the Orange Free State. Five levels of nitrogen and five levels of phosphate, in all combinations, were applied to previously randomised plot positions in each of four blocks and incorporated into the soil. Each block consisted of five rows and five columns, wherein the 25 combinations were randomised. The blocks were positioned as illustrated in Figure 1. The plots are represented by the plus and minus signs and were rectangular with rows running east to west. Thus each experiment may be said to consist of a grid of ten rows by ten columns of plots, in a rectangle.

The experiments, which were carried out over the years 1982 to 1988 in different sites, may also be grouped into four groups according to the levels of N and P applied as shown in Table 1.

In each case, wheat was sown to six rows per plot with a plot seeder. The rows were 5.5m long with 40cm between rows. Prior to harvest the two outer rows of each plot were discarded. Grain yield in kgha^{-1} is the response variable used in this study. Moisture determinations on each trial separately showed no more than 0.5% variation between samples. The trials varied between 10 and 13.5% moisture. There are no missing data.

2. Tea (*Camellia sinensis*) nutritional experiment, 1992/1993, Tzaneen, North Eastern Transvaal

The data from this experiment are included here because of the chemical analyses which were carried out on the soil and leaf samples in addition to recording the plot yield. The data analysis of this trial provides a useful illustration of the spatial analysis.

Treatment design for this experiment consisted of 4 levels of each of the nutrients N, P and K, thus making 64 combinations and these were applied to sixty-four plots which were fully randomised, but there was only one replicate.

Table 1

Wheat nutrition experiments from the Small Grain Institute showing localities, seasons and cultivars

	1982	1983	1984	1985	1986	1987	1988
Bethlehem (BE)	821(i) Betta	831(i) Betta	841(i) Betta	851(ii) Betta	861(iv) Betta	872(iv) Tugela	
		83B(i) Betta	842(iii) Betta		862(iv) Karee		
Clocolan (CL)	821(i) Sche69	831(i) Betta	841(i) Betta	85(ii) Betta	86(iv) Betta	87(iv) Betta	88(ii) Betta
		832(iii) Wilge					
Hertzogville (HE)	82(i) Sche69				86(ii) Sche69		88(iv) Molen
Kroonstad (KR)	82(i) Sche69			85(ii) Betta	86(ii) Betta		88(ii) Molen
Petrusburg (PE)			84(i) Sche69	85(ii) Sche69		87(ii) Molen	88(ii) Molen
Reitz (RE)				85(ii) Betta	86(iv) Betta	87(iv) Betta	88(iv) Betta
Senekal (SE)		832(iii) Betta	841(i) Betta	85(ii) Betta	86(iv) Betta		88(ii) Molen
			842(iii) Karee				
Tweespruit (TW)				85(ii) Sche69	86(ii) Sche		
Wesselsbron (WE)					86(iv) Sche*		88(iv) Molen

* Sche=Scheepers

	(i) Group 1	(ii) Group 2	(iii) Group 3	(iv) Group 4
N Levels (kgha^{-1}):	0,10,20,30,40	0,10,20,30,40	0,15,30,45,60	0,15,30,45,60
P Levels (kgha^{-1}):	0,5,10,15,20	0,8,16,24,32	0,5,10,15,20	0,8,16,24,32

BLOCK I   BLOCK II

```
 +  -  -  -  -  |  -  -  +  -  -
 +  +  +  -  -  |  +  +  +  -  +
 +  +  -  +  +  |  +  +  +  +  -
 +  -  -  -  +  |  -  +  -  -  -
 +  -  -  -  +  |  +  -  +  -  +
 ------------------------------
 -  -  -  -  -  |  -  -  +  -  -
 -  -  -  -  -  |  -  -  -  -  -
 +  +  -  -  +  |  -  -  -  +  +
 +  +  +  +  +  |  +  +  -  +  +
 +  +  +  +  +  |  +  +  +  +  +
```

BLOCK III   BLOCK IV

Figure 1. Field plan of a typical Small Grain Institute experiment. The signs of the residuals also represent the randomised plot positions.

The field plan is given in Figure 2. Each plot consisted of 6 rows of 9 bushes. The response variable in this case is "made tea" in $kgha^{-1}$. Soil analyses before fertilizer application provided data on pH, P, K, Ca, Mg, and the proportion $(Ca + Mg)/K$. Plant analyses consisted of N, P, K, Ca, Mg, Mn, Cu, Zn and the proportion $(Ca + Mg)/K$. Soil analyses after harvest consisted of pH, P, K, Ca, Mg, Na, Al and $(Ca + Mg)/K$.

B. STATISTICAL PROCEDURES
1. Diagnosing the presence of fertility gradients

Establishing whether a fertility gradient of one form or another exists in an experiment is obviously the initial step which will lead to employing a correctional analysis. The following diagnostics are commonly employed:

(i) The residuals according to the customary model of the experimental design are calculated and are written into the randomised field plan of the experiment. Simple inspection of the values may reveal the presence of a fertility gradient because of an obvious grouping of negative values (see Figure 1).

(ii) A contour diagram may reveal the pattern even more clearly as may be seen from the three dimensional diagrams in Figure 3. These were drawn after a polynomial regression had been fitted to the residuals.

(iii) The intraclass correlation coefficient (Snedecor and Cochran, 1980) provides a measure of the average correlation of the ranks of the treatments between the blocks. A low value indicates that treatments take on different ranks from block to block which might be due to a fertility gradient.

(iv) Calculating a reference variance for a particular trial and comparing it to the error mean square (EMS) of the customary analysis. Warren and Mendez (1981, 1982) have investigated the effect of block size and orientation on the experimental error in the analysis of uniformity trials. They define a reference variance as the approximation of the random variation remaining after geographical or positional variation has been accounted for. They point out that, on theoretical grounds, the smallest blocks should have the smallest experimental error and therefore recommend dividing the experiments up into two-plot blocks for the purpose of calculating a reference variance, in fact the experimental error. The ratio of the customary experimental error to the reference variance, called the relative variance, is an indication of the sensitivity of the field trial to spatial effects. By calculating the EMS for two-plot blocks in both directions (horizontally and vertically, in the present case), the effect of block orientation on the spatial variance may also be approximated.

2. Data-analytic procedures

Four procedures have been used with the aim of comparing their usefulness in assessing and reducing spatial variation. The following notation is used in describing the three ensuing models:

Y_{ij} = Observed value for treatment i, block j, or row r and column c.

μ = Mean.

α_i = the i-th treatment effect.

β_j = the j-th block effect.

ρ_r = the r-th row effect.

κ_c = the c-th column effect.

ϵ = residual.

b = regression coefficient of yield on covariate.

a. *Fitting a polynomial response surface*

The model employed in this study is as follows:

$$Y_{i(rc)} = \mu + \alpha_i + \rho_r + \kappa_c + \epsilon_{i(rc)}. \qquad [1]$$

The block effect is not included as it is taken care of by the regression fit.The measurements (e.g., yield of grain) on each plot of a field experiment may be identified by its position in a row and column. Fitting a polynomial regression equation to this data to describe the spatial variation and adjusting the treatment means was proposed as long ago as 1929 by Neyman and coworkers (Kirk et al., 1980). Hinz (1987) has compared this procedure to other methods and concluded that it was often more effective. Kirk et al. (1980) developed a program to fit a greatly extended polynomial to field trial data.

b. *Papadakis Neighbour Analysis (Papadakis, 1937)*

The residuals from the customary analysis are arranged according to the field plan and the following values obtained from each plot:

$$\text{(i) } e'_{rc} = e_{rc} - \tfrac{1}{2}(e_{r(c-1)} + e_{r(c+1)})$$

where r = r-th row and c = c-th column and referred to as PAP_c.

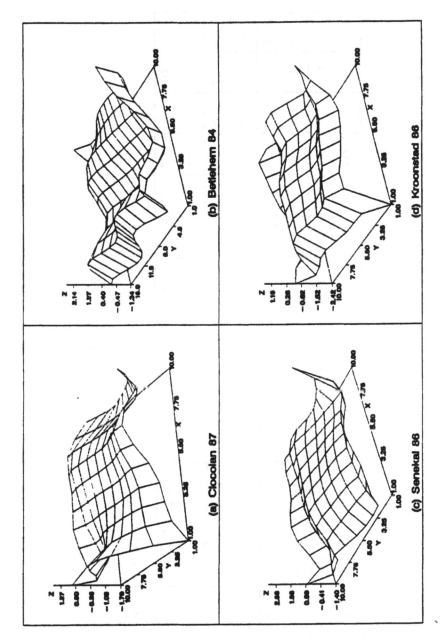

Figure 3. Three dimensional illustrations of response surfaces fitted to residuals

010	021	232	020		011	313	300	221	012	213	103	223	230		031	
-206	60	295	-165		406	282	23	470	93	273	302	491	446		-124	
211	033	123	002	130	331	203	322	210	231	321	030	001	200	013	201	311
-764	-331	-254	111	87	181	-112	22	235	27	-278	-60	487	-308	73	-231	3
233	312	112	101	310	113	133	310	303	222	102	302	110	332	212	121	000
-638	45	-114	-10	-334	-102	-342	-29	42	29	-193	-163	38	-66	-206	14	11
122	032	330	202	220	131	003	100	323	023	111	120	333	320	022	132	
-298	-147	-430	198	-206	119	90	87	-63	-240	-27	240	530	238	-59	452	

Figure 2. Tea Experiment: Field plan with residuals

Two further expansions are:

$$\text{(ii) } e''_{rc} = e_{rc} - \tfrac{1}{2}(e_{(r-1)c} + e_{(r+1)c})$$

for rows and referred to as PAP_r;

$$\text{(iii) } e'''_{rc} = e_{rc} - \tfrac{1}{4}(e_{(r-1)c} + e_{(r+1)c} + e_{r(c-1)} + e_{r(c+1)})$$

for both rows and columns and referred to as PAP_{rc}.

Border plots are accommodated by including the next plot in sequence or basing the PAP_{rc} value on three plots. Either one of the three sets of calculated e_{rc} values are now used as a covariate in the customary model, which for the present data is a randomised complete block, viz.,

$$Y_{ij} = \mu + \alpha_i + \beta_j + b(e''_{rc} - \overline{e}''_{rc}). \qquad [2]$$

c. Nearest Neighbour Analysis (NN) (Schwarzbach, 1984)

This author presents an iterative process whereby the observed values are employed in the detrending process, viz.,
   (i)   Calculate the actual nearest neighbour differences i.e.

$$NND_A = Y_{rc} - \tfrac{1}{2}(Y_{r(c-1)} + Y_{r(c+1)}).$$

   (ii)   Obtain the average of the NND_A values for each treatment.
   (iii) Add the average NND_A to the latest expected value of each treatment.
   (iv) Calculate the nearest neighbour difference of the expected values = NND_E.
   (v) $NND_A - NND_E$
   (vi) Add this difference to the latest expected value of each treatment.
   (vii) Return to (iv) and iterate until an acceptable convergence is obtained.
        The ANOFT program is available for this analysis (Schwartzbach and Betzwar, 1985).

d. Analysis of the 2-way table of residuals by means of the additive main effects and multiplicative interaction model (AMMI)

The experimental design of the wheat trials was such that the randomised field plan consisted of 10 rows and 10 columns (see Figure 1). Whilst all the trials conformed to this general layout, the randomizations of the treatments in all the trials were unique. Residuals written into the randomised field plan, as described in b above, were used as the input data for an analysis according to the additive main effects and multiplicative interaction model (AMMI), viz.,

$$Y_{ge} = \mu + \alpha_g + \beta_e + \Sigma_{n=1}^{N} \lambda_n \gamma_{gn} \delta_{en} + \rho_{ge} \qquad [3]$$

where Y_{ge} = yield of an entry g in environment e

μ = grand mean
α_g = entry mean deviation
β_e = environment mean deviation
N = number of IPCA axes retained in the model
λ_n = singular value for IPCA axis n
γ_{gn} = entry g eigenvector value for IPCA axis n
δ_{en} = environment e eigenvector value for IPCA axis n
ρ_{ge} = residual.

If replications are available, the error may be obtained as follows:

$$\epsilon_{ger} = Y_{ger} - \hat{Y}_{ge}$$

where $\hat{Y}_{ge}$ equals the above equation with the residual ρ_{ge} deleted. The notation given here is the same as that of Gauch (1988, 1992), which refers to genotypes tested in a number of environments. Obviously the rows and columns are coordinates in the present study, and not genotypes and environments, but with the view to maintaining convenient reference to quoted literature, the same notation as that of Gauch (1990) is being used.

The rationale for employing this model to diagnose, as well as to adjust for, fertility gradients is derived from the development of the model for the purpose of analysing genotype-environment interaction in the context of varietal recommendation (Gauch, 1985, 1988, 1990, 1992,1993; Zobel et al., 1988). Thus the model is not suitable only for G x E analyses, but may obviously be employed in the analysis of any two-way table. The abovementioned authors point out that three important concepts must be brought together if the real contribution of the model is to be appreciated. First, the least squares solution of the model results in the data analyses being done in two phases, viz., the additive main effects are removed from the data, leaving residuals which consist of interaction effects and of course, error. The principal component analysis (PCA), which is the multiplicative interaction term in the model, is then carried out on these residuals and if there are replicates, an error term may be estimated. The decomposition of the interaction matrix by PCA results in a partitioning into a number of principal components equal to the lesser of the two values $r-1$ and $c-1$, where r = number of rows and c = number of columns of the table. In the present case, nine principal components may be obtained, which represents the full model. Thus if the overall mean, the row and column effects as well as the scores for all nine principal components were to be substituted into the model given above, the observed values for all the cells would be obtained. The notation for this latter model according to the formal exposition by Gauch (1992) is AMMIF, thus denoting the full model. Obviously, the predicted cell values may be calculated by leaving out all of the principal component scores or by adding in any number up to AMMIF. AMMI0 denotes the basic model with predicted cell means calculated only from row and column effects; AMMI1, AMMI2, etc. denote models which have included the scores of one or more of the principal components. The principal components, which are calculated on the interaction residuals, are termed the IPCA1, IPCA2, etc., to IPCAF for either $(r-1)$ or $(c-1)$ whichever is the smaller.

Clearly, the critical question arises of how many of the IPCA's must be included when employing the AMMI analysis. The necessity of having to make a choice immediately places the whole analysis in a different perspective to that of the conventional ANOVA. Gauch and Zobel (1988) point out that the latter may be described as a "treatment means perspective", for the only information considered relevant to the estimation of the value (e.g., yield) in a cell is the mean from a number of replications.

However, the AMMI analysis may be described as presenting a "modelling perspective", which takes a broader view of the estimation of the value of a cell. The multivariate statistical model applied to the interaction effects relates all the cell values to the cell of interest. Consequently, the information relevant to estimating the value of a cell in row g and column e is the entire data matrix. Not only are the data for row g in column e relevant, but also the data of row g in other columns, the data of other rows in column e and even the data of other rows in other columns. Every datum from the entire data set bears upon the estimate of the value of the cell in row g in column e.

Second, Gauch (1988, 1992) provides clear evidence of how the decomposition of the interaction residuals by PCA results in the capture of most of the pattern which is extant in the interaction, in the early principal components (i.e., those which account for the greatest percentage variance), whilst the later principal components capture less pattern up to the point where they mostly reflect noise.

The choice of a reduced model, restricted to one or two IPCA's for example, would then include the pattern in the predicted cell means, whilst the noise effect would then largely be discarded into a residual. The predicted values from such a model differ from the cell averages in so far that they no longer contain the large noise effect.

Third, an important conceptual as well as practical distinction must be made between the two methods of actually choosing one of the family of models made available by the above-mentioned multivariate approach, viz., an "after the event" or postdictive approach or a predictive approach. In the latter case, the model choice is made by employing the replicates of the genotype x environment values for the purpose of validating the choice of model. One or more replicates are designated for purposes of model fit and the rest for cross validation of the predicted values. All models are fit on the first-mentioned data set and the difference between the predicted cell values and validation set is obtained. The square root of the mean square of these differences (RMSPD) is examined and obviously the model with the smallest value will be the most likely to reflect that which would fit all the data. The general principles of data splitting, as well the specific application of selecting the optimum number of principal components to retain in the model, have been summarised by Crossa et al. (1990). The whole procedure has been included in the computer program MATMODEL which also makes provision for imputing missing values via the expectation maximization algorithm (Gauch and Furnas, 1991).

As mentioned earlier, the residuals denoted in the previous model as ϵ_{ger} where $\epsilon_{ger} = Y_{ger} - \hat{Y}_{ge}$ are arranged according to the randomised field plan. Each residual value now assumes a row and column identity and the AMMI model is now applied to this data as follows:

$$\epsilon_{rc} = \mu' + \alpha'_r + \beta'_c + \Sigma_{n=1}^{N} \lambda'_n \gamma'_m \delta'_{cn} + \rho'_{rc} \qquad [4]$$

where
ϵ_{rc} = residual for row r and column c
μ' = grand mean (zero)
α'_r = row mean deviation
β'_c = column mean deviation
N = number of IPCA axis retained in the model
λ'_n = singular value for IPCA axis n
γ'_m = row eigenvector value for IPCA axis n
δ'_{cn} = column eigenvector value for IPCA axis n
ρ'_{rc} = residual, due to the choice of a reduced model.

In the present case, a model including only one IPCA was chosen and the estimated values resulting from such an analysis therefore represent that part of the residual which is systematic and presumably determined by a fertility gradient, viz.,

$$\hat{\epsilon}_{(AMMI)rc} = \epsilon_{rc} - \rho'_{rc}.$$

What is left is thus largely random error. The systematic estimates $\hat{\epsilon}_{rc}$ are then subtracted from the observed yield values, namely $Y'_{rc} = Y_{rc} - \hat{\epsilon}_{rc}$, and the resultant values reassigned to treatments and blocks and analyzed according to the appropriate design.

III. RESULTS

A. DIAGNOSTICS

The proposal of Warren and Mendez (1981, 1982) to calculate the relative variance (RV) as a diagnostic to indicate which trial is sensitive to spatial effects and therefore may need a form of correctional analysis appears to be useful when carried out on uniformity trial data. They define insensitive trials as having ratios between 1.0 and 1.4, sensitive between 1.4 to 2.04 and highly sensitive overlapping somewhat with 1.43 to 6.42. Since the present experiment is not a uniformity trial, this analysis is applied instead to the residuals obtained by subtracting estimates of treatment effects from the yield data. The relative variances for the trials considered in this study are given in Table 2 and have been grouped according to the level of nutrient as well as locality.

The range of values in this table is considerably less than that quoted above, going from less than unity to greater than 3. Those with ratios of less than one denote error mean squares of the blocks of two that are larger than the RCB error mean square. As the layout of the trials allows both horizontal and vertical blocks-of-two formation, both of these were estimated and the ratios given for both horizontal and vertical. The ratios could be readily grouped into five classes and these classes with the number of trials in each are given in Table 3. In comparison to the values quoted by Warren and Mendez (1981, 1982), these trials, but for a few, appear to mainly fall into the non-sensitive group. The question arises whether the relative variances as calculated on homogeneity trials can be accepted as an absolute standard for these trials where treatment differences are included. The analysis of variance of the AMMI1 model partitions the total sums of squares (TSS) into that for rows, columns and rows x columns as well as for the IPCA1 term. The mean squares for main effects may be tested against the interaction term and the mean square for the IPCA1 term against the residual. The probability of the F ratio for these terms might also be employed as a criterion for judging the sensitivity of a particular experiment. These probabilities are illustrated in Table 2. Absence of a spatial effect would be reflected in non-significant terms of the AMMI analysis. It does not appear that there are especially many non-significant terms in the lower ratio classes.

B. SPATIAL ANALYSIS

The effect of the different methods of spatial analysis on the error term is summarised in Table 4. The error sums of squares (ESS), expressed as a percentage of the total sums of squares (TSS), are provided in the table, for each of the four groups separately. At the bottom of each of the columns representing the different kinds of spatial analyses, the average decrease in percent sums of squares below that for the randomised complete block is given. Only the RCB and AMMI1 analyses were carried out on the 1988 trials.

Overall, the polynomial fit resulted in the least reduction of the percentage error sums of squares, whilst the Papadakis rows and columns and the nearest neighbour analyses were fairly similar. The AMMI1 analysis not only resulted in the greater reduction, but also gave the most consistent reductions.

As a consequence of the greater reduction in the error term, the standard errors of the means (SEM) also exhibited a considerably greater reduction (Table 5). Furthermore, 22 of the 39 trials included in a reanalysis showed a significant N x P interaction term after AMMI1 adjustment. A simple effects analysis of the N x P term aimed at estimating the linear and quadric trends of N for each P level now becomes much clearer for many trials and thus more interpretable (Table 6). This is remarkable because the AMMI analysis was given the spatial layout, not the fertilizer factorial.

C. EVIDENCE THAT THE AMMI1 PREDICTED RESIDUALS CAPTURE A SPATIAL PATTERN WHICH IS CAUSED BY SOIL AND/OR PLANT VARIABLES: FERTILIZER EXPERIMENT WITH TEA (*Camellia sinensis*).

The physical details of this experiment have been described earlier. Yield was measured as the total of a number of harvests throughout the season and expressed as kg of made tea per plot. The field plan of the experiment consisted of four rows by seventeen columns, four of the plots being excluded from the experiment due to unacceptable development.

Table 2
Comparison of relative variance values with significance of AMMI1 ANOVA terms and percentage of the error due to the sum of SS$_{ROWS}$, SS$_{COLS}$ and SS$_{IPCA1}$

	Relative HOR	Variance VERT	ROWS	AMMI1 COLS	IPCA1	TOTAL SS R+C+IPCA1
BE821	1.19	1.06	NS	*	NS	47.8
CL821	1.08	1.03	NS	*	NS	49.0
HE82	0.64	0.62	NS	NS	NS	35.5
KR82	0.88	1.07	NS	NS	*	42.1
BE831	0.60	0.63	NS	***	*	59.8
BE83B	0.86	0.57	***	***	*	72.3
CL831	1.28	0.84	*	NS	**	60.0
CL832	0.99	0.61	**	NS	**	58.1
SE832	3.52	1.87	***	*	NS	72.2
BE841	0.99	1.31	NS	***	*	70.9
CL841	1.48	1.72	**	***	*	63.7
PE84	1.04	1.09	NS	NS	**	56.3
SE841	2.11	2.69	***	***	*	69.5
BE842	0.96	0.72	NS	***	*	59.6
SE842	1.39	1.03	NS	NS	**	56.2
BE851	0.91	0.91	NS	NS	**	55.5
CL85	0.62	0.73	NS	***	*	59.0
KR85	1.31	1.80	**	***	NS	61.9
PE85	1.31	1.12	NS	NS	*	49.0
RE85	0.66	0.76	**	NS	NS	50.2
SE85	1.35	1.39	NS	*	**	58.6
TW85	1.42	1.27	NS	NS	**	53.0
HE86	0.95	1.17	*	**	*	60.2
KR86	1.22	1.57	***	***	NS	61.8
TW86	1.28	1.59	***	*	NS	60.8
BE861	0.63	0.75	NS	NS	*	46.5
BE862	0.73	0.56	*	NS	***	58.0
CL86	0.89	0.86	*	***	*	64.2
RE86	1.49	0.83	***	NS	*	62.4
SE86	0.22	0.29	NS	NS	*	48.8
WE86	1.21	1.10	NS	*	**	59.3
PE87	0.66	1.16	NS	**	NS	51.7
BE872	0.26	0.26	NS	NS	NS	34.4
CL87	1.02	1.19	NS	NS	*	51.3
RE87	1.22	1.28	NS	NS	*	46.8

* = significant at p=0.05
** = significant at p=0.01
*** = significant at p=0.001

Table 2 (continued)
Comparison of relative variance values with significance of AMMI1
ANOVA terms and percentage of the error due to the sum of SS_{ROWS},
SS_{COLS} and SS_{IPCA1}

	Relative HOR	Variance VERT	AMMI1 ROWS	AMMI1 COLS	IPCA1	TOTAL SS R+C+IPCA1
CL88	0.12	0.10	NS	NS	*	49.1
KR88	1.06	0.92	NS	NS	NS	48.4
PE88	0.35	0.31	*	NS	NS	50.5
SE88	0.43	0.51	NS	NS	NS	44.6
HE88	0.22	0.37	*	NS	NS	47.2
RE88	0.84	1.13	NS	*	NS	44.3
WE88	0.75	1.13	***	NS	*	58.8

*   =   significant at p=0.05
** =   significant at p=0.01
*** =   significant at p=0.001

Table 3
Frequency of trials in each of five levels of the relative variance (RV)

RV < 1	Horizontal and vertical	16
	Horizontal only	6
	Vertical only	3
1 ≤ RV ≤ 1.4	Horizontal and vertical	8
	Horizontal only	5
	Vertical only	8
1.4 < RV ≤ 1.8	Horizontal and vertical	2
	Horizontal only	3
	Vertical only	4
1.8 < RV ≤ 2.2	Horizontal and vertical	0
	Horizontal only	1
	Vertical only	1
RV > 2.2	Horizontal only	1
	Vertical only	1

Table 4
Comparison of the ESS as percentage of the TSS, before and after corrective analysis

Trial No	RCB	Polyno-mial	PAP_C	PAP_R	PAP_{RC}	NN	AMMI1
Group 1							
BE821	66.6	70.5	60.7	66.2	63.1	69.7	42.0
BE831	32.8	33.0	29.0	32.4	31.5	35.9	17.8
BE841	51.4	48.0	37.3	45.1	35.4	34.4	25.0
BE83B	26.6	24.0	23.0	15.7	17.6	8.2	20.7
CL821	72.5	55.9	51.7	50.3	51.7	54.4	67.7
CL831	51.0	60.6	49.4	47.6	46.0	42.4	33.4
CL841	72.5	76.9	44.3	41.6	33.8	24.8	43.1
HE82	45.2	52.0	43.0	42.3	42.3	59.0	32.8
KR82	64.7	69.7	64.1	57.5	59.1	88.4	46.9
PE84	59.5	66.9	54.5	59.0	57.7	60.6	32.1
SE841	49.3	49.1	25.9	26.6	20.2	14.4	20.9
Average reduction		-1.3	9.9	9.8	12.2	9.1	19.1
Group 2							
BE851	72.3	72.3	72.3	72.3	72.3	64.9	44.5
CL85	30.3	20.6	23.5	28.6	22.1	17.6	22.8
HE86	58.0	58.7	54.0	57.9	56.6	61.8	34.3
KR85	56.4	54.5	45.0	50.0	45.3	41.6	32.4
KR86	64.7	60.2	69.3	59.8	61.4	52.6	41.5
PE85	76.2	76.6	71.1	69.3	76.0	54.2	57.3
PE87	32.9	30.6	30.4	32.0	32.6	36.0	19.5
RE85	37.7	46.3	37.6	35.8	36.9	34.4	24.6
TW85	70.8	70.9	68.3	68.0	64.6	52.2	45.0
TW86	69.8	71.7	54.8	63.1	57.5	52.2	41.2
CL88	6.1						21.4
KR88	56.7						31.6
PE88	23.6						18.6
SE88	33.1						23.3
Average reduction		0.7	4.3	3.2	4.4	10.2	16.5
Group 3							
CL832	47.7	45.0	46.0	40.5	46.6	31.9	24.1
SE832	75.0	67.5	71.4	43.0	49.5	23.9	30.1
SE842	55.5	57.2	54.7	54.1	53.5	50.7	30.1
Average reduction		2.8	2.0	13.5	9.5	23.9	31.3

Table 4 (continued)
Comparison of the ESS as percentage of the TSS, before and after corrective analysis

			Group 4				
Trial No	RCB	Polyno-mial	PAP_C	PAP_R	PAP_{RC}	NN	AMMI1
BE861	40.9	42.2	36.4	40.1	36.4	31.7	26.0
BE862	42.3	41.8	41.5	50.0	41.9	47.8	16.8
BE872	16.6	16.6	15.6	16.5	16.2	14.6	17.6
CL86	42.4	41.6	30.0	40.1	31.9	34.3	23.4
CL87	57.8	52.5	55.4	57.8	56.0	60.9	33.3
RE87	81.5	80.3	80.7	81.1	79.9		60.9
SE86	16.1	15.6	15.9	16.0	16.1	13.8	17.4
WE86	56.3	55.7	48.8	45.0	41.8	33.7	34.3
HE88	17.9						16.7
RE88	57.8						36.8
WE88	55.0						35.5
Average reduction		0.9	3.7	0.9	4.2	5.1	15.1

Table 5
Percentage reduction in the standard error of the mean

	Group 1	Group 2	Group 3	Group 4
REGR	0	3	INCR	4
PAP+	12	12	6	3
NN	9	6	10	NONE
AMMI1	29	30	44	29

Table 6

Significance of main effects, interactions and trend components before and after AMMI1 adjustment

		N	P	N*P	$N_L P_s$	$N_Q P_s$	$N_l P_l$	$N_Q P_s$	$N_l P_s$	$N_Q P_l$	$N_l P_s$	$N_Q P_s$	$N_l P_l$	$N_Q P_s$	$N_L P_s$	$N_Q P_s$
BE821	U	NS	NS	NS	NS	NS	NS	NS	NS	NS	**	NS	NS	NS	NS	NS
	A	***	NS	•	NS	NS	NS	NS	•	•	***	NS	**	NS	**	NS
CL821	U	**	**	NS	NS	NS	NS	NS	NS	NS	NS	NS	NS	NS	NS	**
	A	•	NS	NS	**	NS	•	NS	NS	NS	NS	NS	NS	NS	**	NS
HE82	U	***	•	NS	NS	NS	NS	NS	NS	•	NS	NS	NS	•	NS	NS
	A	***	***	NS	NS	•	•	•••	**	NS	NS	NS	•	**	**	**
KR82	U	NS	NS	NS	NS	NS	NS	NS	NS	NS	NS	NS	NS	NS	NS	NS
	A	NS	NS	NS	NS	NS	NS	NS	NS	NS	•	NS	NS	NS	NS	NS
BE831	U	***	***	NS	NS	NS	NS	NS	•	NS	NS	NS	NS	NS	NS	NS
	A	NS	***	**	•	NS	NS	NS	NS	•	NS	NS	NS	NS	NS	•
BE83B	U	NS	***	NS	NS	NS	NS	NS	NS	NS	NS	NS	NS	NS	NS	NS
	A	NS	***	**	NS	NS	NS	NS	NS	NS	**	NS	NS	•	NS	•
CL831	U	***	•	NS	**	NS	•	NS	NS	NS	•	NS	NS	•	NS	NS
	A	***	***	NS	**	NS	***	NS	•••	NS	•	NS	NS	NS	NS	**
CL832	U	***	NS	NS	•	NS	•	NS	•••	NS	**	NS	NS	**	••	NS
	A	***	NS	•••	**	NS	•••	NS	•••	••	•••	NS	NS	•••	•••	NS

N_L = linear trend over N levels
N_Q = quadratic trend over N levels

U = unadjusted
A = adjusted

* significant at p=0.05
** significant at p=0.01
*** significant at p=0.001

Table 6

Significance of main effects, interactions and trend components before and after AMMI1 adjustment

		N	P	N°P	N_LP_1	N_QP_1	N_LP_2	N_QP_2	N_LP_3	N_QP_3	N_LP_4	N_QP_4	N_LP_5	N_QP_5
SE832	U	NS	NS	NS	NS	NS	NS	NS	NS	NS	NS	NS	NS	NS
	A	**	NS	NS	NS	NS	NS	NS	NS	NS	NS	NS	NS	**
BE841	U	**	**	NS	NS	NS	**	NS	*	NS	NS	NS	NS	NS
	A	***	***	**	***	NS	***	NS	**	NS	*	NS	***	**
CL841	U	NS	NS	NS	NS	NS	NS	NS	NS	NS	*	NS	NS	NS
	A	***	***	NS	NS	NS	NS	NS	*	NS	**	NS	NS	NS
PE84	U	NS	**	NS	NS	NS	NS	NS	NS	NS	NS	NS	NS	NS
	A	*	***	**	NS	NS	**	NS	NS	*	NS	NS	NS	NS
SE841	U	***	NS	NS	NS	NS	**	NS	**	NS	**	*	*	NS
	A	***	***	***	**	*	***	**	***	NS	**	*	NS	NS
SE842	U	***	*	NS	NS	NS	NS	NS	**	NS	*	*	NS	NS
	A	***	***	***	***	NS	NS	*	***	NS	***	*	NS	**
BE851	U	NS	NS	NS	NS	NS	NS	NS	**	NS	NS	NS	**	NS
	A	**	**	NS	NS	NS	NS	NS	***	NS	NS	NS	NS	NS
CL85	U	NS	***	NS	NS	NS	NS	NS	NS	NS	*	NS	NS	NS
	A	NS	***	NS	NS	NS	NS	NS	NS	NS	NS	NS	NS	NS

N_L = linear trend over N levels U = unadjusted * = significant at p=0.05
N_Q = quadratic trend over N levels A = adjusted ** = significant at p=0.01
*** = significant at p=0.001

Table 6

Significance of main effects, interactions and trend components before and after AMMI1 adjustment

	U/A	N	P	$N{\cdot}P$	N_QP_1	N_LP_1	N_QP_2	N_LP_2	N_QP_3	N_LP_3	N_QP_4	N_LP_4	N_QP_5	N_LP_5	N_QP_6
KR85	U	*	NS	NS	NS	**	NS	NS	NS	NS	NS	NS	NS	NS	NS
	A	**	**	***	NS	**	NS	NS	NS	NS	NS	**	NS	NS	*
PE85	U	NS	NS	NS	*	NS	NS	NS	NS	NS	NS	NS	NS	NS	NS
	A	NS	**	**	**	*	NS	NS	NS	NS	NS	NS	NS	NS	NS
RE85	U	NS	***	NS	NS	NS	NS	NS	NS	NS	NS	NS	NS	NS	NS
	A	*	***	***	*	NS	NS	NS	NS	NS	NS	NS	NS	NS	NS
TW85	U	NS	NS	NS	NS	NS	NS	NS	NS	NS	NS	NS	NS	NS	NS
	A	*	***	**	NS	NS	NS	NS	NS	NS	*	NS	NS	NS	NS
HE86	U	**	NS	NS	NS	NS	***	NS	**	NS	**	NS	**	**	**
	A	***	***	***	*	NS	***	NS	**	NS	***	NS	***	***	***
KR86	U	NS	NS	NS	NS	NS	NS	NS	NS	NS	NS	NS	NS	NS	NS
	A	**	**	*	*	NS	NS	NS	*	NS	**	NS	**	*	NS
TW86	U	NS	*	NS	NS	NS	NS	NS	*	NS	NS	NS	NS	NS	NS
	A	***	***	***	***	NS	*	*	*	*	**	NS	**	**	NS
BE861	U	***	***	NS	**	NS	NS	NS	**	NS	**	NS	**	**	NS
	A	***	***	**	***	NS	NS	NS	**	NS	**	NS	**	**	NS

N_L = linear trend over N levels
N_Q = quadratic trend over N levels

U = unadjusted
A = adjusted

* = significant at p=0.05
** = significant at p=0.01
*** = significant at p=0.001

Table 6
Significance of main effects, interactions and trend components before and after AMMI1 adjustment

		N	P	N•P	$N_Q P_0$	$N_L P_1$	$N_Q P_1$	$N_L P_2$	$N_Q P_2$	$N_L P_3$	$N_Q P_3$	$N_L P_4$	$N_Q P_4$	$N_L P_5$	$N_Q P_5$
BE862	U	***	***	NS	NS	*	**	NS	NS	*	NS	NS	NS	***	***
	A	***	***	**	**	**	***	**	NS	***	**	***	NS	*	***
CL86	U	NS	***	*	NS	***	NS	NS	NS	NS	NS	NS	NS	NS	NS
	A	***	***	***	NS	***	NS	NS	NS	NS	NS	NS	NS	*	NS
SE86	U	NS	***	NS	NS	*	NS	*	NS	NS	NS	NS	*	NS	NS
	A	NS	***	NS	NS	NS	NS	NS	NS	NS	NS	NS	NS	NS	NS
WE86	U	**	NS	**	NS	NS	NS	NS	NS	***	NS	***	NS	NS	NS
	A	***	NS	***	*	NS	NS	NS	*	***	*	***	NS	NS	NS
PE87	U	*	***	*	NS	NS	NS	*	NS	NS	NS	NS	**	NS	NS
	A	**	***	**	NS	NS	NS	***	NS	*	NS	NS	***	**	NS
BE872	U	*	***	NS	NS	NS	NS	NS	*	NS	NS	NS	NS	NS	NS
	A	***	***	NS	NS	NS	*	**	***	NS	NS	*	NS	NS	NS
CL87	U	NS	***	NS	NS	NS	NS	NS	NS	NS	NS	NS	NS	NS	NS
	A	NS	***	NS	NS	NS	NS	NS	NS	NS	NS	NS	NS	NS	NS
RE87	U	NS	NS	NS	NS	NS	NS	NS	NS	NS	NS	NS	NS	NS	NS
	A	NS	NS	NS	NS	NS	NS	NS	NS	NS	NS	NS	NS	NS	NS

N_L = linear trend over N levels　　　U = unadjusted　　　* = significant at p=0.05
N_Q = quadratic trend over N levels　　A = adjusted　　　** = significant at p=0.01
　　　　　　　　　　　　　　　　　　　　　　　　　　　*** = significant at p=0.001

396

Table 6

Significance of main effects, interactions and trend components before and after AMMI1 adjustment

		N	P	$N°P$	N_LP_1	N_QP_1	N_LP_2	N_QP_2	N_LP_3	N_QP_3	N_LP_4	N_QP_4	N_LP_5	N_QP_5
CL88	U	***	**	NS	***	NS	***	NS	***	*	***	NS	***	NS
	A	***	NS	NS	***	NS	***	NS	***	NS	***	NS	***	NS
KR88	U	NS	***	NS	NS	NS	NS	NS	NS	NS	NS	NS	NS	NS
	A	*	***	NS	*	NS	NS	NS	NS	NS	NS	NS	**	NS
PE88	U	***	NS	NS	***	NS	***	NS	***	NS	***	NS	***	NS
	A	***	NS	NS	***	NS	***	NS	***	NS	***	NS	***	*
SE88	U	***	NS	NS	***	NS	***	NS	***	NS	***	NS	***	NS
	A	***	NS	**	***	NS	***	NS	***	NS	***	NS	***	NS
HE88	U	***	NS	**	***	**	***	NS	***	NS	***	**	***	NS
	A	***	NS	***	***	***	***	NS	***	NS	***	*	***	NS
RE88	U	**	NS	NS	NS	NS	**	NS	NS	NS	*	NS	***	NS
	A	***	NS	***	NS	**	***	NS	NS	**	***	NS	***	NS
WE88	U	NS	***	NS	*	NS	NS	NS	NS	NS	NS	NS	NS	NS
	A	NS	***	***	**	**	NS	NS	NS	*	NS	NS	NS	NS

N_L = linear trend over N levels
N_q = quadratic trend over N levels
U = unadjusted
A = adjusted
* = significant at p=0.05
** = significant at p=0.01
*** = significant at p=0.001

Table 7
Tea experiment: Comparison of the unadjusted and adjusted analyses of variance

Analysis before any corrections			Analysis after correction with AMMI1 predicted residuals			Analysis with Mn as a covariate		
Source	DF	P-value	Source	DF	P-value	Source	DF	P-value
N	3	0.0001	N	3	0.0001	N	3	0.0001
P	3	0.0156	P	3	0.0001	P	3	0.0052
K	3	0.3191	K	3	0.0021	K	3	0.2158
N*P	9	0.6072	N*P	9	0.0254	N*P	9	0.4254
N*K	9	0.6532	N*K	9	0.7331	N*K	9	0.4756
P*K	9	0.8685	P*K	9	0.1724	P*K	9	0.7572
						Mn	1	0.0059

Table 8
Tea experiment: Regression of AMMI1 predicted residuals with soil and plant variables

Variable	Measured on	R^2
K	Soil, before application	3%
N	Plant	6%
K		7%
Ca		7%
Mn		23%
pH	Soil, after harvest	7%
Ca		13%
Mg		12%
(Ca+Mg)/K		19%
Mg/K		19%
K+Ca+Mg+Na=S		12%
Al		5%

The basic analysis of variance for this trial is given in Table 7 where the three-factor interaction term is used as error. A subsequent analysis, not relevant to this discussion, was carried out where the non-significant two-factor interaction terms were pooled with the three factor term to obtain an error term, although this did not change the outcome in any way. The residual values for this model were obtained and these were written into the randomised field plan of the experiment. Each residual now takes on a new index which specifies a row and column position on the field plan as shown in Figure 2.

The next step was to obtain an AMMI1 analysis of the two-way table as well as the AMMI1 predicted cell values. As explained earlier, these values embody most of the pattern which may be imbedded in the residuals. When they are subtracted from the original residuals, the difference is virtually only noise, or random error.

Each plot of the tea experiment had been subjected to a soil analysis before the NPK treatments were applied as well as after harvest. Leaf analyses had also been carried out. The soil fertility paradigm would dictate that the spatial pattern which is captured in the AMMI1 predicted residuals should be correlated with some of these variables. Regression analysis (see Table 8) with the AMMI1 predicted residuals as the dependent variable and all the cations separately show clearly that a good association exists between predicted residuals and a number of the cations involved. The model fit for all the regressions enumerated in Table 8 were highly significant. The relatively low R^2 values serve to indicate that a multiple regression model would probably be a better predictor.

Finally, the adjustment of the yield data by subtracting the predicted residuals results in the analysis of variance given in Table 7. It may be seen that more terms are now significant. When the cation Mn is employed as a *covariate*, it exhibits a highly significant regression with yield, but does not reduce the error to the same extent as in the case of AMMI1 predicted residuals. Common experience indicates that the action of soil variables, which also serve to cause a spatial pattern, may range from a fairly clear dominance of one variable, through increasing complexity of interactions of a number of variables to the point where it may be difficult to pinpoint the offending individual variables, with all of these affected to some extent by environmental factors such as precipitation. The present results indicate that the AMMI1 predicted residuals serve to integrate these effects, separating them from the random noise and presenting them as an interpretable effect.

IV. DISCUSSION

Enough published evidence seems to have accumulated to indicate that an appreciable number of field experiments are affected by some systematic variables which cannot be accounted for by classical blocking alone. The fact that the arable soils of South Africa are inherently very heterogeneous, both on a macro as well as on a micro scale, is simply accepted as common knowledge, supported by a number of surveys which have been conducted by researchers, often with the view to planning their experiments. However, the accepted methodology by many agricultural researchers, if not the majority, is a reliance on circumstantial or even anecdotal evidence about the homogeneity of the site where the experiment is to be carried out and the expectation that replication and classical blocking will take care of any soil heterogeneity which may be present.

Apart from the necessity of keeping very clear and detailed plans of the disposal of experiments on the fields of experimental stations as well as on cooperating farmers' fields, additional information regarding the field experiment is needed. In the first place a form of diagnostic is needed to help in the decision of whether a correctional spatial analysis is needed or not. Secondly, a way to accumulate experience is needed so that a site may, in time, be classified as to its sensitivity to large blocks and therefore spatial effects. Such information would obviously be very useful in planning experiments. Finally, the researcher needs to have feedback on the effectiveness of his choice of blocking so as to improve his skill in planning the experiments (Warren and Mendez, 1981).

Two diagnostic methods have been applied to the experiments included in this study, namely the customary variance as a ratio of the reference variance calculated on the observed yield values, and the AMMI1 model which was applied to the residuals. In this instance the AMMI model is used in a postdictive sense. Thus the analysis is carried out and the mean squares for rows and columns are tested against the rows-columns interaction, whilst the IPCA1 is tested against the residual.

Whilst it is theoretically expected that an analysis based on blocks of two would provide the smallest possible error mean square, eliminating all but random variation (Warren and Mendez, 1981), this does not happen in practice. In fact the ratios range from smaller than unity to about 3.5. Ratios of less than one would indicate that the blocks of two are less efficient for the elimination of systematic variation than the existing blocking structure. However, this criterion may also be influenced by the "treatments" involved in the analysis, for the error term in the blocks of two analysis may, by chance, be large because of greater block by entry interaction.

As pointed out earlier, the sums of squares of the AMMI1 analysis having been determined from standardised residuals represent a percentage of the error term. Thus the total of SS for rows, columns and IPCA1 would represent the percentage of the error term due to pattern (Table 4). These percentages show some agreement with the relative variance, though limited to the low and high ends of the ranges of their values.

It is suggested that both these criteria should be calculated for any particular trial. The ratio certainly provides a diagnostic of the efficiency of blocking both as to the size as well as the orientation of the blocks. However, the presence of systematic variation appears to be more clearly reflected in the AMMI1 analysis of the standardised residuals.

Four methods of corrective analysis have been applied to the wheat fertilizer trials considered here. The response surface model which was fit to this data is fairly simple and is based on the row and column coordinates of the yield data in the field. In their application of this model, Kirk et al. (1980) extended the model to a higher order polynomial and developed a computer program to fit the model which included a criterion to prevent overfitting. The relatively modest reduction in error sums of squares obtained in this study is possibly due to the fact that a low order polynomial was fit to all the experiments, whereas a higher order might have described the systematic effect more accurately. However, Tamura et al. (1988) point out that model mis-specification in this kind of analysis, which they term trend analysis, will result in biased estimates of both the treatment and error sums of squares. Both the type of treatment effect and the overall magnitude of treatment variability are likely to affect the power of detection of the treatment effects. They further point out that the chance of selecting an inappropriate response surface model in trend analysis will be greater when the placement of treatments is confounded with systematic field variation. It was with the aim of avoiding overfitting that a lower order polynomial was fit.

The Papadakis method aims at developing a covariate from the residuals which will incorporate the spatial effect. Covariance analysis may then be used to remove the spatial effect. In the case of the present data, the covariate could be calculated along rows or columns or by incorporating the four plots around the target plot. The last-named method is recommended and two iterations of the covariance are usually considered to be necessary (Bartlett, 1938, 1978). From the results in Table 6, it appears that the calculation of the covariate via columns or rows may differ considerably, but those calculated via the "target" method generally favour the lower of the two. The covariance was only run once on each experiment. Whilst this method, or modifications of this method, has by no means fallen into disuse (Magnussen, 1993), the results from an extensive investigation by Wilkinson et al. (1983) into the Papadakis method indicate that it has severe drawbacks. They found that it is inefficient in the sense that it does not estimate all the trend, especially when large trends are present in the data. These authors show that after the covariance analysis with the Papadakis covariate, the error is still upwardly biased and a further iteration of the covariance using the treatment estimates from the previous iteration to redefine the nearest-neighbour covariate for the current resulted in a serious underestimation of the error and obviously a serious upward bias in the treatment F ratio. The source of inefficiency in the Papadakis method was found to be the prior correction for treatment effects when forming the nearest neighbour covariate.

Wilkinson suggested that treatment effects might be estimated more efficiently by equating treatment totals of the simpler NN-adjusted values

$$Y_r - (\tfrac{1}{2}(Y_{r-1} + Y_{r+1}) \times b) \qquad [5]$$

to their treatment totals, initially with b=1 and then with a value of b for optimal efficiency determined from a variance ratio for trend effects in the data.

Thus the difference between the yields of a plot and the average of its nearest neighbours on either side is calculated. The treatment totals of these differences are then obtained. They are then readjusted with a value for b which is obtained from a variance component for trend effects.

The results from this kind of analysis, using the algorithm published by Schwarzbach (1984) and Schwarzbach and Betzwar (1985, 1986), are presented in Table 4. The overall reduction in the ESS is somewhat greater than that of the Papadakis method and the polynomial fit.

The remarkable power of the additive main effects and multiplicative interaction model (AMMI1) to reveal the structure in data of two-way tables is the motivation for employing this model in a spatial analysis context. The analysis of a particular trial was carried out as follows:

(i)   The model for the RCB design

$$Y_{ij} = \mu + \alpha_i + \beta_j + \epsilon_{ij}.$$

(ii) The AMMI1 analysis of the residuals, viz.,

$$\epsilon_{rc} = \mu' + \alpha'_r + \beta'_c + \Sigma_{n=1}^{N} \lambda'_n \gamma'_m \delta'_{cn} + \rho'_{rc}.$$

(iii) Adjustment of the Y_{ij} by the predicted residuals $\hat{\epsilon}_{(AMMI1)rc}$ and reanalysis according to the model in (i).

Note that the AMMI1 analysis is applied to the residuals, i.e., after the Y_{ij} had been adjusted for treatment and block effects.

The EMS of an ANOVA on yield data which has been standardised to a mean of zero and standard deviation of 1, where the standard deviation is calculated as the root mean square of the error term, provides an error term equal to 1. An AMMI1 analysis of the residuals thus allows an estimate of the proportion of the EMS due to pattern which is equated to systematic variation in the trial. This proportion appears to have some value as a diagnostic for the presence of spatial effects.

Whilst the reduction in error sums of squares after AMMI1 adjustment varies from experiment to experiment, the overall reduction in ESS is clearly greater than that of the other methods, resulting in considerably smaller SEMs. This provides a sharpened focus on the differences in treatment means as well as on the trend analysis (Table 6).

Perhaps the major value of the application of the AMMI model to the study of the residuals of maize nutrition trials is the multivariate and predictive perspective inherent in the application. By confining the model to one IPCA, predicted residual values for all the cells are obtained wherein most of the pattern is captured. Quite often the row and column effects are large, but equally often the IPCA1 accounts for a large part of the rows-columns interaction (Table 4). The multivariate model relates all the data in the matrix to the cell of interest. Thus in a spatial analysis context, the effects of soil characteristics which are responsible for systematic variation, which may gradually increase their contribution to the effect over the experiment, are none the less integrated into the calculated value. The results from the analysis of the tea experiment are therefore very relevant in this context for they provide evidence that the AMMI1 predicted residuals reflect systematic spatial distribution of soil characteristics in the experiment. After the usual procedure of obtaining predicted AMMI1 cell values had been carried out, a regression analysis with the individual cation concentrations as the independent variable and the AMMI1 predicted residuals as dependent variable was carried out. The soil variables obtained from chemical analysis before the fertilizer treatments were applied showed no association, except perhaps with K although this was very tenuous. The variables obtained from the leaf analyses showed a highly significant regression with the AMMI1 residuals, especially Mn. The acidity related cations measured in an analysis after harvest were also highly significantly associated with the residuals.

Thus it appears that the pattern which has been captured in the tea experiment by the AMMI1 model is a reflection of the systematic variation in the site of the experiment due to, amongst others, certain measurable cations. The R^2 values of the regressions are of interest in that, at best, these cations explain only some 20% of the variation in the prediction of the AMMI1 residuals. It appears that a more comprehensive search for soil variables (e.g., physical characteristics) is necessary if other important sources of spatial variation are to be uncovered.

This finding is substantiated by the work of Ball et al. (1993), who showed that the systematic distribution of organic carbon in one field was consistent with the systematic distribution of residuals for grain yield.

These considerations obviously have a bearing on the statistical analysis of the results from crop nutrition experiments for the purpose of developing recommendations. These analyses usually involve multiple regression models wherein the predictors are both the applied fertilizers as well as variables emerging from soil analyses. However, if some of the latter variables are responsible for a systematic effect in the experiment, the treatments may be confounded with cations already present and so result in a distortion of the regression results.

The significance of the plant analyses as explanatory variables for the AMMI1 residuals hints at a further dynamism in experiments with chemical fertilizers, namely the possibility that the treatments might have interacted with some soil characteristic which helped create the systematic effect. If this is a true state of affairs, planning of such experiments would have to be carried out with greater care if spurious conclusions are to be avoided.

These results are encouraging for using AMMI to describe spatial patterns and clarify treatments effects. However, on balance it must be noted that AMMI does not model two-dimensional spatial patterns explicitly. For example, rows 1, 2, and 3 are just different rows for AMMI, whereas for explicitly spatial models like NN, row 2 is between rows 1 and 3. More research is needed to further understand AMMI's response to various spatial patterns, and to compare its performance relative to other models. Simulated as well as real data sets would be useful in reaching an accurate and balanced assesment.

ACKNOWLEDGMENTS

We wish to thank Mrs. M.M. Saunders and Mrs. L. Morey for their valuable and valued technical assistance in the preparation of this paper. We also appreciate insightful suggestions from Z. Huang.

REFERENCES

Atkinson, A.C. 1969. The use of residuals as a concomitant variable. *Biometrika* 56: 33-41.

Ball, S.T., D.J. Mulla, and C.F. Konzak. 1993. Spatial heterogeneity affects variety trial interpretation. *Crop Science* 33: 931-935.

Bartlett, M.S. 1938. The approximate recovery of information from field experiments with large blocks. *Journal of Agricultural Science* 28: 418-427

Bartlett, M.S. 1978. Nearest neighbour models in the analysis of field   experiments. *Journal of the Royal Statistical Society B* 40(2), 147-174.

Brownie, C., D.T. Bowman, and J.W. Burton. 1993. Estimating spatial variation in analysis of data from yield trials; a comparison of methods. *Agronomy Journal* 85: 1244-1253.

Cochran, W.G., and G.M. Cox. 1957. *Experimental designs.* John Wiley and Sons, Inc., New York.

Colwell, J.D. 1978. *Computations for studies of soil fertility and fertilizer requirements.* Commonwealth Agricultural Bureau, Farnham Royal, Slough, England.

Crossa, J., H.G. Gauch, und R.W. Zobel. 1990. Additive main effects and multiplicative interaction analyses of two international maize cultivar trials. *Crop Science* 30: 493-500.

Ellis, P.R., G.H. Freeman, B.D. Dowker, J.A. Hardman, and G. Kingswell. 1987. The influence of plant density and position in field trials designed to evaluate the resistance of carrots to carrot fly (*Psila rosae*) attack. *Annals of Applied Biology* 111: 21-31.

Freeman, G.H. 1973. Statistical methods for the analysis of genotype-environment interactions. *Heredity* 31: 339-354.

Gauch, H.G. 1985. *Integrating additive and multiplicative models for analysis of yield trials with assessment of predictive success.* Mimeo 85-7. Dept. of Soil, Crop and Atmospheric Sciences, Cornell University, Ithaca, New York.

Gauch, H.G. 1988. Model selection and validation for yield trials with interaction. *Biometrics* 44: 705-715.

Gauch, H.G. 1990. Full and reduced models for yield trials. *Theoretical and Applied Genetics* 80: 153-160.

Gauch, H.G. 1992. *Statistical analysis of regional yield trails: AMMI analysis of factorial designs.* Elsevier, Amsterdam.

Gauch, H.G. 1993. Prediction, parsimony and noise. *American Scientist* 81: 468 - 478.

Gauch, H.G., and R.W. Zobel. 1988. Predictive and postdictive success of statistical analyses of yield trials. *Theoretical and Applied Genetics* 76: 1-10.

Gauch, H.G., and R.E. Furnas. 1991. Statistical analysis of yield trials with MATMODEL. *Agronomy Journal* 83: 916-920.

Gomez, K.A., and A.A. Gomez. 1984. *Statistical procedures for agricultural research.* 2nd Edition. John Wiley and Sons, Inc., New York.

Hinz, P.N. 1987. *Nearest neighbour analysis in practice.* Iowa State Journal of Research, 62(2), 199-217.

Kempton, R.A. (ed). 1984. *Spatial methods in field experiments.* Biometric Society Workshop, University of Durham, 1984.

Kempton, R.A. 1990. Developments in the statistical analysis of field experiments. Paper presented at the 1st African Biometric Conference, Nairobi, Kenya.

404

Keppel, G. 1982. *Design and analysis. A researchers handbook.* 2nd Edition. Prentice-Hall, Englewood Cliffs, New Jersey.

Kirk, H.J., F.L. Haynes, and R.J. Monroe. 1980. Application of trend analysis to horticultural field trials. *Journal of the American Society of Horticultural Science* 105: 189-193.

Magnussen, S. 1993. Bias in genetic variance estimates due to spatial auto-correlation. *Theoretical and Applied Genetics* 86: 349-355.

Mercer, W.B., and A.D. Hall. 1911. The experiment error of field trials. *Journal of Agricultural Science* 4, 107-132.

Nelson, L.A. 1987. The role of response surfaces in soil test calibration. p 31-40. *In* Soil Testing, SSSA Special Publication 21. Soil Science Society of America, Inc., Madison, Wisconsin.

Papadakis, J.S. 1937. Methode statistique pour des experiences sur champ. *Bulletin Institute Amelioration des Plantes à Salonique,* No 23.

Pearce, S.C. 1978. Discussion of the paper by Bartlett, M.S. *Journal of the Royal Statistical Society B* 40(2), 147-174.

Pearce, S.C. 1983. *The agricultural field experiment.* John Wiley and Sons, Chichester, England.

Pearce, S.C., and C.S. Moore. 1976. Reduction of experimental error in perennial crops, using adjustment by neighbouring plots. *Experimental Agriculture* 12: 267-272.

Pearce, S.C., G.M. Clarke, G.V. Dyke, and R.V. Kempson. 1988. *Manual of crop experimentation.* Charles Griffin and Co. Ltd., London.

Pichot, C. 1993. Iterated nearest neighbour analysis of field experiments: 1993. *Agronomie* 13: 109-119.

Scharf, P.C., and M.M. Alley. 1993. Accounting for spatial field variability in field experiments increases statistical power. *Agronomy Journal* 85: 1254-1256.

Schwarzbach, E. 1984. A new approach in the evaluation of field trials. *Vortschrit der Pflanzüchtung* 6: 249-259.

Schwarzbach, E., and W. Betzwar. 1985. Evaluation of field trials with the nearest neighbour analysis using the ANOFT program. *In:* Genetics and breeding of *triticale.* Edited by: M. Bernard and S. Bernard. Institut National de la Recherche Agronomique, Paris, France.

Schwarzbach, E., and W. Betzwar. 1986. Nearest neighbour analysis of non-replicated standard plot trials in Monte Carlo simulations. Proceedings of the Eucarpia meeting of the Cereal Section on rye, 11-13 June, 1985. Part II, p601 - 604. Svalov, Sweden.

Snedecor, G.W., and W.G. Cochran. 1980. *Statistical methods.* 7th Edition, Iowa State University Press, Ames, Iowa.

Stroup, W.W., P.S. Baenziger, and D.K. Mulitze. 1994. Removing spatial variation from wheat trials: a comparison of methods. *Crop Science* 34: 62-66.

Sumner, M.E. 1987. Field experimentation: Changing to meet current and future needs. *In:* Soil Testing. SSSA Special Publication No. 21: 119-131, Soil Science Society of America, Madison, Wiscosin.

Tamura, R.N., L.A. Nelson, and G.C. Naderman. 1988. An investigation of the validity and usefulness of trend analysis for field plot data. *Agronomy Journal* 80: 712-718.

Warren, J.A., and I. Mendez. 1981. Block size and orientation, and allowance for positional effects, in field experiments. *Experimental Agriculture* 17: 17-24.

Warren, J.A., and I. Mendez. 1982. Methods for estimating background variation in field experiments. *Agronomy Journal* 74: 1004-1009.

Wilkinson, G.N. 1984. Nearest neighbour methodology for design and analysis of field experiments. Invited paper for proceedings of 12th International Biometrics Conference, Japan 1984: 69-79

Wilkinson, G.N., S.R. Eckert, T.W. Hancock, and O. Mayo. 1983. Nearest neighbour (NN) analysis of field experiments. *Journal of the Royal Statistical Society* B 45: 151-211.

Zobel, R.W., M.J. Wright, and H.G. Gauch. 1988. Statistical analysis of a yield trial. *Agronomy Journal* 80: 388-393.

Appendix

BIBLIOGRAPHY ON GENOTYPE-BY-ENVIRONMENT INTERACTION

Jean-Baptiste Denis,[1] Hugh G. Gauch, Jr.,[2] Manjit S. Kang.[3] Fred A. Van Eeuwijk,[4] and Richard W. Zobel[2]

I. INTRODUCTION

Genotype-by-environment interaction (GEI) offers significant challenges and opportunities to agronomists and breeders, as the chapters in this volume indicate. It is gratifying for researchers in this field to see such a dramatic increase in recent years in the number and quality of papers on GEI. However, it is not easy to keep up with this expanding literature. One problem is that many important developments on interaction are published in statistical journals not read routinely by many agricultural researchers, and relevant statistical applications appear in an even wider assortment of medical, engineering, and other journals. Another problem is that merely the agricultural literature itself is enormous, appearing in various languages and published in numerous countries. Some important items are published as departmental mimeographs or as chapters in books that may not come to some researchers' attention.

To help researchers handle this enormous and diffuse literature on GEI in particular and interaction in general, two GEI bibliographies have been developed. One is a basic bibliography with 83 references presented in this chapter. The other is an extensive online bibliography accessible through electronic mail using Gopher. The online GEI database is to be updated in an ongoing manner. Authors of GEI papers are invited to send copies to insure inclusion in the database.

II. BASIC PRINTED BIBLIOGRAPHY

A helpful bibliography on GEI, with 129 references, was published by Aastveit and Mejza (1992). Also, the books by Kang (1990) and Gauch (1992) and the reviews by Denis and Vincourt (1982), Cox (1984), and Crossa (1990) contain many references.

---

[1] Institut National de la Recherche Agronomique, Laboratoire de Biométrie, Route de Saint Cyr, F 78026 Versailles Cedex, FRANCE.
[2] Soil, Crop, and Atmospheric Sciences, 1021 Bradfield Hall, Cornell University, Ithaca, New York 14853, UNITED STATES.
[3] Department of Agronomy, 104 Madison B. Sturgis Hall, Louisiana State University, Baton Rouge, Louisiana 70803-2110, UNITED STATES.
[4] DLO-Centre for Plant Breeding and Reproduction Research, P.O. Box 16, NL-6700 AA Wageningen, THE NETHERLANDS.

This chapter offers a basic bibliography on GEI with 83 references. Obviously, there is a considerable subjective element in deciding which papers are most important and relevant. Doubtless, many excellent papers are not included here. Nevertheless, this brief bibliography may be useful as a point of departure. Some factors considered in selecting these few references are historical importance, statistical innovation, agricultural interpretation of the genotypic and environmental causes of interaction and the heritability and repeatability of stability or interaction indices, and citations leading into the GEI literature with special interest in review articles. This printed bibliography is deliberately brief because its role is complemented by a longer bibliography described next.

III. EXTENSIVE ONLINE BIBLIOGRAPHY

An extensive online bibliography on GEI is being updated in an ongoing manner and is accessible through electronic mail using Gopher. The main agricultural and statistical journals are scanned routinely to add new items. Also, authors of GEI papers are invited to send four copies to any author of this chapter to ensure consideration for addition to the database. Copies of articles already accepted for publication but not yet published are also welcome, provided that the author is willing to send a preprint (or at least an abstract) to any researcher who notes the database entry and requests a preprint from the author. By this means, new results can be disseminated more rapidly.

The GEI bibliographic database is structured as a straight ASCII file. To enhance consistency with the major bibliographic services, the record structure follows BIOSIS (BioSciences Information Service, 2100 Arch Street, Philadelphia, Pennsylvania 19103), which is typical of all these services.

To access the GEI database on Internet, the Gopher server is RHIZO.CIT.CORNELL.EDU on port 70. On World Wide Web, GOPHER://RHIZO.CIT.CORNELL.EDU:70/ is the URL. Researchers without access to electronic mail may instead request a floppy disk copy of the bibliographic database by sending $25 in US funds to cover costs of materials and handling to: Hugh Gauch, SCAS, 1021 Bradfield Hall, Cornell University, Ithaca, New York 14853. The check should be payable to Cornell University. Be sure to specify the desired disk size (5 1/4 or 3 1/2 inches) and format (IBM or McIntosh) or else the default applies of a 3 1/2 inch 1.44 MB format for IBM compatible computers.

REFERENCES

Aastveit, A.H., and H. Martens. 1986. ANOVA interactions interpreted by partial least squares regression. *Biometrics* 42:829-844.
Aastveit, A.H., and S. Mejza. 1992. A selected bibliography on statistical methods for the analysis of genotype x environment interaction. *Biuletyn Oceny Odmian* 25:83-97.
Abou-el-Fittouh, H.A., J.O. Rawlings, and P.A. Miller. 1969. Classification of environments to control genotype by environment interactions with an application to cotton. *Crop Science* 9:135-140.

Allard, R.W., and A.D. Bradshaw. 1964. Implications of genotype-environment interactions in applied plant breeding. *Crop Science* 4:503-508.

Annicchairico, P. 1992. Cultivar adaptation and recommendation from alfalfa trials in northern Italy. *Journal of Genetics and Breeding* 46:269-277.

Bailey, R.A. 1983. Interaction. *In* Encyclopedia of Statistical Sciences, Volume 4, S. Kotz and N.L. Johnson (Eds.). John Wiley, New York, New York, pages 176-181.

Baker, R.J. 1988. Test for crossover genotype-environmental interactions. *Canadian Journal of Plant Science* 68:405-410.

Baril, C.P. 1992. Factor regression for interpreting genotype-environment interaction in bread wheat trials. *Theoretical and Applied Genetics* 83:1022-1026.

Basford, K.E. 1982. The use of multidimensional scaling in analysing multi-attribute genotype response across environments. *Australian Journal of Agricultural Research* 33:473-480.

Basford, K.E., P.M. Kroonenberg, I.H. de Lacy, and P.K. Lawrence. 1990. Multiattribute evaluation of regional cotton variety trials. *Theoretical and Applied Genetics* 79:225-234.

Berke, T.G., P.S. Baenzinger, and R. Morris. 1992. Chromosomal location of wheat quantitative trait loci affecting stability of six traits, using reciprocal chromosome substitutions. *Crop Science* 32:628-633.

Bradu, D., and K.R. Gabriel. 1978. The biplot as a diagnostic tool for models of two-way tables. *Technometrics* 20:47-68.

Braun, H.-J., W.H. Pfeiffer, and W.G. Pollmer. 1992. Environments for selecting widely adapted spring wheat. *Crop Science* 32:1420-1427.

Breese, E.L. 1969. The measurement and significance of genotype-environment interactions in grasses. *Heredity* 24:27-44.

Brown, K.D., M.E. Sorrells, and W.R. Coffman. 1983. A method for classification and evaluation of testing environments. *Crop Science* 23:889-893.

Calinski, T., S. Czajka, and Z. Kaczmarek. 1987. A model for the analysis of a series of experiments repeated at several places over a period of years -I: Theory. *Biuletyn Oceny Odmian* 12:7-33.

Ceccarelli, S. 1989. Wide adaptation: How wide? *Euphytica* 40:197-205.

Chadoeuf, J., and J.-B. Denis. 1991. Asymptotic variances for the multiplicative interaction model. *Journal of Applied Statistics* 18:331-353.

Charcosset, A., J.-B. Denis, M. Lefort-Buson, and A. Gallais. 1993. Modelling interaction from top-cross design data and prediction of F1 hybrid value. *Agronomie* 13:597-608.

Cherney, J.H., and J.J. Volenec. 1992. Forage evaluation as influenced by environmental replication: A review. *Crop Science* 32:841-846.

Chuang, C. 1982. Empirical Bayes methods for a two-way multiplicative-interaction model. *Communications in Statistics, Theory and Methods* 11(25):2977-2989.

Cooper, M., and G.L. Hammer (ed.). 1995. *Plant adaptation and crop improvement.* CAB International, Wallingford, UK, ICRISAT, Patancheru, 502 324, India, and IRRI, Manila, Philippines.

Cornelius, P.L., D.A. Van Sanford, and M.S. Seyedsadr. 1993. Clustering cultivars into groups without rank-change interactions. *Crop Science* 33:1193-1200.

Corsten, L.C.A., and J.-B. Denis. 1990. Structuring interaction in two-way tables by clustering. *Biometrics* 46:207-215.

Cox, D.R. 1984. Interaction. *International Statistical Review* 52:1-31.

Crossa, J. 1990. Statistical analyses of multilocation trials. *Advances in Agronomy* 44:55-85.

Crossa, J., and P.L. Cornelius. 1993. Recent developments in multiplicative models for cultivar trials. *In* International Crop Science I., D.R. Buxton et al. (Eds.). Crop Science Society of America, Madison, Wisconsin, pages 571-577.

Crossa, J., P.N. Fox, W.H. Pfeiffer, S. Rajaram, and H.G. Gauch. 1991. AMMI adjustment for statistical analysis of an international wheat yield trial. *Theoretical and Applied Genetics* 81:27-37.

Denis, J.-B. 1988. Two way analysis using covariates. *Statistics* 19:123-132.

Denis, J.-B. 1991. Adjustements de modèles linéaires sous contraintes linéaires avec données manquantes. *Revue de Statistique Appliquée* 39:5-24.

Denis, J.-B., and P. Vincourt. 1982. Panorama des méthodes statistiques d'analyse des interactions génotype x milieu. *Agronomie* 2:219-230.

Fabian, V. 1991. On the problem of interactions in the analysis of variance. *Journal of the American Statistical Association* 86:362-375.

Federer, W.T., and B.T. Scully. 1993. A parsimonious statistical design and breeding procedure for evaluating and selecting desirable characteristics over environments. *Theoretical and Applied Genetics* 86:612-620.

Finlay, K.W., and G.N. Wilkinson. 1963. The analysis of adaptation in a plant-breeding programme. *Australian Journal of Agricultural Research* 14:742-754.

Fisher, R.A., and W.A. Mackenzie. 1923. Studies in crop variation. II. The manurial response of different potato varieties. *Journal of Agricultural Science, Cambridge* 23:311-320.

Freeman, G.H. 1973. Statistical methods for the analysis of genotype-environment interactions. *Heredity* 31:339-354.

Freeman, G.H. 1985. The analysis and interpretation of interaction. *Journal of Applied Statistics* 12:3-10.

Gabriel, K.R. 1978. Least squares approximation of matrices by additive and multiplicative models. *Journal of the Royal Statistical Society, Series B* 40:186-196.

Gauch, H.G. 1992. Statistical Analysis of Regional Yield Trials: AMMI Analysis of Factorial Designs. Elsevier, Amsterdam.

Gollob, H.F. 1968. A statistical model which combines features of factor analytic and analysis of variance techniques. *Psychometrika* 33:73-115.

Hanson, W.D. 1964. Genotype-environment interaction concepts for field experimentation. *Biometrics* 20:540-552.

Hardwick, R.C., and J.T. Wood. 1972. Regression methods for studying genotype-environment interactions. *Heredity* 28:209-222.

Helms, T.C. 1993. Selection for yield and stability among oat lines. *Crop Science* 33:423-426.

Hill, J. 1975. Genotype-environment interactions — a challenge for plant breeding. *Journal of Agricultural Science, Cambridge* 85:477-493.

Jalaluddin, M., and S.A. Harrison. 1993. Repeatability of stability estimators for grain yield in wheat. *Crop Science* 33:720-725.

Johnson, D.E., and F.A. Graybill. 1972. An analysis of a two-way model with interaction and nonreplication. *Journal of the American Statistical Association* 67:862-868.

Johnson, D.E., and F.A. Graybill. 1972. Estimation of σ^2 in a two-way classification model with interaction. *Journal of the American Statistical Association, Theory and Methods* 67:388-394.

Kang, M.S. (Ed). 1990. Genotype-by-Environment Interaction and Plant Breeding. Louisiana State University Agricultural Center, Baton Rouge, Louisiana.

Kang, M.S., and H.N. Pham. 1991. Simultaneous selection for high yielding and stable crop genotypes. *Agronomy Journal* 83:161-165.

Kempton, R.A. 1984. The use of biplots in interpreting variety by environment interactions. *Journal of Agricultural Science, Cambridge* 103:123-135.

Kempton, R.A., and M. Talbot. 1988. The development of new crop varieties. *Journal of the Royal Statistical Society, Series A* 151:327-341.

Knight, R.K. 1970. The measurement and interpretation of genotype-environment interactions. *Euphytica* 19:225-235.

Krishnaiah, P.R., and M.G. Yochmowitz. 1980. Inference on the structure of interaction in two-way classification model. *In* Handbook of Statistics, P.R. Krishnaiah (Ed.), volume 1. North-Holland, New York, New York, 973-994.

Lin, C.S., and M.R. Binns. 1991. Genetic properties of four types of stability parameter. *Theoretical and Applied Genetics* 82:505-509.

Lin, C.S., M.R. Binns, and L.P. Lefkovitch. 1986. Stability analysis: Where do we stand? *Crop Science* 26:894-900.

Mandel, J. 1961. Nonadditivity in two-way analysis of variance. *Journal of the American Statistical Society* 56:878-888.

Mandel, J. 1971. A new analysis of variance model for non-additive data. *Technometrics* 13:1-18.

Mandel, J. 1993. The analysis of two-way tables with missing values. *Applied Statistics* 42:85-93.

Nachit, M.M., M.E. Sorrels, R.W. Zobel, H.G. Gauch, R.A. Fischer, and W.R. Coffman. 1992. Association of environment variables with sites' mean grain yield and components of genotype-environment interaction in durum wheat II. *Journal of Genetics and Breeding* 46:369-372.

Perkins, J.M. 1972. The principal component analysis of genotype-environmental interactions and physical measures of the environment. *Heredity* 29:51-70.

Piepho, H.-P. 1993. Use of the maximum likelihood method in the analysis of phenotypic stability. *Biometrical Journal* 35:815-822.

Prabhakaran, V.T., and J.P. Jain. 1994. *Statistical techniques for studying genotype-environment interactions.* South Asian Publishers Pvt. Ltd., New Delhi, India.

Rao, V., I.E. Henson, and N. Rajanaidu (Eds.). 1993. Genotype-Environment Interaction Studies in Perennial Tree Crops. Oil Palm Research Institute of Malaysia, Kuala Lumpur, Malaysia.

Romagosa, I., and P.N. Fox. 1993. Genotype x environment interaction and adaptation. *In* Plant Breeding: Principles and Prospects, M.D. Hayward, N.O. Bosemark, and I. Romagosa (Eds.). Chapman and Hall, London, pages 373-390.

Romagosa, I., P.N. Fox, L.F. García del Moral, J.M. Ramos, B. García del Moral, F. Roca de Togores, and J.L. Molina-Cano. 1993. Integration of statistical and physiological analyses of adaptation of near-isogenic barley lines. *Theoretical and Applied Genetics* 86:822-826.

Russel, T.S., and R.A. Bradley. 1966. One-way variances in a two-way classification. *Biometrika* 45:111-129.

Saidon, G., and G.B. Schaalje. 1993. Evaluation of locations for testing dry bean cultivars in western Canada using statistical procedures, biological interpretation and multiple traits. *Canadian Journal of Plant Science* 73:985-994.

Schlesselman, J.J. 1973. Data transformation in two-way analysis of variance. *Journal of the American Statistical Association* 68:369-378.

Seyedsadr, M., and P.L. Cornelius. 1992. Shifted multiplicative models for nonadditive two-way tables. *Communications in Statistics, Simulation and Computation* 21:807-832.

Shukla, G.K. 1972. Some statistical aspects of partitioning genotype-environment components of variability. *Heredity* 29:237-245.

Snijders, C.H.A., and F.A. Van Eeuwijk. 1991. Genotype x strain interactions for resistance to Fusarium head blight caused by *Fusarium culmorum* in winter wheat. *Theoretical and Applied Genetics* 81:239-244.

Van Eeuwijk, F.A. 1992. Interpreting genotype-by-environment interaction using redundancy analysis. *Theoretical and Applied Genetics* 85:89-100.

Van Eeuwijk, F.A. 1995. Multiplicative interaction in generalized models. *Biometrics* (in press).

Van Eeuwijk, F.A., and A. Elgersma. 1993. Incorporating environmental information in an analysis of genotype by environment interaction for seed yield in perennial ryegrass. *Heredity* 70:447-457.

Van Oosterom, E.J., D. Kleijn, S. Ceccarelli, and M.M. Nachit. 1993. Genotype-by-environment interactions of barley in the Mediterranean region. *Crop Science* 33:669-674.

Van Sanford, D.A., T.W. Pfeiffer, and P.L. Cornelius. 1993. Selection index based on genetic correlations among environments. *Crop Science* 33:1244-1248.

Wallace, D.H., J.P. Baudoin, J. Beaver, D.P. Coyne, D.E. Halseth, P.N. Masaya, H.M. Munger, J.R. Myers, M. Silbernagel, K.S. Yourstone, and R.W. Zobel. 1993. Improving efficiency of breeding for higher crop yield. *Theoretical and Applied Genetics* 86:27-40.

Westcott, B. 1986. Some methods of analysing genotype-environment interaction. *Heredity* 56:243-253.

Wood, J.T. 1976. The use of environmental variables in the interpretation of genotype-environment interaction. *Heredity* 37:1-7.

Yates, F., and W.G. Cochran. 1938. The analysis of groups of experiments. *Journal of Agricultural Science, Cambridge* 28:556-580.

Yau, S.K. 1995. Regression and AMMI analyses of genotype × environment interactions: An empirical comparison. *Agronomy Journal* 87:121-126.

Zavala-Garcia, F., P.J. Bramel-Cox, and J.D. Eastin. 1992. Potential gain from selection for yield stability in two grain sorghum populations. *Theoretical and Applied Genetics* 85:112-119.

Zobel, R.W., M.J. Wright, and H.G. Gauch. 1988. Statistical analysis of a yield trial. *Agronomy Journal* 80:388-393.

INDEX

Abiotic, i, 55-56, 361
Accuracy, 86-88, 98-111, 114-115, 144,
    354, 373
Active loci, 341
Adaptability, 293, 309-310, 323, 329-330
Adaptation, 10, 54-55, 59, 93, 96, 309-
    310, 313-315, 320-321, 329,
    331, 333-335
  broad, 54-57, 60, 70, 77, 82, 87, 94,
    309, 312, 315, 321, 328, 335
  specific, 54-57, 60, 82, 94, 97, 310,
    313, 315, 335
Adapted genotype(s), i, 10
Additive effects, 128, 140-141
Additive main effects and multiplicative
   interaction, i-ii, 37, 85-115, 146, 152,
      171-172, 201-203, 209-211,
      227, 233-234, 324, 327, 329,
      334, 362, 383, 385-389, 392-
      399, 401-402
  choice as model form, 223-225
  hypothesis tests, 202-208, 226-227
  least squares estimates, 202
  PRESS statistic, 224
  relation to Tukey-Mandel tests, 224-
    225
  shrinkage estimators, 211-217, 219,
    228-229
  software, 230
  unreplicated trials, 226-229
Additive model, 20, 23, 28, 103
Agronomic concept, 6
Alfalfa, 89, 93, 96, 99-100, 103
Alternative hypothesis, 3, 242, 249
AMMI
  See Additive main effects and
  multiplicative interaction
Analysis of variance, 16, 20, 27-28, 54,
    57-60, 70, 74, 76, 79, 81, 85-
    86, 89, 91-92, 113, 131-133,
    146, 162, 186, 190-191, 237-
    238, 243, 257, 261-262, 327,
    344-345, 348, 355, 359- 361,
    365, 388-389, 397-398, 401
ANOVA
  See Analysis of variance
AOV
  See Analysis of variance
Artificial covariates, 18-19, 43

Balanced data, 2, 86
Barley, 89, 97, 99, 102, 124, 316-317,
    320, 323, 331-332, 334

Bayes estimates, 211
Best linear unbiased estimators, 6-7, 132
Best linear unbiased predictors, 6, 40,
    54, 60, 109, 132, 134- 137,
    145-146, 211-212, 214-217
Biadditive mixed models, 44-45
Bibliography, ii, 405
Biotic, i, 55
BLUE
  See Best linear unbiased estimators
BLUP
  See Best linear unbiased predictors
Breeding program(s), i, 52, 55-56, 87,
    93, 96, 106, 123, 151, 156,
    158, 273, 309-312, 321, 324,
    331-332, 362, 374
Buffering
  ability, 314
  individual, 313-315
  population, 313-315
Building block method, 196-197

Cabbage, 89, 95
Canola, 95
Citrus, 89
Cluster analysis, ii, 135, 145-146, 177-
    178, 190, 244, 251, 263-264,
    267, 311-312, 325, 357, 363
Clustering genotypes, ii, 175-176
CM
  See Conventional method
Coefficient of concordance, 238, 244,
    247, 251, 263, 267
Coheritability, 313
COI
  See Crossover interaction
Combining ability, 140, 331
COMM
  See Completely multiplicative model
Completely multiplicative model, 200-
    203, 211-213, 215-217, 226,
    230
  choice as model form, 223-225
  hypothesis tests, 202-208, 226
Computer simulation, ii, 341, 345
Concurrent regression, 214-215
Conventional method, 1, 3
Corn
  See Zea mays L.
Correlated response, i, 312-313, 316-
    317, 319-322, 329, 331, 335
Co-segregation, 124-126, 144, 146
Covariate(s), 16, 19, 21-23, 25-26, 33,
    35, 37, 39-40, 42-44, 98, 400

412

416

416